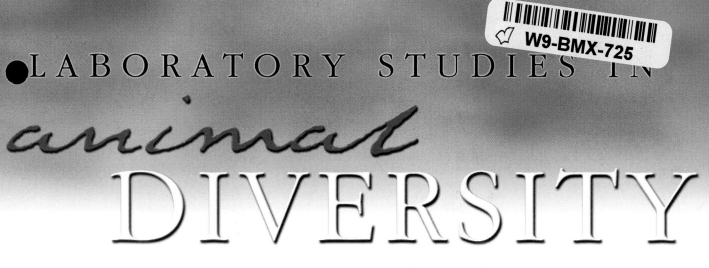

LABORATORY STUDIES IN
animal
DIVERSITY

THIRD EDITION

CLEVELAND P. HICKMAN, JR.

**Professor Emeritus
Washington and Lee University**

LEE B. KATS

**Professor
Pepperdine University**

Original Artwork by
WILLIAM C. OBER, M.D. and CLAIRE W. GARRISON, R.N.

Boston   Burr Ridge, IL   Dubuque, IA   Madison, WI   New York   San Francisco   St. Louis
Bangkok   Bogotá   Caracas   Lisbon   London   Madrid
Mexico City   Milan   New Delhi   Seoul   Singapore   Sydney   Taipei   Toronto

McGraw-Hill Higher Education

A Division of The **McGraw-Hill** *Companies*

LABORATORY STUDIES IN ANIMAL DIVERSITY, THIRD EDITION

Published by McGraw-Hill, a business unit of The McGraw-Hill Companies, Inc., 1221 Avenue of the Americas, New York, NY 10020. Copyright © 2003, 2000, 1995 by The McGraw-Hill Companies, Inc. All rights reserved. No part of this publication may be reproduced or distributed in any form or by any means, or stored in a database or retrieval system, without the prior written consent of The McGraw-Hill Companies, Inc., including, but not limited to, in any network or other electronic storage or transmission, or broadcast for distance learning.

Some ancillaries, including electronic and print components, may not be available to customers outside the United States.

♻ This book is printed on recycled, acid-free paper containing 10% postconsumer waste.

1 2 3 4 5 6 7 8 9 0 QPD/QPD 0 9 8 7 6 5 4 3 2

ISBN 0-07-251883-9

Publisher: *Margaret J. Kemp*
Senior developmental editor: *Donna Nemmers*
Marketing manager: *Heather K. Wagner*
Senior project manager: *Susan J. Brusch*
Production supervisor: *Kara Kudronowicz*
Coordinator of freelance design: *David W. Hash*
Cover designer: *Kaye Farmer*
Cover photo: © *Tui De Roy/Minden Pictures*
Senior photo research coordinator: *John C. Leland*
Photo research: *Billie Porter*
Senior media project manager: *Tammy Juran*
Media technology producer: *Renee Russian*
Compositor: *Precision Graphics*
Typeface: *10/12 Garamond*
Printer: *Quebecor World Dubuque Inc.*

COVER PHOTO: Sally Lightfoot crabs (*Grapsus grapsus*), vivid splashes of color on a rugged shoreline of the Galápagos Islands, share a lava rock with three marine iguanas (*Amblyrhynchus cristatus*). Unlike any other lizard in the world, the unique Galápagos marine iguanas are totally dependent on the sea for their living, diving for their diet of marine algae. Marine iguanas, although described as ugly by some visitors, appear starkly beautiful to others in this exotic setting. Around the iguanas the nimble Sally Lightfoot crabs scamper over the rocks with surprising agility, while feeding principally on algae. It was here in Galápagos that the young English naturalist Charles Darwin observed the strange animals and plants that later contributed to his theory of evolution by natural selection.

The credits section for this book begins on page 266 and is considered an extension of the copyright page.

Some of the laboratory experiments included in this text may be hazardous if materials are handled improperly or if procedures are conducted incorrectly. Safety precautions are necessary when you are working with chemicals, glass test tubes, hot water baths, sharp instruments, and the like, or for any procedures that generally require caution. Your school may have set regulations regarding safety procedures that your instructor will explain to you. Should you have any problems with materials or procedures, please ask your instructor for help.

www.mhhe.com

BRIEF CONTENTS

PART ONE

ACTIVITY OF LIFE

EXERCISE 1
Ecological Relationships of Animals   2

EXERCISE 2
Introduction to Animal Classification   7

PART TWO

THE DIVERSITY OF ANIMAL LIFE

EXERCISE 3
The Microscope   16

EXERCISE 4
Protozoan Groups   24

EXERCISE 5
The Sponges   49

EXERCISE 6
The Radiate Animals   57

EXERCISE 7
The Acoelomate Animals   72

EXERCISE 8
The Pseudocoelomate Animals   88

EXERCISE 9
The Molluscs   98

EXERCISE 10
The Annelids   112

EXERCISE 11
The Chelicerate Arthropods   127

EXERCISE 12
The Crustacean Arthropods   134

EXERCISE 13
The Uniramia Arthropods: Myriapods and Insects   146

EXERCISE 14
The Echinoderms   165

EXERCISE 15
Phylum Chordata: A Deuterostome Group   180

EXERCISE 16
The Fishes—Lampreys, Sharks, and Bony Fishes   188

EXERCISE 17
Class Amphibia   207

EXERCISE 18
Class Reptilia   224

EXERCISE 19
Class Aves   229

EXERCISE 20
Class Mammalia   232

CONTENTS

preface vi

PART ONE

ACTIVITY OF LIFE

EXERCISE 1

Ecological Relationships of Animals 2

Exercise 1: A study of population growth, with application of the scientific method 2

EXERCISE 2

Introduction to Animal Classification 7

Use of a taxonomic key for animal identification 7
Key to the major animal taxa 9

PART TWO

THE DIVERSITY OF ANIMAL LIFE

EXERCISE 3

The Microscope 16

Exercise 3A: Compound light microscope 16
Exercise 3B: Stereoscopic dissecting microscope 22

EXERCISE 4

Protozoan Groups 24

Classification 24
Exercise 4A: Amebas 25
Exercise 4B: Phylum Chlorophyta—*Volvox*, Phylum Euglenozoa—*Euglena*, and *Trypanosoma* 33

Exercise 4C: Phylum Apicomplexa, Class Gregarinea, Class Coccidea—*Gregarina* and *Plasmodium* 41
Exercise 4D: Phylum Ciliophora—*Paramecium* and other ciliates 43

EXERCISE 5

The Sponges 49

Classification: Phylum Porifera 49
Exercise 5: Class Calcarea—*Sycon* 49

EXERCISE 6

The Radiate Animals 57

Classification: Phylum Cnidaria 57
Exercise 6A: Class Hydrozoa—*Hydra, Obelia,* and *Gonionemus* 58
Exercise 6B: Class Scyphozoa—*Aurelia,* a "true" jellyfish 71
Exercise 6C: Class Anthozoa—*Metridium,* a sea anemone, and *Astrangia,* a stony coral 71

EXERCISE 7

The Acoelomate Animals 72

The acoelomate phyla 72
Classification: Phylum Platyhelminthes 72
Exercise 7A: Class Turbellaria—the planarians 73
Exercise 7B: Class Trematoda—the digenetic flukes 78
Exercise 7C: Class Cestoda—the tapeworms 81
Experimenting in Zoology: Planaria regeneration experiment 84

EXERCISE 8

The Pseudocoelomate Animals 88

Exercise 8A: Phylum Nematoda—*Ascaris* and others 88
Exercise 8B: A brief look at some other pseudocoelomates 94

EXERCISE 9

The Molluscs 98

Classification: Phylum Mollusca 98
Exercise 9A: Class Bivalvia (= Pelecypoda)—the freshwater clam 99
Exercise 9B: Class Gastropoda—the pulmonate land snail 107
Exercise 9C: Class Cephalopoda—*Loligo,* the squid 109

EXERCISE 10

The Annelids 112

Classification: Phylum Annelida 112
Exercise 10A: Class Polychaeta—the clamworm 113
Exercise 10B: Class Oligochaeta—the earthworms 115
Exercise 10C: Class Hirudinea—the leech 123
Experimenting in Zoology: Behavior of the medicinal leech, *Hirudo medicinalis* 125

EXERCISE 11

The Chelicerate Arthropods 127

Classification: Phylum Arthropods 127
Exercise 11: The chelicerate arthropods—the horseshoe crab and garden spider 128

EXERCISE 12

The Crustacean Arthropods 134

Exercise 12: Subphylum Crustacea—the crayfish (or lobster) and other crustaceans 134
Experimenting in Zoology: The Phototactic Behavior of *Daphnia* 144

EXERCISE 13

The Uniramia Arthropods: Myriapods and Insects 146

Exercise 13A: The myriapods—centipedes and millipedes 146
Exercise 13B: The insects—the grasshopper and the honey bee 148
Exercise 13C: The insects—the house cricket 155
Exercise 13D: Collection and classification of insects 158
Key to the principal orders of insects 159

EXERCISE 14

The Echinoderms 165

Classification: Phylum Echinodermata 165
Exercise 14A: Class Asteroidea—the sea stars 166
Exercise 14B: Class Ophiuroidea—the brittle stars 171
Exercise 14C: Class Echinoidea—the sea urchin 173
Exercise 14D: Class Holothuroidea—the sea cucumber 176

EXERCISE 15

Phylum Chordata: A Deuterostome Group 180

What Defines a Chordate? 180
Classification: Phylum Chordata 180

Exercise 15A: Subphylum Urochordata—*Ciona*, an ascidian 181
Exercise 15B: Subphylum Cephalochordata—amphioxus 184

EXERCISE 16

The Fishes—Lampreys, Sharks, and Bony Fishes 188

Exercise 16A: Class Cephalaspidomorphi (= Petromyzontes)—the lampreys (ammocoete larva and adult) 188
Exercise 16B: Class Chondrichthyes—the cartilaginous fishes 192
Exercise 16C: Class Osteichthyes—the bony fishes 198
Experimenting in Zoology: Aggression in Paradise Fish, *Macropodus opercularis* 202
Experimenting in Zoology: Analysis of the multiple hemoglobin system in *Carassius auratus,* the common goldfish 204

EXERCISE 17

Class Amphibia 207

Exercise 17A: Behavior and adaptations 207
Exercise 17B: The skeleton 210
Exercise 17C: The skeletal muscles 211
Exercise 17D: The digestive, respiratory, and urogenital systems 216
Exercise 17E: The circulatory system 219

EXERCISE 18

Class Reptilia 224

Exercise 18: The painted turtle 224

EXERCISE 19

Class Aves 229

Exercise 19: The pigeon 229

EXERCISE 20

Class Mammalia 232

Exercise 20A: The skeleton 233
Exercise 20B: The muscular system 237
Exercise 20C: The digestive system 245
Exercise 20D: The urogenital system 251
Exercise 20E: The circulatory system 255

Appendix: Sources of Living Material and Prepared Microslides 263

Credits 265

Index 266

PREFACE

Laboratory Studies in Animal Diversity offers students hands-on experience in learning about the diversity of life. It provides students the opportunity to become acquainted with the principal groups of animals and to recognize the unique anatomical features that characterize each group as well as the patterns that link animal groups to each other. Although this manual was written to accompany a particular textbook, *Animal Diversity,* it can easily be adapted to use with any other introductory zoology text and with a variety of course plans. Every effort has been made to provide clear instructions and enough background material to create interest and an understanding of the subject matter. Many illustrations complement the written word.

New to the Third Edition

- Project exercises are placed with certain chapters in units entitled "Experimenting in Zoology." A new project exercise uses molecular techniques to explore questions that are important to our understanding of zoology and evolution. Some of these exercises can be completed within a single laboratory period; others are followed for a longer period. In all project exercises the student follows experimental procedures, records and analyzes quantitative data, and draws conclusions from the results. Many instructors will want their

students to gain additional experience by writing a laboratory report in which the student states the objectives, methods followed, results obtained, and conclusions that can be drawn from the results. The Experimenting in Zoology exercises are: Regeneration in *Planaria* (Exercise 7); Behavior of the medicinal leech, *Hirudo medicinalis* (Exercise 10); The Phototactic behavior of *Daphnia* (Exercise 12); Aggression in paradise fish, *Macropodus opercularis* (Exercise 16); Analysis of the multiple hemoglobin system in *Carassius auratus,* the common goldfish (Exercise 16).

- We have made the exercises more interactive with questions placed throughout the text, and with spaces provided for students to write down their responses and observations. This "active learning" approach involves students in the exercise and encourages them to think about the information as they read. Some questions may require students to consult their textbook for the answers. In some exercises we have placed questions in the figure legends, to be answered in the spaces provided when the student consults the figure. Examples of this interactive approach are found in Exercises 15 through 20.

- In several exercises we rewrote or expanded the "Where Found" sections to provide more interesting information

on the biology of the organism or group. These enhanced introductions are found 6B (*Aurelia*), 7B (*Clonorchis*), 8A (*Ascaris* and pinworms), 9A (bivalves), 11A (horseshoe crab), 14 (sea stars, brittle stars, sea urchins, sea cucumbers), and 15A (tunicates).

- A complete new set of full-color illustrations drawn from frog specimens was prepared for Exercise 17 on the frog. Many existing illustrations throughout the manual were converted to full color.

Supplements

- New to this edition is a supplement to Exercise 2, "Taxonomic Identification of Organisms to Species," which is found on this manual's website. This interactive exercise, prepared by Louise Wootton of Georgian Court College, leads the student through the construction of a dichotomous key, an exercise in phenetic analysis, and the construction of a cladogram. This interactive cladistics exercise may be found at www.mhhe.com/zoology (click on this book's cover).

- McGraw-Hill's **Digital Zoology CD-ROM and Student Workbook** by Jon Houseman is an interactive guide to the specimens and materials that you study in your laboratory and lecture sessions. This easy-to-use CD-ROM provides laboratory modules containing illustrations, photographs, annotations of the major structures of organisms, interactive quizzes, and video clips. Students will also find interactive cladograms within lab modules, along with links to interactive synapomorphies of the various animal groups. Key terms are linked to an interactive glossary. The accompanying student workbook and website provide additional study tips, exercises, and phyla characteristics. *Digital Zoology* is the perfect complement to a zoology lab manual to ensure the best possible results in your zoology lab.

There are many aids for the student in this laboratory manual. Throughout the exercises, working instructions are clearly set off from the descriptive material. Classifications, where appropriate, are included with the text, together with a "pie" diagram showing the relative sizes of the classes in a phylum. Function is explained along with anatomy. Topic headings help the student mentally organize the material. Met-

ric tables and definitions are placed on the inside front and back covers for convenient use. Much of the artwork was designed to assist the student with difficult dissections.

Acknowledgments

We are indebted to the reviewers whose many suggestions were essential in guiding our revision for this edition.

Carollyn Boykins-Winrow
   Elizabeth City State University
John A. Byers
   University of Idaho
Gerald L. DeMoss
   Morehead State University
Patricia M. Dorris
   Saint Leo College
Sharon C. McDonald
   Henry Ford Community College
Charles R. Moser
   California State University, Sacramento
Sarah H. Swain
   Middle Tennessee State University
Louise Wootton
   Georgian Court College

We especially thank Louise Wootton who provided the exercise on

different approaches to taxonomic classification: *Taxonomic Identification of Organisms to Species,* found on our web site at www.mhhe.com/zoology.

The authors express their appreciation to the editors and support staff at McGraw-Hill Higher Education who guided this revision and were a pleasure to work with. Special thanks are due Marge Kemp, Publisher, and Donna Nemmers, Senior Developmental Editor, who guided this manual throughout its development. Susan Brusch, Senior Project Manager, guided the project expertly through production. Although we make every effort to bring to you an error-free manual, errors of many kinds inevitably find their way into a book of this scope and complexity. We will be grateful to readers who have comments or suggestions concerning content to send their remarks to Donna Nemmers, Senior Developmental Editor, McGraw-Hill Publishers, 2460 Kerper Boulevard, Dubuque, IA 52001. Donna may also be contacted by e-mail: donna_nemmers@mcgraw-hill.com, or through this textbook's website at www.mhhe.com/zoology.

LABORATORY SAFETY PROCEDURES

1. Keep your work area uncluttered. Unnecessary books, backpacks, purses, etc. should be placed somewhere other than on your desktop.
2. Avoid contact with embalming fluids. Wear rubber or disposable plastic gloves when working with preserved specimens.
3. Wear eyeglasses or safety glasses to protect your eyes from splattered embalming fluid.
4. Keep your hands away from mouth and face while in the laboratory. Moisten labels with tap water, not your tongue.
5. Sponge down your work area and wash all laboratory instruments at the end of the period.
6. Wash your hands with soap and water at the end of the laboratory period.

GENERAL INSTRUCTIONS

Equipment

Each student will need to supply this equipment:

Laboratory manual and textbook
Dissecting kit containing scissors, forceps, scalpel, dissecting needles, pipette (medicine dropper), probe, and ruler, graduated in millimeters
Drawing pencils, 3H or 4H
Eraser, preferably kneaded rubber
Colored pencils—red, yellow, blue, and green
Box of cleansing tissues
Loose-leaf notebook for notes and corrected drawings

The department will furnish each student with all other supplies and equipment needed during the course.

Aim and Purpose of Laboratory Work

The zoology laboratory will provide your "hands-on" experience in zoology. It is the place where you will see, touch, hear, smell—but perhaps not taste—living organisms. You will become acquainted with the major animal groups, make dissections of preserved or anesthetized specimens to study how animals are constructed, ask questions about how animals and their parts function, and gain an appreciation of some of the architectural themes and adaptations that emphasize the unity of life.

General Instructions for Laboratory Work

Prepare for the Laboratory. Before coming to the laboratory, read the entire exercise to familiarize yourself with the subject matter and procedures. Read also the appropriate sections in your textbook. Good preparation can make the difference between a frustrating afternoon of confusion and mistakes and an experience that is pleasant, meaningful, and interesting.

Follow the Manual Instructions Carefully. It is your guide to exploring and understanding the organisms or functions you are investigating. Its instructions have been written with care and with you in mind, to help you do the work (1) in logical sequence, (2) with economy of time, and (3) to arouse a questioning attitude that will stimulate interest and curiosity.

Use Particular Care in Making Animal Dissections. A glossary of directional terms used in dissections will be found inside the back cover. The object in dissections is to separate or expose parts or organs so as to see their relationships. Working blindly without the manual instructions may result in the destruction of parts before you have had an opportunity to identify them. **Learn the functions** of all the organs you dissect.

Record Your Observations. Keep a personal record in a notebook of everything that is pertinent, including the laboratory instructor's preliminary instruction and all experimental observations. Do not record data on scraps of paper with the intention of recopying later; record directly into a notebook. The notes are for your own use in preparing the laboratory report later.

Take Care of Equipment. Glassware and other apparatus should be washed and dried after use. Metal instruments in particular should be thoroughly dried to prevent rust or corrosion. Put away all materials and equipment in their proper places at the end of the period.

Tips on Making Drawings

You need not be an artist to make laboratory drawings. You do, however, need to be **observant.** Study your specimen carefully. Your simple line drawing is a record of your observations.

Before you draw, locate on the specimen all the structures or parts indicated in the manual instructions. Study their relationships to each other. Measure the specimen. Decide where the drawing should be placed and how much it must be enlarged or reduced to fit the page (read further for estimation of magnification). Leave ample space for labels.

When ready to draw, you may want first to rule in faint lines to represent the main axes, and then sketch the general outlines lightly. When you have the outlines you want, draw them in with firm dark lines, erasing unnecessary sketch lines. Then fill in details. Do not make overlapping, fuzzy, indistinct, or unnecessary lines. Indicate differences in texture and color by stippling. Stipple deliberately, holding the pencil vertically and

making a neat round dot each time you touch the paper. Placing the dots close together or farther apart will give a variety of shading. Avoid line shading unless you are very skilled. Use color only when asked for it in the directions.

Label the Drawing Completely. Print labels neatly in lowercase letters and align them vertically and horizontally. Plan the labels so that there will be no crossed label lines. If there are to be many labels, center the drawing and label on both sides.

Indicate the magnification in size beneath the drawing, for instance, "×3" if the drawing is three times the length and width of the specimen. In the case of objects viewed through a microscope, indicate also the magnification at which you viewed the subject, for example, 430× (43× objective used with a 10× ocular).

Estimating the Magnification of a Drawing

A simple method for determining the magnification of a drawing is to find the ratio between the size of the drawing and the actual size of the object you have drawn. The magnification of the drawing can be expressed in this formula:

$$\times = \frac{\text{Size of drawing}}{\text{Size of object}}$$

If your drawing of the specimen is 12 cm (120 mm) long, and you have estimated the specimen to be 0.8 mm long, then × = 120 ÷ 0.8, or 150. The drawing, then, is ×150, or 150 times the length of the object drawn.

This same formula will hold good whether the drawing is an enlargement or a reduction. If, for example, the specimen is 480 mm long, and the drawing is 120 mm, then × = 120/480, or 1/4.

STATEMENT ON THE USE OF LIVING AND PRESERVED ANIMALS IN THE ZOOLOGY LABORATORY

Congress has probably received more mail on the topic of animal research in universities and business firms than on any other subject. Do humans have the right to experiment on other living creatures to support their own medical, pharmaceutical, and commercial needs? A few years ago, Congress passed a series of amendments to the Federal Animal Welfare Act, a body of laws covering animal care in laboratories and other facilities. These amendments have become known as the three R's: **r**eduction in the number of animals needed for research; **r**efinement of techniques that might cause stress or suffering; and **r**eplacement of live animals with simulations or cell cultures whenever possible. As a result, the total number of animals used each year in research and in commercial product testing has declined steadily as scientists and businesses have become more concerned and more accountable. The animal rights movement, largely comprising vocal anti-vivisectionists, has helped to create an awareness of the needs of laboratory research animals and has stretched the resources and creativity of the researchers to discover cheaper and more humane alternatives to animal experimentation.

However, computers and cell cultures—the alternatives—can only simulate the effects on organismal systems of, for instance, drugs, when the principles are well acknowledged. When the principles are themselves being scrutinized and tested, computer modeling is insufficient. Nor can a movie or computer simulation match the visual and tactile comprehension of anatomical relationships provided by direct dissection of preserved or anesthetized animals. Medical and veterinarian progress depends on animal research. Every drug and every vaccine that you and your family have ever taken has first been tested on an animal. Animal research has wiped out smallpox and polio; has pro-vided immunization against diseases previously common and often deadly, such as diphtheria, mumps, and rubella; has helped create treatments for cancer, diabetes, heart disease, and manic depression; and has been used in the development of surgical procedures such as heart surgery, blood transfusions, and cataract removal.

Animal research has also benefited other animals for veterinary cures. The vaccine for feline leukemia that could threaten the life of your cat, as well as the parvo vaccine given to your puppy, were first introduced to other cats and dogs. Many other vaccinations for serious animal diseases were developed through animal research; for example, rabies, distemper, anthrax, hepatitis, and tetanus.

The animal models used by the artist for the illustrations in the exercises of this laboratory manual, and the animals you will dissect in this laboratory course, were prepared for educational use following strict humane procedures. No endangered species have been used. No living vertebrate organisms will be harmed in this laboratory setting. Invertebrate animals that are to be dissected while alive are anesthetized before the procedure. The experiments selected are unoffensive, are respectful of the integrity of the animal's evolutionary contributions, and often require only close observation. The experiments closely follow the tenets of the scientific method, which cannot dictate ethical decisions but can provide the structure for common sense. Do not be wasteful. Share the animals with the other students as often as possible. At the same time, you are encouraged to observe the live animal in its natural setting and its relationships to other species, for only in this manner will you gain a full appreciation of the unique evolutionary position and special structure and systems of each animal.

PART ONE

Activity of Life

Ecological Relationships of Animals

Exercise 1*

A Study of Population Growth, with Application of the Scientific Method

EXERCISE 1

A Study of Population Growth, with Application of the Scientific method
The scientific method
Application of the scientific method to the study of populations
Experimental Procedures

One goal of this course is to introduce students to the methods scientists use to gather knowledge. In this project you will apply the "scientific method" to the problem of determining what regulates animal populations.

The Scientific Method

People, scientists included, often acquire knowledge by applying a two-stage process, **conjecture** followed by **confirmation**—although few of us think of it this way. Conjecture consists of generating a general explanation of how the world is constructed, and is often based on general observations. Confirmation tests the validity of this conjecture. We all use such a method in our everyday lives; for example, we speculate on the quality of a future music concert based on our observations of recorded music, then subsequently confirm, reject, or modify that speculation based on our experiences while attending the concert.

Scientists employ a similar method, although with a lot more rigor, in attempting to discover new facts. An idealized form of the method used by scientists is known as the **scientific method,** and is broken down into four steps.

1. **Observation.** Observations may be based on direct examination of a system, or based on something we read, or even be the result of discussions with others about a process or concept. Such observations frequently stimulate questions about why species exhibit certain traits, why internal organs interact the way they do, or what the advantages of a particular body shape might be, or the role of certain genes in a particular process.

2. **Hypothesis formulation.** Formulating an hypothesis is like saying "Let's suppose . . ." Its objective is to explain, by induction, the observation. Typically several alternative hypotheses are formulated, each a possible and reasonable explanation for the observation. Weeding out these hypotheses is the role of the third and fourth steps.

3. **Prediction.** Predictions are deduced from hypotheses, and often will be based on some knowledge of the organisms or concepts being studied. They follow the form of "If hypothesis A is true, then I predict the following pattern." Predictions must be generated such that one set of hypotheses predicts one result, but alternative hypotheses predict another result. A prediction is worthless if it can be made for all of the hypotheses under consideration. Testing whether a prediction holds true is the way in which one or more hypotheses can be rejected, thus reducing the number of hypotheses still under consideration.

4. **Testing of predictions.** The final step is to design a test so that a prediction, if incorrect, can confidently be rejected. Tests are accomplished using observations or experimental manipulations. The confidence with which we can make such a rejection is quantified by the use of inferential statistics, which we shall not discuss. However, note that our confidence in a result increases with the use of (1) treatments known as controls, in which all variables except for the one manipulated

---

* Exercise written by James C. Munger, Department of Biology, Boise State University, Boise, Idaho; and Richard S. Inouye, Department of Biological Sciences, Idaho State University, Pocatello, Idaho.

are held the same, and (2) several replicates of each treatment, to give assurance that the observed result was due to conditions of the treatment and not simply due to variation among individuals. With skill (and perhaps some luck) all but one hypothesis will have been rejected. The unrejected hypothesis, however, is not proven to be true. Hypotheses can never be fully accepted; they can only be rejected (what lonely lives they must lead).

The next step is to repeat this process. With the results in hand from the tests of previous predictions, it is possible to fine-tune the hypothesis, then set about testing the new one. Here is an example of this repetitive process.

Initial observation:	Roommate breaks dish
Generalization (hypothesis):	Roommate breaks everything
Prediction:	Will wreck borrowed car
Test/observation	Didn't wreck car
New generalization/ hypothesis:	Only breaks dishes
Prediction:	Won't break borrowed camera
Test:	Does break camera
New generalization/ hypothesis:	Roommate breaks small objects

And so on . . .

The unrejected hypothesis at each stage is our best guess as to how the world works. If a hypothesis withstands repeated tests and has great explanatory value, it may be elevated to the level of a scientific theory. Note that a scientific theory is not an untested hypothesis, but is instead as close as scientists will come to calling a hypothesis proven. The theory of evolution is an example.

Whether a hypothesis is accepted or rejected, the observations made while testing the hypothesis frequently lead to more hypotheses, more predictions, more tests, more observations, and so on. It is often said scientific investigation raises more questions than it answers; it is this aspect of science that, to many people, is most exciting.

Application of the Scientific Method to the Study of Populations

If you were a scientist, you would make your own observations, formulate your own hypotheses, derive your own predictions, and perform your own tests of those predictions. In a classroom situation, however, there are certain constraints, as you will see. We will apply the scientific method to the study of population growth.

Step 1—Observation

In 1798 a British economist, Thomas Malthus, published an essay in which he observed that populations do not grow indefinitely, but often tend to stay at relatively stable numbers. We can make similar observations: if we look around us, we do not see populations of organisms forever growing—instead they are relatively stable.

Step 2—Hypothesis Formulation

Why might this occur? Again we can look to Malthus, this time for one possible explanation (a hypothesis) as to why populations do not grow indefinitely. Malthus reasoned that if a population had unlimited resources, it would grow exponentially to infinite size. However, since no population grows to infinite size, resources must be limiting. He therefore hypothesized that a limitation of resources is the cause of limited population growth. We can depict these two possible conditions graphically. If resources are unlimited, the population should grow **exponentially.**

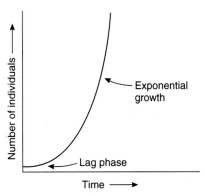

If resources are limited but in relatively constant supply, the population will experience **logistic growth,** rapidly growing at first and then eventually reaching an equilibrium, known as the carrying capacity. Logistic growth occurs in situations where resources are renewed (such as plankton reaching a barnacle), or because they are not consumed (such as nest sites, which can be used again and again).

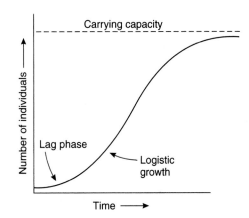

A third type of population growth occurs when resources are consumed but not renewed. This sort of population growth would occur in a test tube bacterial culture in which the population increases until all the nutrients are consumed, then crashes.

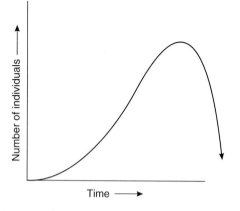

Now we can erect our first hypothesis regarding what limits population growth:

H₁: **Limited food limits population size.**

However, there are a number of factors that could limit population size, such as predation, climate, disease, limited nest sites, and intraspecific strife, including cannibalism. Can you think of others? We can cast these factors as this list of hypotheses:

H₁: **Limited food regulates population size.**
H₂: **Predation regulates population size.**
H₃: **Disease regulates population size.**
H₄: **Limited nest sites regulate population size.**
H₅: **Intraspecific strife regulates population size.**

Note that this list is not exhaustive. Also note that in this case the hypotheses are not mutually exclusive, that is, more than one may be true in a particular population; this is especially true if we consider a wide range of species.

Step 3—Prediction Derivation

What predictions logically follow from these hypotheses? The best predictions are those that (1) allow the investigator to decide between two or more competing hypotheses, and (2) are straightforward to test.

For the purposes of discussion, we will focus on our hypotheses as they apply to the setup to be used in this exercise: a population of flour beetles eating flour, living in flour, laying eggs in flour, all contained in a small jar. In this system, we can discount the possibility of one hypothesis, predation, because we will not allow predators into the system.

A prediction that follows from H₁ is that if we limit food availability, we expect a smaller population to result (less flour, fewer beetles; more flour, more beetles). But do the other hypotheses make different predic-

tions? If nest sites are limited, then adding more food will increase the availability of nest sites, giving the same prediction: more flour, more beetles. If intraspecific strife (e.g., cannibalism) is limiting, what will adding more flour do? It will give the beetles more room to hide, meaning fewer encounters, and more to eat, meaning less hunger; both mean less cannibalism. Again, more flour, more beetles. And if disease is limiting, more flour means fewer encounters among beetles and less disease transmission. Again, more flour, more beetles.

However, what if we were to vary the amount of food available while holding constant the total volume available for the beetles to roam? If H₁ were true, more food would lead to more beetles. But if H₃ or H₄ or H₅ (but not H₁) were true, then more food would have no effect on beetle numbers, so long as the total volume was constant. So this prediction allows us to distinguish among competing hypotheses.

Step 4—Test of Predictions

Next, we need to create an experiment that will allow us to vary food without varying volume. One way to accomplish this is to put various amounts of food into jars, then add an inert filler (such as vermiculite) to maintain constant volume.

Step 5—Repeat the Process

When you look at the results from your experiment, you can consider what modifications to make to your hypotheses and what new predictions you could use to test your new hypothesis.

Experimental Procedures

We will start cultures of *Tribolium confusum* (a flour beetle) with the same initial population size, but varying amounts of food and varying amounts of space. Near the end of the term, we will count the number of larvae, pupae, and adults in each container and compare age distribution and resulting densities.

T. confusum develops from egg to adult in about 28 days as follows: egg stage 5 days, larval stages 17 days, pupal stage 6 days (Figure 1-1). The average life span of adult beetles is roughly 200 days.

☞ **Work in groups of four students each. Each group should prepare:**

A. One low-density, high-food jar, containing 50 g of resource (95% whole wheat flour; 5% brewer's yeast)

B. One medium-density, medium-food jar, containing 10 g of resource

C. One high-density jar, low-food jar, containing 3 g of resource

D. One low-density, medium-food jar, containing 10 g of resource with filler (such as vermiculite

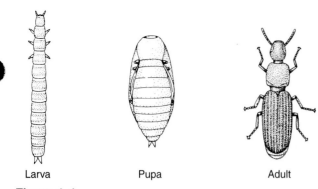

| Larva | Pupa | Adult |

Figure 1-1

Appearance of *Tribolium confusum* at larval, pupal, and adult stages.

screened to standardize the size) added to bring the total volume to that of jar A.

E. One low-density, low-food jar, containing 3 g of resource with filler added to bring the total volume to that of jar A.

Next, sort through the culture that has been provided to you. Be careful not to damage the animals. Use a fine brush or small spatula to push them around. Each group should sort and count 250 healthy looking adults, and put 50 into each of the half-pint Mason canning jars. We will assume that within each group of 50 there are plenty of both males and females. Cover jar with the precut wire mesh (window screen), then a piece of paper towel, then screw on the top ring. Place the jars in a cabinet with a light (for warmth; about 30° C) and an open container of water (for humidity).

You and your lab instructor may decide to try other variations on this experimental setup. For example, you might try to see if temperature is important. How would you do this?

Why don't we use just one jar at each density for the whole class? If there were only one jar, it would be difficult to say, because of biological variation and experimental error, that the results from that jar are representative of any jars the class might start with the same density. For example, what if only the jar your class had used for its medium-density culture had previously contained a toxic chemical? That could invalidate your results. However, if each group of four students prepares one jar at each density, then a class of 24 students would have a total of six jars at the same density (known as **replicates**). If all six give approximately the same result, then we can have substantial confidence that those jars are representative of all jars at that density.

During the term, make occasional observations on beetle behavior, and record those observations in your notebook. For example, do the beetles live on top of the flour or within it? Do the beetles congregate or space themselves out?

At the end of the experiment, sort through each jar and count live adults, dead adults, pupae, and large (final instar) larvae. For the purposes of this experiment we will count neither the eggs nor the early instar larvae. Compile data for the whole class and calculate averages for each age class for each jar. Construct an age distribution for each treatment.

This is a time to try your hand at "playing with the data." Look for interesting patterns such as correlations between variables. Try to think of patterns you would expect to occur if the hypothesis is correct.

Questions

1. Comparison between which jars will test what hypotheses? For example, what hypothesis can be tested by comparing jars A, D, and E? What would you conclude if they had the same densities of beetles?

2. What other hypotheses can you formulate to explain the observation that populations do not grow indefinitely? What predictions can you derive from these hypotheses and what tests could you perform?

3. Two life stages (eggs and pupae) are not mobile, and so are particularly vulnerable to cannibalism. Do you see evidence of this when comparing jars, for example, jar A with jar C?

4. An attitude commonly encountered in undergraduate science labs is that if you did not get the result the instructor expected, then the experiment "did not work." What do you think of this view?

5. Given the results thus far, what would be your next step in this study, if your goal is to understand what determines the abundance of flour beetles?

Written Report

For your report prepare the suggested graphs and write a summary statement of the experimental approach and an explanation of the results. Answer any of the preceding questions that your instructor may request.

Alternatively, your instructor may want you to follow the format of a scientific paper for your report: introduction, materials and methods, results, discussion, and literature cited. These questions are designed to bring up possible topics that might be included in the report. Be sure to think about your results and look at your data in original ways before writing the report.

References

Hasting, A., and R. Costanino. 1987. Cannibalistic egg-larva interactions in *Tribolium:* an explanation for the oscillations in population numbers. American Naturalist **130:**37–52.

Ho, F., and P. Dawson. 1966. Egg cannibalism by *Tribolium* larvae. Ecology **47:**318–322.

Lloyd, M. 1968. Self regulation of adult numbers by cannibalism in two laboratory strains of flour beetles (*Tribolium castaneum*). Ecology **49:**245–259.

Lutherman, C., E. Miller, and T. Park. 1939. Studies in population physiology, IX. The effect of imago population density on the duration of larval and pupal stages of *Tribolium confusum* Duval. Ecology **20:**365–373.

Park, T. 1932. Studies in population physiology: the relation of numbers to initial population growth in the flour beetle *Tribolium confusum* Duval. Ecology **13:**172–181.

Park, T. 1933. Studies in population physiology, II. Factors regulating initial growth of *Tribolium confusum* populations. Jour. Exper. Zoology **65:**17–42.

Peters, M., and P. Barbosa. 1977. Influence of population density on size, fecundity, and developmental rate of insects in culture. Ann. Rev. Entomol. **22:**431–450.

Rich, E. R. 1956. Egg cannibalism and fecundity in *Tribolium*. Ecology **37:**109–120.

Stevens, L. 1989. The genetics and evolution of cannibalism in flour beetles. Evolution **43:**169–179.

Young, A. 1970. Predation and abundance in populations of flour beetles. Ecology **51:**602–619.

Introduction to Animal Classification

Taxonomy involves the scientific naming of organisms and the grouping or classifying of them with reference to their exact position in the kingdoms of life. Scientific names are always latinized and are recognized internationally. This tends to prevent confusion, for whereas one animal may be called by several different common names in different geographical areas, its scientific name is the same the world over.

Hundreds of thousands of different species of animals have been classified on the basis of shared derived anatomical characters and, more recently, by biochemical procedures such as DNA hybridization. Most of these species tend to fall into certain large groups because of similarities in structural organization. These primary groups are known as **phyla** (singular, **phylum**). Members of a phylum share certain distinctive characteristics that set members of that group apart from all other members of the animal kingdom.

On the basis of differences within a phylum, it is subdivided into smaller groups called **classes;** classes are further subdivided into **orders;** orders into **families;** families into **genera** (singular, **genus**); and genera into **species** (singular and plural). In large groups other categories, such as superclass, suborder, infraorder, and subfamily, also exist.

A **species** is a distinctive kind of living thing. The species, or scientific, name is a **binomial,** that is, it consists of two parts, the genus name and the species epithet. We call this two-name system the Linnaean system of **binomial nomenclature.** The human species is *Homo sapiens; Homo* ("a man") is the genus and *sapiens* ("mighty" or "wise") is the species epithet, actually an adjective that modifies the genus name. The genus name can be used alone when one is referring to a group of species included in that genus, such as *Rana* (a large genus of frogs) or *Felis* (a genus of cats including wild and domestic species). The specific epithet, however, would be meaningless if used alone because the same epithet may be used in combination with different genera. The domestic cat is designated *Felis domestica; domestica* used alone is without significance, since it is a commonly used epithet that identifies no particular organism. Therefore, the species

---

Use of a Taxonomic Key
for Animal Identification
How to use a taxonomic key
Key to the Major Animal Taxa

---

epithet must always be preceded by the genus name. However, you can abbreviate the genus name when it is used in a context in which it is understood. *Felis domestica* might then be designated *F. domestica.*

In some cases, where geographical varieties or races of a species exist, three names may be used, in which case the last name indicates the **subspecies.** When three names are thus used, the method is called **trinomial nomenclature.** For example, one race (subspecies) of the long-tailed salamander, *Eurycea longicauda longicauda,* is classified as follows:

Phylum Chordata
  Subphylum Vertebrata
    Class Amphibia
      Order Urodela
        Family Plethodontidae
          Genus *Eurycea*
            Species *Eurycea longicauda*
              Subspecies *Eurycea longicauda longicauda*

Note that all except the species and subspecies names are capitalized; species and subspecies names begin with lowercase letters. Genus, species, and subspecies names are printed in italics or are underlined when written or typed.

Here is a brief exercise in classification that shows you how to use a taxonomic key to "run down" or "key out" the classification of an animal when neither its common nor its scientific name is known.

Use of a Taxonomic Key for Animal Identification

How does one identify an unknown specimen? One way is by direct comparison to specimens in a museum reference collection. However, few biologists have ready access to such collections. Even with such access, most nonspecialists would find this a tedious approach, involving working through thousands of museum specimens.

A practical alternative is to use a taxonomic key. A key is a convenient tabular device that enables us to identify a specimen by comparing it feature by feature with alternatives given in key couplets. Keys may be designed to identify species, genera, families, orders, or any other taxon.

In this exercise, you will identify specimens of animals (and animal-like protists) representing several different phyla and classes. Your instructor may assign this key or may substitute a different key based on forms common to your area. An alternative is the key to insect orders found in Exercise 13.

Once you have made an identification, it is important to verify its accuracy by consulting one or more references containing a text description of the distinctive characters of the species or group in question and a drawing or photograph of the species or of representatives of the group to which the species belongs.

☞ Select a specimen and then, using the appropriate key and the following instructions, identify the specimen and record the phylum or class on pp. 12 and 13. Verify the identification in one or more of the reference books provided by the instructor. Identify as many of the specimens as requested by the instructor. Return all specimens to the proper trays on the service desk.

How to Use a Taxonomic Key

A two-choice system serves as the basis of a **dichotomous key.** In the dichotomous key, two contrasting alternatives are offered at once, so you can choose the one that fits your specimen. At the end of the choice you will find a reference number to the next set of alternatives to be considered. Again make a decision and proceed in the same manner until you arrive at the scientific name of the animal or the taxon to which it belongs.

This key also has the capacity for reverse use, so that you can retrace your steps if you make a mistake. In each couplet the number in parentheses refers to the number of the couplet from which that couplet was reached.

Keep in mind that individual variations exist; keys are based on the average, or "typical," adult specimen, whereas your specimen may be immature or somewhat abnormal. It is often very helpful to examine more than one specimen of a species or group, if available, when a particular descriptive character proves troublesome.

Written Report

✍ When you have identified and verified a specimen from one of the trays, record its individual number and the taxon to which it belongs in the proper box on pp. 12 and 13. Give the reference source from which you verified the taxon. In giving references, always list the author's surname first, followed by the author's initials, the year of publication, title of book or article, and publisher. Repeat for each animal identified.

Key to the Chief Phyla and Classes of the Animal Kingdom

Here is a simple key to the more common phyla and classes of animals (and animal-like Protista). The key, for the most part, uses external characters that can be visualized without dissection. It is designed for use with adult specimens. Like most keys, this key is utilitarian in the sense that the animal groups are not arranged in perfect phylogenetic sequence, and the characters used in the key may have no particular phylogenetic significance for the taxon. They are simply the characters that provide the best assurance of correct identification.

Key to the Major Animal Taxa

The numbers in parenthesis refer back to the couplets from which these couplets were reached, making it possible to work backward if a selection is wrong, following the path of the choices made.

1a   Body with numerous pores; body radially symmetrical or irregular, with one or more large openings (oscula); no mouth or digestive tract . . . . **Phylum Porifera** (sponges)

1b   Body without numerous pores or oscula; usually symmetrical in outline; mouth and digestive tract usually present 2

2a (1b) Body with radial or biradial symmetry 3

2b   Body with bilateral symmetry 12

3a (2a) Body mostly soft and gelatinous; cylindrical, umbrella shaped, or somewhat spherical; body parts usually in divisions of four, six, or eight 4

3b   Body usually hard and spiny or with leathery skin; body parts in divisions of five; tentacles branched when present; usually with tube feet 8

4a (3a) Body cylindrical or umbrella shaped; mouth or rim of umbrella usually encircled with unbranched tentacles; nematocysts (stinging cells) present; mostly marine . . . . **Phylum Cnidaria** (jellyfish, hydroids, corals, sea anemones, and relatives) 5

4b   Symmetry biradial; one pair of tentacles (not encircling mouth) or none; eight radial rows of ciliated comb plates . . . . **Phylum Ctenophora** (comb jellies, sea walnuts)

5a (4a) Body a gelatinous medusa in bell or umbrella shape; free-swimming 6

5b   Body a cylindrical polyp, usually sessile or attached; single or colonial; tentacles surrounding the mouth 7

6a (5a) Medusa small and possessing a velum; usually four to eight radial canals . . . . **Class Hydrozoa** (hydromedusae)

6b   Medusa usually large (2 to 20 cm or more) and lacking a velum; fringed oral lobes; highly branched radial canals; scalloped margins . . . . . . **Class Scyphozoa** (true jellyfish)

7a (5b) Polyps typically small, often in branching colonies with more than one type of polyp; gastrovascular cavity not divided by septa. . . . . **Class Hydrozoa** (hydroids)

7b   Polyps typically large (>1 cm diameter); gastrovascular cavity divided by septa . . . . **Class Anthozoa** (sea anemones and corals)

8a (3b) Body without arms; body with hard endoskeleton of calcareous plates, or with soft leathery skin 9

8b   Body with branched or unbranched arms; body with hard endoskeleton of calcareous plates. 10

9a (8a) Body with rigid endoskeleton; body globular or flattened, with movable spines. . . . . **Class Echinoidea** (sea urchins, sea biscuits, sand dollars)

9b   Saclike body with leathery skin; body elongated in mouth-to-anus axis; mouth surrounded by branching tentacles. . . . . **Class Holothuroidea** (sea cucumbers)

10a (8b) Body with five movable, branched, and feathery arms; stalked or free-swimming; mouth and anus on oral surface, which is directed upward. . . . . **Class Crinoidea** (sea lilies, feather stars)

10b   Body with unbranched arms (except in some brittle stars); oral surface directed downward 11

11a (10b) Arms not sharply set off from central disc; ambulacral grooves with tube feet on ventral side of each arm. . . . . **Class Asteroidea** (sea stars)

11b   Arms sharply set off from central disc; no ambulacral grooves. . . . . . . **Class Ophiuroidea** (brittle stars)

12a (2b) No gill slits on pharynx, no internal skeleton (skull or vertebrae) of cartilage or bone 13

12b   Lateral gill slits in pharynx present, or body with internal skeleton of cartilage or bone, or both; nerve cord dorsal and single 35

13a (12a) Body slender, wormlike or leaflike, with no body segments, lateral appendages or fins, or shell 14

13b   Body not as above; if body wormlike then having appendages or segments or both 20

14a (13a) Body flat and soft, rarely cylindrical. 15

14b   Body narrowly cylindrical with hard or tough cuticle. 18

15a (14a) Body flattened dorsoventrally; no anus, no proboscis . . . . **Phylum Platyhelminthes** (flatworms). 16

15b   Body long, soft, and highly contractile; long eversible proboscis present above mouth; digestive tract with anus; mostly free-living . . . . **Phylum Nemertea** (ribbon worms)

16a (15a) Mouth and digestive tract present 17

continued

16b   No mouth or digestive tract; body of scolex and usually numerous pseudosegments (proglottids); body increasing in size posteriorly, usually long and ribbonlike . . . . **Class Cestoda** (tapeworms)

17a (16a) No attachment organs, no suckers around mouth; ciliated epidermis . . . . **Class Turbellaria** (planarians)

17b   Attachment organs present, some with hooks at posterior end; suckers present around mouth; nonciliated epidermis; body leaflike or slender in shape . . . . **Class Trematoda** (digenetic flukes)

18a (14b) Lateral lines present . . . . . . **Phylum Nematoda** (roundworms)

18b   Lateral lines absent 19

19a (18b) Body extremely slender and elongate . . . . **Phylum Nematomorpha** (horsehair worms)

19b   Anterior retractile proboscis armed with spines . . . . **Phylum Acanthocephala** (spiny-headed worms)

20a (13b) Body slender and torpedo shaped, with lateral and caudal fins; mouth with bristles or spines; planktonic . . . . **Phylum Chaetognatha** (arrow worms)

20b   Body not as above 21

21a (20b) Body soft and unsegmented; body enclosed in shell, or body with ventral muscular foot or both; some with fleshy arms or tentacles 22

21b   Body segmented (often wormlike) or with jointed appendages, or both 27

22a (21a) Shell of two valves arranged in dorsal and ventral position to each other; ventral shell usually larger than dorsal; stalk or peduncle for attachment . . . . **Phylum Brachiopoda** (lamp shells)

22b   Shell single, or of two lateral valves, or of eight dorsal plates; or shell absent or reduced and internal; ventral muscular foot or fleshy arms or tentacles present; body unsegmented . . . . **Phylum Mollusca** (molluscs) 23

23a (22b) No prehensile arms with suckers; eyes small or absent 24

23b   Head large and well developed with two large eyes and foot modified into 8 or 10 prehensile arms with suckers . . . . **Class Cephalopoda** (nautiluses, squids, cuttlefishes, and octopuses)

24a (23a) Shell of eight dorsal plates; radula present . . . . **Class Polyplacophora** (chitons)

24b   Single shell, or shell of two valves, or shell absent or internal 25

25a (24b) Shell of two lateral valves with ligamentous hinge; muscular foot present; head reduced; no tentacles or radula . . . . **Class Bivalvia** (bivalves)

25b   Shell of one piece or absent or internal 26

26a (25b) Shell tubular and open at both ends; head absent; mouth with tentacles and radula . . . . **Class Scaphopoda** (tooth shells)

26b   Shell usually coiled or spiraled (uncoiled or absent in some); head with radula, one or two pairs of tentacles and one pair of eyes . . . . **Class Gastropoda** (snails, slugs, nudibranchs, tectibranchs)

27a (21b) Body wormlike and segmented throughout; setae, parapodia, or both often present; no jointed appendages . . . . . . **Phylum Annelida** (segmented worms) 28

27b   Segmented body encased in firm exoskeleton of chitin; jointed appendages . . . . **Phylum Arthropoda** (arthropods) 30

28a (27a) Setae present on each segment; segments distinct; suckers absent 29

28b   Setae absent; no parapodia; segments indistinct and with many annuli; clitellum present; suckers present anteriorly and posteriorly or posteriorly only . . . . **Class Hirudinea** (leeches)

29a (28a) Many setae on each segment; parapodia or fleshy lateral appendages present (may be reduced); clitellum absent . . . . **Class Polychaeta** (marine segmented worms)

29b   Few setae per segment; no parapodia; clitellum present . . . . **Class Oligochaeta** (earthworms; freshwater segmented worms)

30a (27b) Paired antennae present 31

30b   Antennae absent; body of cephalothorax and . . abdomen; segmentation often obscured 34

31a (30a) Two pairs of antennae; appendages mostly biramous and specialized for different functions; many with gills; head, thorax, and abdomen present but head and at least part of thorax fused . . . . **Subphylum Crustacea** (crustaceans)

31b   One pair of antennae . . . . **Subphylum Uniramia** 32

32a (31b) Head, thorax, and abdomen distinct; three pairs of legs on thorax; one or two pairs of wings often present . . . . **Class Insecta** (insects)

32b   Body elongate, 15 or more pairs of jointed legs 33

continued

Key to the Major Animal Taxa—*continued*

33a (32b) Each segment with one pair of legs; dorsoventrally flattened. . . . . **Class Chilopoda** (centipedes)

33b Each segment with two pairs of legs; subcylindrical body . . . . **Class Diplopoda** (millipedes)

34a (30b) Four pairs of walking legs; head completely fused with thorax; no wings . . . . **Class Arachnida** (spiders, scorpions, ticks)

34b Five pairs of walking legs; lateral compound eyes present . . . . **Class Merostomata** (horseshoe crabs)

35a (12b) Short and wormlike; body divided into proboscis, collar, and trunk . . . . **Phylum Hemichordata** (acorn worms)

35b Body not wormlike 36

36a (35b) Cranium and brain absent 37

36b Cranium and brain present. 38

37a (36a) Adults saclike and sedentary; body covered with a test and with two siphons at one end . . . . **Subphylum Urochordata** (tunicates)

37b Body lance shaped; lateral musculature in conspicuous V-shaped segments; notochord and dorsal nerve cord extend length of body . . . . **Subphylum Cephalochordata** (amphioxus)

38a (36b) Body fishlike 39

38b Body not fishlike 42

39a (38a) Without true fins, scales, or paired fins. . . . . **Superclass Agnatha** (hagfishes, lampreys). 40

39b With jaws and (usually) paired appendages; notochord replaced by vertebrae . . . . **Superclass Gnathostomata** (jawed fishes, all tetrapods) 41

40a (39a) Suctorial mouth with horny teeth; seven pairs of gill pouches. . . . . **Class Cephalaspidomorphi** (lampreys)

40b Terminal mouth with four pairs of tentacles; 5 to 15 pairs of gill pouches. . . . . **Class Myxini** (hagfishes)

41a (39b) Skeleton cartilaginous; ventral mouth; placoid scales or no scales; five to seven pairs of gills, each in a separate pharyngeal cleft . . . . **Class Chondrichthyes** (sharks, rays, skates)

41b Skeleton mostly bony; body primarily fusiform but variously modified; cycloid, ganoid, or ctenoid scales; terminal mouth with many teeth . . . . **Class Actinopterygii** (ray-finned fishes)

42a (38b) Skin with horny epidermal scales, sometimes with bony plates; paired limbs, usually with five toes, or limbs absent; no gills . . . . **Class Reptilia** (snakes, turtles, lizards, crocodilians)

42b Skin without scales 43

43a (42b) Epidermis with feathers on body and scales on legs; forelimbs (wings) adapted for flying; toothless horny beak . . . . **Class Aves** (birds)

43b No feathers, wings, or beak 44

44a (43b) Skin naked, often moist, slimy, and sometimes warty . . . . **Class Amphibia** (frogs, toads, salamanders, caecilians)

44b Body covered with hair (reduced in some); integument with sweat, sebaceous, and mammary glands. . . . . **Class Mammalia** (mammals)

Name_____

Date_____

Section_____

Use of a Key for Animal Identification

In each box, print the number of a specimen and taxon (phylum or class) or, if other animal keys are used, the species (genus name and species epithet) as required by the instructor. Enter the complete reference consulted to verify the identification.

A _____ _____

Reference

B _____ _____

Reference

C _____ _____

Reference

D _____ _____

Reference

E _____ _____

Reference

F _____ _____

Reference

G _____ _____

Reference

H _____ _____

Reference

I _____ _____

Reference

J _____ _____

Reference

K _____ _____

Reference

L _____ _____

Reference

LAB REPORT

PART TWO

The Diversity of Animal Life

The Microscope

For a biologist the compound microscope is probably the most important tool ever invented. It is indispensable not only in biology but also in the fields of medicine, biochemistry, and geology; in industry; and even in crime detection and many hobbies. Yet even though the microscope is one of the most common tools in the biologist's laboratory, too frequently it is used without any effective understanding of its construction and operation. The results may be poor illumination, badly focused optics, and misleading interpretations of what is (barely) seen.

EXERCISE 3A

Compound Light Microscope
Understanding the parts and operation of the microscope
Getting acquainted with your microscope
Taking control of your microscope
Magnification in the microscope
How to measure size of microscopic objects
Exercises with the compound light microscope

EXERCISE 3B

Stereoscopic Dissecting Microscope
Exercises with the dissecting microscope

EXERCISE 3A
Compound Light Microscope

The compound light microscope may be either monocular or binocular, with either vertical or inclined oculars.

☞ Use both hands to carry a microscope. Grasp it firmly by the arm with one hand and support the **base** with the other. Carry it in a fully upright position.

Understanding the Parts and Operation of the Microscope

If you are not familiar with the parts of the microscope, please study Figures 3-1 and 3-2.

The **image-forming optics** consist of (1) a set of **objectives** screwed into a **revolving nosepiece,** and (2) a **body tube,** or head, with one or two **oculars (eyepieces).**

Each **objective** is a complex set of tiny lenses that provide most of the magnification. Your microscope may have two, three, or four objectives, each with its magnification, or power, engraved on the side. For example, if the objective magnifies an object 10 times, the magnification is said to be 10 diameters and is commonly written simply as 10×. Most microscopes include a **scanning objective** (3.5× or 4.5×), a **low-power objective** (10×), and a **high-power objective** (40×, 43×, or

45×). Some microscopes also carry an **oil-immersion objective** (95×, 97×, or 100×), which must always be used with a drop of oil to form a liquid bridge between itself and the surface of the slide being viewed.

☞ Revolve the nosepiece and note the clicking sound when an objective swings into place under the tube. Enter here the magnifications of the objectives on your microscope: _____ _____ _____ Does your microscope have an oil-immersion lens? _____ If so, what is its magnification? _____

The lenses in the **ocular** further magnify the image formed by the objective. The ocular most often used is the 10×. A 6× or a 15× ocular may also be provided. Often a pointer is mounted into the ocular. Enter here the magnification of the ocular on your microscope: _____

☞ If there is a pointer in your microscope, rotate the ocular and note the movement of the pointer. Remember that if you look through a binocular scope with only one eye you may miss the pointer. Always use both eyes.

Directly beneath the **stage** of most microscopes is a **substage condenser,** a system of enclosed lenses that concentrates the light on the specimen above. The condenser may have a knob that allows you to move the condenser up and down.

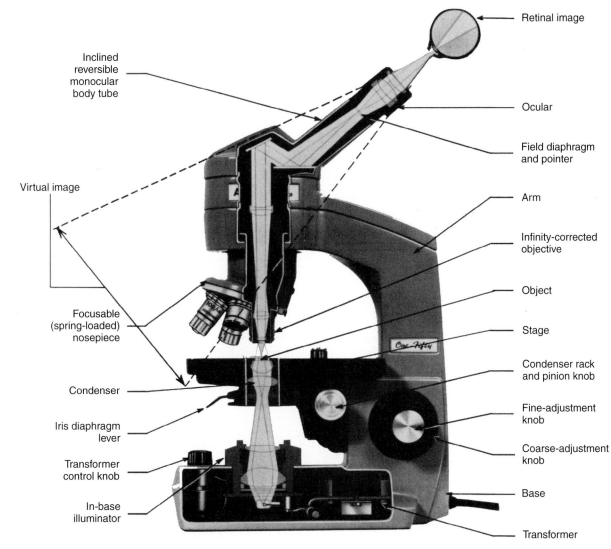

Figure 3-1

Optical and mechanical features of a compound microscope.

Beneath the condenser, many microscopes have a built-in, low-voltage **substage illuminator.** Microscopes lacking a substage illuminator employ an adjustable **reflecting mirror** that reflects natural light or light from a microscope lamp up into the optical system. The **concave surface** of the mirror is used with natural light or with a separate microscope lamp when there is no condenser on the microscope. The **plane (flat) surface** of the mirror is used with a substage condenser.

☞ Turn on the substage illuminator. If your microscope lacks one, use a microscope lamp or position your microscope and mirror to take advantage of natural light. With the low-power objective in place, adjust the mirror to bring a bright, evenly distributed circle of light through the lens.

An adjustable iris diaphragm under the stage is used to reduce glare caused by stray light from the illuminator.

☞ Raise or lower the substage condenser to near its uppermost limit. Then close down the iris diaphragm until all glare is gone (but don't close so far that dark halos appear around objects).

On a straight microscope, you raise and lower the body tube with two sets of adjustment knobs. Use the **coarse-adjustment knob** for low-power work and for initial focusing. Use the **fine-adjustment knob** for final adjustment.

On a microscope with an inclined tube you focus by *moving the stage* rather than the body tube. If you are using an inclined microscope, read "lower the stage" when the directions say "raise the objective." In either case the distance between objective lens and the object is increased.

☞ Turn the coarse-adjustment knob and note how it moves the body tube (or stage). Find out which way to turn the knob to raise and lower the tube.

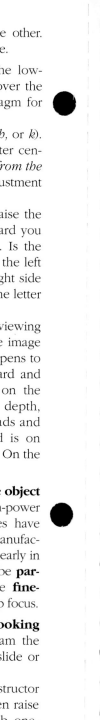

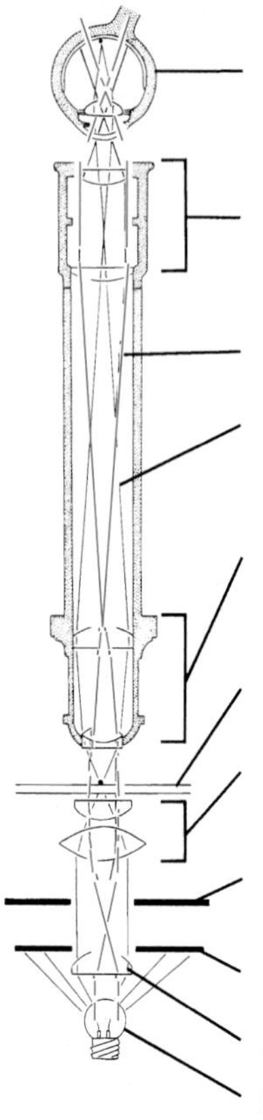
Your eye
Eyepiece (ocular) normally 10× magnification, may contain pointer to assist in locating objects in field of view.
Illuminating rays
Image-forming rays
Objective—two, three, or four objectives of different magnification are mounted in a rotating nosepiece: objective magnification times ocular magnification gives total magnification
Specimen stage
Condenser lenses concentrate light on specimen; correct position is at or just below its uppermost position
Condenser iris diaphragm controls amount of light passing through the specimen
Lamp diaphragm (also called the field stop) controls amount of light reaching condenser
Lens of lamp
Low-voltage lamp

Figure 3-2
Optical path of light through a microscope.

Never use the coarse-adjustment knob when a high-power objective is in place. Turn the fine-adjustment knob. This moves the tube so slightly that you cannot detect it unless you are examining an object through the ocular. The fine-adjustment knob works the same as the coarse-adjustment knob. To focus downward, turn the knob in the same direction as you would to focus down with the coarse adjustment. Practice this. **Always use the fine adjustment when the high-power objective is in place.**

Getting Acquainted with Your Microscope

You need not wear glasses when using the microscope unless they correct a severe astigmatism. Nearsightedness and farsightedness can be corrected by adjusting the microscope.

Clean the lenses of both ocular and objectives, when necessary, by wiping them gently with a clean sheet of lens paper. **Never touch lenses with anything except clean lens paper.**

Keep both eyes open while using a monocular microscope. If this seems difficult at first, hold a piece of paper over one eye while viewing the object with the other. Should one eye become tired, shift to the other one.

How to Focus with Low Power. Turn the low-power (10×) objective until it clicks in place over the aperture. Adjust the condenser and iris diaphragm for optimal illumination as already described.

Obtain a slide containing the letter *e* (or *a, h,* or *k*). Place it, coverslip up, on the stage with the letter centered under the objective lens. *While watching from the side,* lower the objective with the coarse adjustment until it is close to the slide surface.

Now look through the ocular and slowly raise the objective by turning the coarse adjustment toward you until the object on the slide is in sharp focus. Is the image upside down? Is it reversed; that is, does the left side of the letter appear on the right, and the right side on the left? On a separate sheet of paper, draw the letter as it appears.

Shift the slide very slightly to the right while viewing it through the ocular. In what direction does the image move? Move the slide away from you. What happens to the image? Turn the fine-adjustment knob toward and then away from you and observe the effect on the image. To gain experience with interpreting depth, obtain a slide with three different colored threads and focus with low power. Which colored thread is on top? ____________ In the middle? ____________ On the bottom? ____________

How to Focus with High Power. Focus the object first with low power; then slowly rotate the high-power objective into position. If the microscope lenses have been constructed in a particular way by the manufacturer, the object in focus with low power will be nearly in focus under high power. Such lenses are said to be **parfocal** with respect to each other. Now turn the **fine-adjustment knob** to bring the specimen into sharp focus.

Never use the coarse adjustment while looking at an object under high power; you may ram the objective into the slide. This may damage the slide or ruin the lens of the objective.

If the microscope is not parfocal (your instructor will tell you), focus first with low power and then raise the tube by turning the coarse-adjustment knob one-half turn. **With your eye at the level of the stage** carefully swing the high-power objective into place, raising the tube a little further if the objective touches the slide. Now, **still watching the high-power objective from the side,** lower it slowly to about 1 mm from the cover glass. Then, **looking through the ocular,** raise the tube with the fine adjustment until the object is in focus. Do this several times to acquire skill in focusing.

If you cannot find anything at all with high power, it may be that, because of the small size of the high-power field, the object is not in the field. Turn back to lower power, adjust the slide so that the object is in the very center of your field of view, and then return to high power. If you still cannot find the object, it may be that it is too far out of focus to be seen. Use the fine-adjustment knob. You will have to learn by experience whether to rack (turn) up or down to bring the slide into sharp focus.

Because light decreases when you switch to high power, you must adjust the light with mirror and/or iris diaphragm.

While viewing the object, keep your hand on the fine adjustment and constantly focus up and down. This enables you to see detail throughout the depth of the object.

How to Use the Oil-Immersion Objective. Occasionally a project or demonstration exercise requires using an oil-immersion objective. This is an objective in the $90\times$ to $100\times$ range. Because the resolving power of this objective is so great, you must use oil to bridge the separation between slide and objective.

To use oil immersion, bring the specimen into focus first with the low-power and then with the high-power objective. Carefully center the point of interest in the field of view; then rotate the nosepiece to move the high-power objective off to one side. Place a single drop of immersion oil on the coverslip at the point where the objective will come into position. Now move the oil-immersion objective carefully into position, **watching from the side** to be certain that the lens clears the coverslip. The oil should now form a bridge between lens and coverslip. **Carefully** adjust the fine focus to bring the specimen into focus. Adjust the iris diaphragm or substage condenser to increase light as required.

When finished, clean the lens with a lens tissue wetted with xylol and then with a lens tissue wetted with Kodak Lens Cleaner. **Never** use alcohol, which will dissolve the cement around the lens system. If the lens is to be used again soon (within a day or two), it is best not to clean the lens face. Residual oil will not harm the lens unless it is allowed to harden over a long period without use.

Taking Control of Your Microscope

1. **Proper lighting** is the first requirement for happy microscopy. Too much light is as bad as too little (beginners usually tend to use too much). Transparent objects are often clearer in reduced light. Reduce light by closing down the iris diaphragm, **not** by lowering the substage condenser.

2. **Focus with eyes relaxed.** The image appears to your eye as though it were about 10 inches away, but the eye should be relaxed as though it were viewing an image in the distance. Look up periodically and train your eyes on something across the room. Then, if you keep your eyes relaxed, you should not have to readjust your focus very much when looking through the microscope. If you get a headache, the chances are you are trying to look *into* the microscope rather than *through* it.

3. **If you are using a binocular microscope,** it is important (a) to adjust the distance between the microscope's ocular to match the distance between your own pupils, and, (b) to adjust the oculars for a sharp focus. If the focus is not sharp you may have to focus each eye separately. The microscope will have one fixed and one adjustable ocular. Adjust focus for the fixed ocular first to suit that eye, then adjust the other ocular (usually by rotation) until focus is sharp for both eyes.

4. **Find the correct eye distance** from the oculars, one that affords a full view of the field. Keep relaxed, hold your head steady, and enjoy the view.

5. **Where's the dirt?** If spots or smudges appear in the field of vision, it may be dirt on the ocular, on the slide, or on the objective. To find out which, first rotate the ocular; if the spots move, the ocular needs to be cleaned (in laboratory, avoid using eye makeup, which may smear on the ocular surface). Move the slide; if the spots move with it, clean the slide. If after cleaning ocular and slide the spots persist, it is probably a dirty objective lens. Use only special lens paper to clean lenses. The slide may be cleaned very gently with a soft damp cloth or damp cleansing tissue. If after cleaning everything you still see spots or smudges, try moving the condenser up or down a little. Still not satisfied? You may need help from the assistant. Some people may see "floaters" that drift across the field of vision while viewing a brightly illuminated object. These are defraction images of red blood cell "ghosts" in the vitreous humor of the eye. While annoying at first, one can usually learn to ignore them.

6. **Keep one hand on the fine adjustment** and constantly focus up and down. This is the only way to see everything.

7. **If the fine adjustment refuses to turn,** the knob has reached its range limit. To correct this, give the fine-adjustment knob several turns in the opposite direction, and then refocus with the coarse adjustment.

8. **Be friendly to your microscope.** It must be kept dust free, so return it **carefully** to its box or cupboard when finished with it. Before putting it away, put the low-power objective in place and raise the tube a little. Be sure not to leave a slide on the stage.

The prepared slides you will be using in the laboratory were made at the cost of great skill and patience and are quite expensive to buy. They are fragile and should be handled with great care. Your instructor may provide a demonstration of how slides are made.

Magnification in the Microscope

How much your microscope will magnify depends on the power of the combination of lenses you are using. Your microscope is probably equipped with a 10× ocular, which magnifies the object 10 times in diameter. Other oculars may magnify 2×, 5×, or 20×. The objectives may be designated, respectively, 3.5× (scanning objective), 10× (low-power objective), and 45× (high-power objective). The total magnifying power is determined by multiplying the power of the objective by the power of the ocular. Here are examples of the magnification of certain combinations:

Ocular	Objective	Magnification
5×	3.5×	17.5 diameters
5×	10×	50 diameters
10×	3.5×	35 diameters
10×	10×	100 diameters
10×	45×	450 diameters
10×	90×	900 diameters

For handy reference, enter in the table, the total magnifying power for lens combinations on *your* microscope.

How to Measure Size of Microscopic Objects

It is often important to know the size of an organism or object you are viewing through the microscope because size may be a diagnostic characteristic. Or, if you are looking for a particular species of protozoan in a mixed culture and know that the species is usually about 400 μm long, it saves time to know just how large 400 μm will appear at either low or high power. Alternative methods for measuring object sizes are described here.

Measuring Objects with Transparent Ruler Calibration. With this simple method, the viewer measures the diameter of the field of view and then estimates the proportion of the field occupied by the object. This method is not as accurate as the alternative described in the next section, but often an approximation of size is all that the viewer requires.

☞ With scanning objective in position, place a transparent ruler on the microscope stage and focus on its edge so that you can see the scale. Move the ruler right or left so that one of the vertical millimeter lines is just visible at the edge of the circular field of view. Count the number of millimeter lines spanning the field; you will probably have to estimate the last decimal fraction of a millimeter. Enter diameter in mm here _____ Now convert this figure to micrometers by multiplying by 1000. Enter this value here _____ This is the diameter in micrometers of the field for the scanning lens.

Diameters of the fields of view for your low-power and high-power objectives cannot be measured directly with the transparent ruler because of the high magnification. To calculate the field diameter of the low-power objective, multiply the diameter of the scanning lens field by the magnifying power of the scanning objective and divide by the magnifying power of the low-power objective:

$$\frac{\text{Magnification of scanning objective}}{\text{Magnification of low-power objective}}$$

$$\times \text{ Diameter of scanning objective field}$$

$$= \text{ Diameter of low-power field}$$

Make this calculation for your microscope and enter the value here _____ . It usually ranges between 1.5 and 1.6 mm (1500 and 1600 μm).

Similarly, the diameter of the high-power field is determined as:

$$\frac{\text{Magnification of the low-power objective}}{\text{Magnification of the high-power objective}}$$

$$\times \text{ Diameter of low-power field}$$

$$= \text{ Diameter of high-power field}$$

Magnifying powers for microscope number _____						
Magnification for:	**Ocular**		**Objective**		**Magnification**	
Scanning lens	_____	×	_____	=	_____	diameters
Low-power lens	_____	×	_____	=	_____	diameters
High-power lens	_____	×	_____	=	_____	diameters
Oil-immersion lens	_____	×	_____	=	_____	diameters

Make this calculation for your microscope and enter the value here _____ It usually ranges between 0.36 and 0.42 mm (360 and 420 µm). Now enter these values in the next table for quick reference.

☞ Place a hair from your head and from your eyebrow on a slide. Examine with low and high power and measure their diameters. Write the diameters in micrometers here: eyebrow _____ head _____ Are they the same? Compare diameters with others in the class.

Measuring Objects with Ocular and Stage Micrometer Calibration. An **ocular micrometer** can be fitted into the microscope's ocular. It is a disc on which is engraved a scale of (usually) either 50 or 100 units. These units are arbitrary values that always appear the same distance apart no matter which objective is used in combination with the eyepiece. Therefore, the ocular micrometer cannot be used to measure objects until it has been calibrated with a **stage micrometer.**

The stage micrometer resembles an ordinary microscope slide but bears an engraved scale on its upper surface, usually 1 or 2 mm long, divided into 0.1 and 0.01 mm divisions.

Place the stage micrometer on the microscope stage and focus on the engraved scale with the low-power objective. Both scales should appear sharply defined. Rotate the eyepiece until the two scales are parallel. Now move the stage micrometer to bring the 0 line of the stage scale in exact alignment with the 0 marking of the ocular scale. The scales should be slightly superimposed (Figure 3-3).

To calibrate the ocular scale for this objective, use the longest portion of the ocular scale that can be seen to coincide precisely with a line on the stage scale. For example, suppose that 70 units on the ocular scale equal 0.24 mm on the stage scale. Then:

$$70 \text{ ocular units} = 0.24 \text{ mm, and}$$

$$1 \text{ ocular unit} = \frac{0.24}{70} = 0.0034 \text{ mm, and}$$

$$1 \text{ mm} = 1000 \text{ µm}$$

Therefore:

$$1 \text{ ocular unit} = 0.0034 \text{ mm} = 3.4 \text{ µm}$$

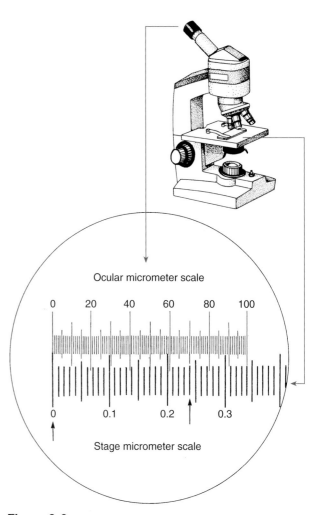

Ocular micrometer scale

Stage micrometer scale

Figure 3-3
Calibrating an ocular micrometer.

Repeat the calibration procedure for the scanning lens and the high-power lens. For handy reference, enter in the next table the micrometer values you have calculated.

To measure any object available to you in the laboratory, it is only necessary to multiply the number of divisions covered by the specimen or part thereof by the micrometer value you have determined (3.4 µm in this example). Note that the micrometer value applies only to the objective with which the calibration was made.

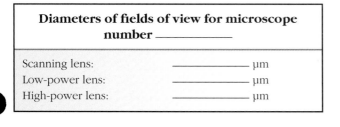

Diameters of fields of view for microscope number _____	
Scanning lens:	_____ µm
Low-power lens:	_____ µm
High-power lens:	_____ µm

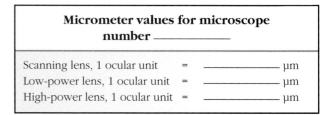

Micrometer values for microscope number _____		
Scanning lens, 1 ocular unit	=	_____ µm
Low-power lens, 1 ocular unit	=	_____ µm
High-power lens, 1 ocular unit	=	_____ µm

Exercises with the Compound Light Microscope

Try these or any similar exercises with the compound microscope.

1. Examine crystals of salt or sugar. Sketch what you see. Estimate size of the crystals.

2. Examine the letter *e* made on onionskin paper by two different typewriters and see if you can find individual differences between the two typewriters. Sketch.

3. Prepare a temporary wet mount (Figure 3-4). Select a clean slide and coverslip. Mount a piece of insect wing or appendage, a bit of feather, or some similar object on the slide. Add a drop of distilled water with a pipette. Hold the coverslip at an angle with one edge on the slide and the other held up with a dissecting needle. Slowly lower the raised edge onto the drop of water. This helps prevent air bubbles from forming. Examine with low power and sketch what you see in the field of vision. Switch to high power without moving the slide, and draw the portion now visible.

4. Make a wet mount using a drop of water from a solution in which gum arabic has been dissolved and shaken up. With low or scanning power, look for large and small air bubbles, focusing up and down on them until you are sure you will always recognize an air bubble. This may later prevent you from interpreting a bubble as a small organism or other small object. What is the characteristic outline of an air bubble?

5. Mount a drop of pond water, cover, and look for living organisms in it. As you watch movement of the swimming animals, remember that movement as well as size is magnified by the microscope. Sketch some of the organisms. On your drawing state estimated size of each animal and magnification of your drawing (on a separate paper).

EXERCISE 3B
Stereoscopic Dissecting Microscope

The stereoscopic dissecting microscope (Figure 3-5) is as indispensable to the laboratory as is the compound microscope. It enables you to study objects too large or too thick for the compound microscope. It furnishes a three-dimensional view of objects at a very low power (5× to 50×, depending on the microscope). The image is not inverted, and there is ample space for manipulation and dissection under the lens. The microscope stage can be illuminated either by **transmitted light** (light passing through the object from below) or by **reflected light** (light illuminating the object from above and being reflected by the object into the microscope). Which type of lighting is used with the compound microscope? _____ Some dissecting microscopes have a substage mirror or a substage lamp. Focusing is done with very little adjustment.

Place an object on the stage and illuminate with reflected light. Looking through the oculars, adjust them to fit the distance between your eyes so that you can

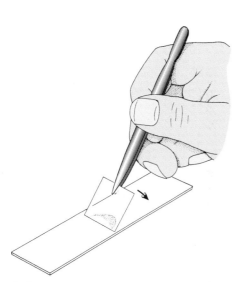

Figure 3-4
Preparation of a wet mount.

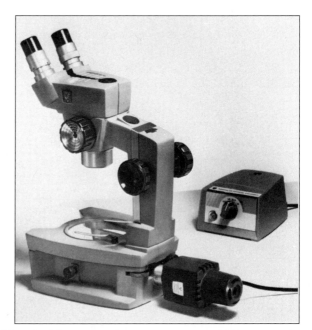

Figure 3-5
Stereoscopic dissecting microscope with illuminator.

look comfortably through both oculars at once. You should see a single field of vision. If focus is not sharp you may have to focus for each eye separately. The microscope will have one fixed and one adjustable ocular. Adjust focus for the fixed ocular first to suit that eye, then adjust the other ocular (usually by rotation) until the focus is sharp for both eyes.

Move the object on the stage away from you. Which way does the image move?

Move the object to one side. Which way does the image move? How does this compare with the compound microscope?

Exercises with the Dissecting Microscope

Try these or any similar exercises with the dissecting microscope and **keep a record** of the results of different lighting and background effects with different types of material. This will save time for you in later studies.

1. Examine some pond water. How many kinds of organisms can you see? Are they more distinct on a white or a black background?

2. Examine a prepared slide of the whole mount of a fluke or tapeworm or a similar slide. Try it with reflected lighting, moving the light source about for the best effects. Then try transmitted lighting. If the microscope lacks substage lighting, try placing the slide over a small microscope lamp with a piece of writing paper between slide and lamp. Which method gives better illumination?

3. Examine the surface of a preserved sea star or gills of a crayfish. Study the material first without water; then study it submerged in a finger bowl or glass of water. In which preparation do you see the most detail? Why? Try the gills both with top lighting and with transmitted light. Can both methods be used with the sea star? Why?

Be sure to make good use of the dissecting microscopes that are available for your use in the laboratory. You will find them invaluable.

Protozoan Groups
Amebas
Phylum Chlorophyta
Phylum Euglenozoa
Phylum Apicomplexa
Phylum Ciliophora

In the protozoa all functions of life are performed within the limits of a single plasma membrane. There are no organs or tissues, but there is division of labor within the cytoplasm, where various complex **organelles** are specialized to carry out specific tasks. Each organelle is a small aggregate of macromolecules arranged in a definite manner that fits it for its specific function. Specialized organelles function as skeletons, modes of locomotion, sensory systems, conduction mechanisms, defense mechanisms, and contractile systems.

Among the different phyla there are various levels of organization, depending on the complexity of their structural plan. Because protozoa are not divided into cells, they are said to belong to the **protoplasmic level of organization.**

Protozoa are often erroneously referred to as "simple" organisms. There are no simple organisms. Many are exceedingly complex, the most elaborately organized of all known cells. They are widespread ecologically, being found in fresh, marine, and brackish water and in moist soils. Some are free-living, others live as parasites or in some other symbiotic relationship.

Classification

The traditional phylum Protozoa contained four classes: flagellates, amebas, sporozoans (an important parasitic group including malarial organisms), and ciliates. An enormous amount of information on protozoan structure, life histories, and physiology accumulated, and the Society of Protozoologists published a new classification of protozoa in 1980, recognizing seven separate phyla, the most important of which were the Sarcomastigophora (flagellates and amebas), Apicomplexa (sporozoans and related organisms), and Ciliophora (ciliates). However, analyses of sequences of bases in genes, primarily those

---

Classification

EXERCISE 4A

Amebas
Amoeba and others
Projects and demonstrations

EXERCISE 4B

Phylum Chlorophyta—*Volvox*
Volvox

Phylum Euglenozoa—*Euglena* and *Trypanosoma*
Euglena
Trypanosoma
Projects and demonstrations

EXERCISE 4C

Phylum Apicomplexa, Class Gregarinea, Class Coccidea—*Gregarina* and *Plasmodium*
Gregarina
Plasmodium

EXERCISE 4D

Phylum Ciliophora—*Paramecium* and Other Ciliates
Paramecium
Other ciliates
Project

---

encoding the small subunit of ribosomal RNA, but also those encoding several proteins, have revolutionized our concepts of phylogenetic affinities and relationships, not only of protozoan groups, but all eukaryotes. According to some scientists, if we retain classical kingdoms such as animals (Metazoa), fungi, and plants, we must recognize no fewer than *12 additional* kingdoms.[1] In some cases, however, molecular data are available for very few (or only one) species in a group; this seems slender evidence upon which to establish a kingdom of living organisms. Thus we will generally follow the system of phyla in comprehensive recent monographs, such as Hausmann and Hülsmann (1996). Ameboid organisms (former Sarcodina) fall into numerous lineages, but for the sake of simplicity, we will discuss these together under an informal heading "Amebas."

Amebas (an informal grouping for former members of Sarcodina to simplify presentation). Amebas move by pseudopodia or locomotive protoplasmic flow without discrete pseudopodia; flagella, when present, usually restricted to development or other temporary stages; body naked or with external or internal test or skeleton; asexual reproduction by fission; sexuality, if present, associated with flagellated or, more rarely, ameboid gametes; most free living.

> **Rhizopodans** Locomotion by lobopodia, filopodia (thin pseudopodia that often branch but do not rejoin), or by protoplasmic flow without production of discrete pseudopodia. Examples: *Amoeba, Endamoeba, Difflugia, Arcella, Chlamydophrys.*
>
> **Granuloreticulosans** Locomotion by reticulopodia (thin pseudopia that branch and often rejoin [anastomose]); includes foraminiferans. Examples: *Globigerina, Vertebralima.*
>
> **Actinopodans** Locomotion by axopodia; includes radiolarians. Examples: *Actinophrys, Clathrulina.*

Phylum Euglenozoa (yu-glen-a-zo′a) (Gr. *eu-,* good, true, + *glene,* cavity, socket, + *zōon,* animal). With cortical microtubules; flagella often with paraxial rod (rodlike structure accompanying axoneme in flagellum); mitochondria with discoid cristae; mucleoli persist during mitosis.

> **Subphylum Euglenida** (yu-glen′i-da). With pellicular microtubules that stiffen pellicle.
>
> > **Class Euglenoidea** (yu-glen-oyd′e-a) (Gr. *eu-,* good, true, + *glene,* cavity, socket, + *-ōideos,* form of, type of). Two heterokont flagella (flagella with different structures) arising from apical reservoir; some species with light-sensitive stigma and chloroplasts. Example: *Euglena.*
>
> **Subphylum Kinetoplasta** (ky-neet′o-plas′ta) (Gr. *kinetos,* to move + *plastos,* molded, formed). With a unique mitochondrion containing a large disc of DNA; paraxial rod.
>
> > **Class Trypanosomatidea** (try-pan′o-som-a-tid′e-a) (Gr. *trypanon,* a borer, + *soma,* the body). One or two flagella arising from pocket; flagella typically with paraxial rod that parallels axoneme; single mitochondrion (nonfunctional in some forms) extending length of body as tube, hoop, or network of branching tubes, usually

with single conspicuous DNA-containing kinetoplast located near flagellar kinetosomes; Golgi body typically in region of flagellar pocket, not connected to kinetosomes and flagella; all parasitic. Examples: *Leishmania, Trypanosoma.*

Phylum Chlorophyta (klor-off′i-ta) (Gr. *chloros,* green, + *phyton,* plant). Unicellular and multicellular algae; photosynthetic pigments of chlorophyll *a* and *b,* reserve food is starch (characters in common with "higher" plants: bryophytes and vascular plants); all with biflagellated stages; flagella smooth and of equal length; mostly free-living photoautotrophs. Examples: *Chlamydomonas, Volvox.*

Phylum Apicomplexa (a′pi-com-plex′a) (L. *apex,* tip, + *complex,* twisted around, + *a,* suffix). Characteristic set of organelles (apical complex) at anterior end in some stages; cilia and flagella usually absent; all species parasitic.

> **Class Gregarinea** (gre-ga-ryn′e-a) (L. *gregarius,* belonging to a herd or flock). Mature gamonts (individuals that produce gametes) large, extracellular; gametes usually alike in shape and size; zygotes forming oocysts within gametocysts; parasites of digestive tract of body cavity of invertebrates, life cycle usually with one host. Examples: *Monocystis, Gregarina.*
>
> **Class Coccidea** (kok-sid′e-a) (Gr. *kokkos,* kernel, grain). Mature gamonts small, typically intracellular; life cycle typically with merogony, gametogony, and sporogony; most species in vertebrates. Examples: *Cryptosporidium, Cyclospora, Eimeria, Toxoplasma, Plasmodium, Babesia.*

Phylum Ciliophora (sil-i-of′o-ra) (L. *cilium,* eyelash, + Gr. *phora,* bearing). Cilia or ciliary organelles present in at least one stage of life cycle; usually two types of nuclei; binary fission across rows of cilia; budding and multiple fission also occur; sexuality involving conjugation, autogamy, and cytogamy; heterotrophic nutrition; mostly free-living; contractile vacuole typically present. (This is a very large group, now divided into three classes and numerous orders.) Examples: *Paramecium, Colpoda, Tetrahymena, Stentor, Blepharisma, Epidinium, Vorticella, Euplotes, Didinium.*

EXERCISE 4A
Amebas

Amoeba
Rhizopoda
> Class Lobosa
> > Species *Amoeba proteus*

Where Found

The amebas may be naked or enclosed in a shell. The naked amebas, which include the genera *Amoeba* and

---

[1] Baldauf, S. L., A. J. Roger, I. Wenk-Siefert, and W. F. Doolittle. 2000. A kingdom-level phylogeny of eukaryotes based on combined protein data. Science **290:** 972–977

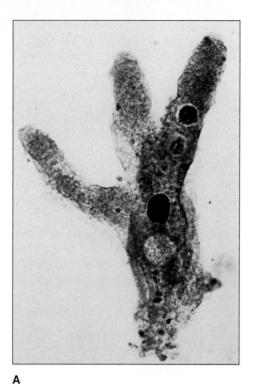

A

B

Advancing pseudopodia

Hyaline cap

Endoplasm

Ectoplasm

Pseudopodium being withdrawn

Food vacuoles

Plasmalemma

Contractile vacuole

Nucleus

Figure 4-1
Ameba. **A,** Whole mount of a living specimen. **B,** Interpretive drawing.

Pelomyxa, live in both fresh water and seawater and in the soil. They are bottom dwellers and must have a substratum on which to glide. *A. proteus*** is a freshwater species that is usually found in slow-moving or still-water ponds. They are often found on the underside of lily pads and other water plants. They feed on algae, bacteria, protozoa, rotifers, and other microscopic organisms.

Study of Live Specimens

☞ In the center of a clean slide, the instructor will place a drop of culture drawn from the bottom of the undisturbed culture bottle. Adjust the iris diaphragm of the microscope to provide *subdued light.* With *low power* explore the contents of the slide, at first *without a coverslip.*

You may see some masses of brownish or greenish plant matter, and there will probably be some small ciliates or flagellates moving about. The ameba, in contrast, is gray, rather transparent, irregularly shaped, and finely granular in appearance (Figure 4-1). If it is not apparently moving, watch it a moment to see if the granules are in motion or if the shape is slowly changing. If you do not find a specimen after several minutes of careful examination, ask the instructor to check your slide.

** The genus name (*Amoeba;* from the Greek meaning "change") may be abbreviated when used in a context in which it is understood. Recall, however, that the species is binomial (*Amoeba proteus*) and that the species epithet (*proteus*) has no taxonomic meaning by itself (see p. 7).

☞ After you have found some specimens and observed their locomotion, carefully cover with a coverslip. You will need to support the coverslip to prevent crushing the amebas as the water evaporates. Place two short lengths of hair or thread or two pieces of broken coverslip, one on each side of the drop of culture, before adding the coverslip. *Never use high power without a coverslip.* From time to time add a drop of culture water to the edge of the coverslip to replace water lost by evaporation.

Some types of substage microscope lamps will warm the stage enough to kill the ameba. Turn off the lamp when not actually viewing the specimen.

General Features

The outer cell membrane is the **plasmalemma** (*plasma,* form, + *lemma,* skin). Electron micrographs show the plasmalemma is fringed with fine, hairlike projections (too small to be seen with the light microscope), which are thought to be involved in the adhesion of the cell surface to the substratum or to nutrient particles and to aid in the capture and intake (ingestion) of food.

The cytoplasm enclosed by the plasmalemma is differentiated into a thin, peripheral rim of stiff **ectoplasm** and an inner, more fluid, **endoplasm.** The ectoplasm looks clear, even glassy, because it lacks the subcellular organelles that are present in abundance in the endoplasm (Figure 4-1B). As a result the clear ectoplasm is often referred to as a **hyaline cortex,** or "glassy rind."

Locomotion. The ameba moves and changes shape by thrusting out **pseudopodia** ("false feet"), which are extensions of the cell body. The ameba creeps forward when a protruding pseudopodium grasps a surface and a motile force pulls the rest of the cell forward. The advancing end of the pseudopodium is called the **hyaline cap.** As the endoplasm flows into the hyaline cap, it fountains out into the cortex and is converted into stiff ectoplasm, thus extending the sides of the pseudopodium much like a tube or sleeve. As the pseudopodium lengthens, the ectoplasm at the temporary "tail end" converts again into streaming endoplasm to replenish the forward flow. At any time the action can be reversed, the endoplasm streaming back, the tube shortening, and another pseudopodium forming elsewhere.

Within the pseudopod is a fibrous network of actin, one of the major proteins of the muscle of more derived organisms. Actin molecules form a rigid polymer meshwork in the clear ectoplasm that gives the pseudopod its strength and form. The alternate rapid assembly and disassembly of this actin meshwork are in some way responsible for the propulsive force of ameboid movement, although the molecular details are not yet fully understood.

Observe the formation of pseudopodia. Does the ameba have a permanent anterior and posterior end? Can you observe the change from endoplasm to ectoplasm, and vice versa? Does the ameba move steadily in one direction? Does more than one pseudopodium ever start at once? How is a pseudopodium withdrawn? What happens when the ameba meets an obstruction? Tap the coverslip gently with the tip of a pencil or probe. Does the ameba respond?

Drawings

On separate paper make a series of five or six outline sketches, each about 2 cm in diameter, to show the changes in shape that occur in an ameba in a period of 1 to 10 minutes. Sketch rapidly and use arrows freely to indicate the direction of flow of the cytoplasm. Indicate the magnification of the drawings.

Feeding. Note the **food vacuoles,** which are particles of food surrounded by water and enclosed in a membrane. Cells and unicellular animals that engulf foreign particles are called **phagocytes,** and this type of ingestion is known as **phagocytosis** (Gr. *phagein,* to eat, + *kytos,* hollow vessel) (Figure 4-2). Phagocytosis involves encircling the prey by pseudopodia to form a food cup. Subsequently the prey, along with a quantity of water, becomes completely enclosed by cytoplasm to

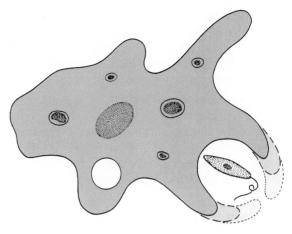

Figure 4-2
Feeding by ameba. Successive positions of the pseudopodia are shown as the food cup forms and closes around the prey organism.

form a food vacuole. The contents are digested by hydrolytic enzymes secreted into the food vacuoles by tiny membrane-bound organelles called **lysosomes.**

Undigested end products can be eliminated (egested) at any point along the plasmalemma.

Cells also take in fluid droplets and minute food particles by a process of channel formation called **pinocytosis** (Gr. *pinein,* to drink, + *kytos,* hollow vessel). A demonstration method is given on p. 29.

Osmoregulation. Look for the **contractile vacuole,** a clear bubble containing no particles. Note that this bubble gradually increases in size by accumulation of fluid and then ruptures and disappears. This organelle rids the ameba of excess water that has been taken in along with food vacuoles or acquired by osmosis. The contractile vacuole is surrounded by a network of continuous membranous channels that are populated by proton pumps. These pumps create an osmotic gradient that moves water into the vacuole. The vacuole is then pushed against the plasmalemma, where it ruptures and empties to the outside. Note that the vacuole, as it enlarges, tends to be located in the temporary posterior end of the moving ameba.

Time the appearance and disappearance of the contractile vacuole in the culture medium. Then place a drop of distilled water at the edge of the coverslip and draw the water through to the other side using a piece of absorbent paper. Repeat. The culture fluid now has been largely replaced with distilled water. After a minute or so, make another recording of vacuole discharge and compare with the first. Has the discharge rate increased?

Nucleus. Locate the **nucleus.** It is disc shaped, often indented, finely granulated, and somewhat refractive to light. You can distinguish it from the contractile vacuole because the latter is perfectly spherical, increases in size, and finally disappears. The nucleus is usually carried along in the cytoplasm and is often found near the center of the animal. As it turns over and over, it sometimes appears oval and sometimes round.

Reproduction. It is possible, although not too likely, that you might find a specimen in division. If you do, call it to the attention of your instructor. The amebas reproduce asexually by a type of mitotic cell division known as **binary fission.** The life cycles of some sarcodines are much more complex than that of *Amoeba*.

Study of a Stained Slide

☞ On a stained slide locate the structural features that you have studied in the living specimen. Note especially the **nucleus**, the **plasmalemma**, the **ectoplasm**, the **endoplasm**, the **contractile vacuole**, and the **food vacuoles.**

Written Report

✎ Answer the questions on p. 32.

Other Amebas

☞ Examine stained slides or specimens from living cultures of as many of the listed sarcodines are available in your laboratory (Figures 4-3 and 4-4).

Parasitic Amebas

A number of species of *Entamoeba* (Figure 4-3) are found in humans and other vertebrates. *E. gingivalis* lives in the mouth and feeds on bacteria around the base of the teeth. *E. histolytica* lives chiefly in the large intestine, where it causes dysentery by feeding on body tissues and red blood cells. *E. histolytica* exists in two phases, the trophozoite, or active feeding phase, and the encysted stage, in which nuclear divisions occur. The mature cyst usually contains four small nuclei. After ingestion by a suitable host, the multinucleated ameba emerges from the cyst and undergoes a series of fissions resulting in uninucleate daughter amebas.

Endamoeba is a similar genus found in cockroaches and termites.

Some Shelled Amebas

Protozoa employ many different kinds of materials for building their shells, or tests (Figure 4-4). *Arcella*, which occurs in bogs or swamps where much vegetation exists, has a hemispherical hat-shaped skeleton (up to 260 μm in diameter) made up of siliceous or chitinous material set in a base of polymerized proteins. Small finger-

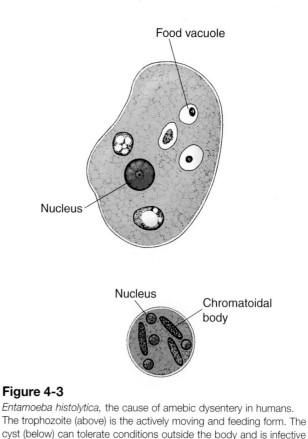

Figure 4-3
Entamoeba histolytica, the cause of amebic dysentery in humans. The trophozoite (above) is the actively moving and feeding form. The cyst (below) can tolerate conditions outside the body and is infective to the new host. Note the four small nuclei and chromatoidal bodies.

like pseudopodia extend through an opening in the flat underside of the skeleton.

Difflugia is often found in leaf-choked puddles or in delicate aquatic vegetation. This protozoan has an inverted flask-shaped skeleton (up to 500 μm long) made up of sand grains cemented together with the polymerized protein base, sometimes with the addition of diatom shells or sponge spicules.

Foraminiferans, such as *Globigerina* (Figure 4-4), are marine sarcodines that secrete a skeleton of one or more chambers, usually calcareous and sometimes incorporating silica, sand, or sponge spicules. Long, delicate, feeding pseudopodia extend through pores in the skeleton. As the foraminiferan grows, it adds new chambers to the shell. On death their skeletons join the ooze of the ocean bottom. Enormous limestone deposits were formed over millions of years by the skeletons of these animals.

Actinopod Amebas

Heliozoans are spherical, primarily freshwater sarcodines with long, slender, feeding pseudopodia radiating out in all directions in a sunburst arrangement (hence

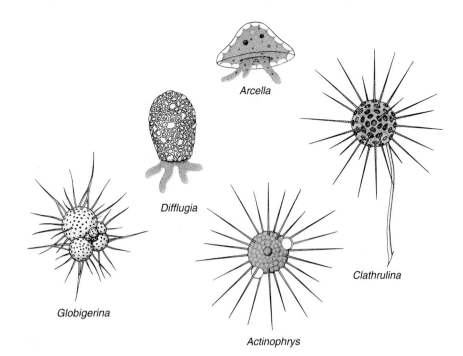

Figure 4-4

Some ameba protozoa. *Arcella* and *Difflugia* are shelled amebas with lobopodia. *Globigerina* is a foraminiferan. *Actinophrys* and *Clathrulina* are actinopod amebas with long, slender feeding pseudopodia called axopodia.

the common name, "sun animals"). In *Clathrulina* the pseudopodia project from perforations in a latticelike organic capsule (Figure 4-4). *Actinophrys,* another heliozoan, lacks a capsule, but the cytoplasmic surface has a frothy, vacuolated appearance. Heliozoans feed on other protozoa and certain algae that, on contact, become attached to the pseudopodia and then paralyzed. These forms occur in ponds, usually floating among aquatic plants.

Radiolarians are marine actinopod amebas that secrete about themselves a transparent skeleton of silica, and then thrust out slender pseudopodia through pores in the skeleton. They float in surface plankton, and when they die, their skeletons become a part of the ocean bottom ooze. Examine a prepared slide showing a variety of shell types.

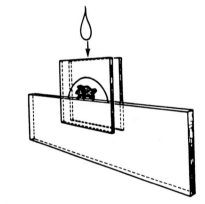

Figure 4-5

Preparation for studying a side view of ameba.

Drawings

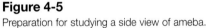 Sketch any of the above amebas or their skeletons on p. 31.

Projects and Demonstrations

1. *Viewing the locomotion of ameba from the side.* With a toothpick smear a thin line of petrolatum (Vaseline) on each side of a microscope slide, then attach coverslips with the edges extended as shown in Figure 4-5. Place a drop of culture on the edge of the slide between the coverslips. Tilt the microscope 90° and clamp the slide on the stage with the spring clips. You will need to tape small spacers such as pieces of cardboard or plastic between the slide and the microscope stage to prevent the lower coverslip from pressing against the stage. Observe the ameba as it crawls on the edge of the slide.

2. *Demonstration of pinocytosis.* Pinocytosis, or "cell drinking," may be demonstrated in *Amoeba* by placing the amebas in a solution of 0.125 M NaCl in 0.01M phosphate buffer at a pH of 6.5 to 7.0.

Channels should begin to form in 2 or 3 minutes and continue for several minutes.

3. *Demonstrating the contractile vacuole (water-expulsion vesicle).* Add a drop of 10% nigrosine to a drop of *Amoeba* culture on a slide, add a coverslip, and examine with high power.

4. *Demonstration of osmoregulation.* How would the osmoregulatory requirements of marine protozoa compare with those of freshwater forms? This might be demonstrated by placing a drop of rich *Amoeba* culture in each of four deepwell slides. Fill three of the wells with culture water that has been made up to 2%, 4%, and 6% saline (by adding 2, 4, or 6 g of salt/100 ml of water), and the fourth with plain culture water. After a few minutes place a drop from the bottom of each well on a slide and examine. Time the rate of discharge of the contractile vacuoles in specimens from each osmotic concentration. If there is a difference, can you explain it?

Phylum _____

Subphylum _____

Genus _____

Name _____

Date _____

Section _____

Location Sketches of *Amoeba*

Other Amebas

The Ameba

1. What is the average size of the specimens on your slide? _____

2. Does the ameba react to stimuli? _____ What reactions have you observed? _____
   _____
   _____

3. What is the main function of the contractile vacuole? _____

4. What was the vacuole's discharge rate with the ameba in culture medium?
   _____ Discharge rate in distilled water? _____ Explain any
   observed difference in discharge rate.
   _____
   _____
   _____

5. How does the ameba respire? _____
   _____
   _____

6. How does the ameba reproduce? _____
   _____

7. If you observed ingestion or egestion occurring, or if you could identify the contents of any of the food vacuoles, describe what you saw.
   _____
   _____
   _____
   _____
   _____

8. Distinguish between phagocytosis and pinocytosis.
   _____
   _____
   _____

9. How would you describe and explain ameboid movement? (Include answers to questions on p. 44.)
   _____
   _____
   _____
   _____
   _____
   _____
   _____

EXERCISE 4B
Phylum Chlorophyta—*Volvox*
Phylum Euglenozoa—*Euglena* and *Trypanosoma*

Phyla Euglenozoa and Chlorophyta, previously members of the Subphylum Mastigophora, are characterized by having one or more flagella—undulating, whiplike organelles that move these protozoa efficiently through their fluid environment. We recognize two major groups of flagellates. The phytoflagellates, or "plant-flagellates," are pigmented species such as *Euglena* and *Volvox* that possess chlorophyll and consequently can synthesize carbohydrates from carbon dioxide and water like any green plant. They are autotrophic (some phytoflagellates are heterotrophic as well, able to gather nutrients by pinocytosis).

The zooflagellates, or "animal-flagellates," are colorless, lack photosynthetic pigments, and obtain their nutrients by absorbing them through the plasma membrane or by engulfing prey in food vacuoles. Thus they are all heterotrophic. Many are free-living but some, such as *Trypanosoma,* parasitize humans and other animals.

Volvox

Phylum Chlorophyta
    Order Volvocida
        Family Volvocidae
            Genus *Volvox*
                Species *Volvox globator*

Where Found

Volvox (*volvere,* to roll) is a beautiful, large, spherical colonial genus often found with other protozoa in stagnant pools and ponds. It is an important component of summer plankton "blooms" that appear in nitrogen-rich ponds, pools, and ditches.

General Features

Colonies of *Volvox* may reach a diameter of 1.5 mm (most are smaller) and are easily visible to the unaided eye. These colonies are especially interesting because they represent a transition between the single-celled protozoa and the many-celled metazoans. In *Volvox* we see the beginning of **cell differentiation,** resulting in the division of labor among cells—an important step toward the metazoan body plan. They are not metazoans, however, because there is no tissue differentiation. This group also illustrates an important stage in the **development of sex.**

Study living or preserved colonies and stained slides. For living specimens, use a pipette to transfer a colony (visible to the naked eye) to a ringed or depression slide and cover with a cover glass. Study with *low power* and focus up and down to get all the details. Be especially careful if you use high power to avoid damage to slides or objective.

Each spherical colony is composed of a variable number of one-celled individuals called **zooids** (zō′oid; *zōon,* life). There may be a few hundred or many thousands of zooids in a colony, arranged on the surface of a gelatinous ball and connected to each other by fine **protoplasmic strands.** The zooids are differentiated into **somatic cells** and a smaller number of **reproductive cells.** The somatic cells are quite similar to other flagellate animals and make up most of the colony. They handle nutrition, locomotion, and response to stimuli for the entire colony. Each somatic cell contains **chloroplasts** (which give the colony its green color), a **stigma** for light sensitivity, and a pair of **flagella** for locomotion.

Locomotion. If you have living colonies, observe the locomotion. Do you notice that one end usually goes foremost? This is the anterior pole, and the anterior zooids are so highly specialized that they are unable to reproduce. Some of the colonies on your slide will contain smaller **daughter colonies** revolving about inside the gelatinous center of the mother colony (Figure 4-6).

Reproduction. *Volvox* reproduces both sexually and asexually. During spring and summer, *Volvox* reproduces asexually by repeated division of the zooids, which form miniature daughter colonies. These eventually escape by rupture of the mother colony. In the fall the asexual colonies develop **sex cells.** Some zooids enlarge to form **macrogametes** (eggs). Other zooids divide repeatedly to form packets of spindle-shaped **microgametes** (sperm). The sperm escape and fertilize the eggs, which become **zygotes.** Each zygote secretes a spiny cyst wall around itself, providing protection during winter after the mother colony dies. In spring, the zygotes break out of the cysts to give rise by cell division to new asexual colonies (Figure 4-6).*

Written Report and Drawings

Answer questions on p. 40 and sketch one of the colonies for your report.

---

*Several invertebrate groups use the strategy of employing asexual reproduction during the warm months of the year when rapid reproduction and colonizing new habitats is more important than genetic variability furnished by sexual reproduction. Sexual reproduction is then used during the unstable conditions of winter.

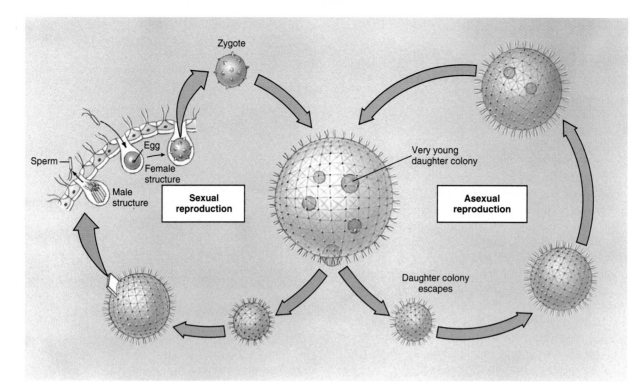

Figure 4-6
Life cycle of *Volvox,* a colonial flagellate.

Euglena

Phylum Euglenozoa
  Subphylum Euglenida
    Class Euglenoidea
      Order Euglenida
        Genus *Euglena*
          Species *Euglena gracilis* (or *Euglena viridis*)

Where Found

Euglenoids are most common in still pools and ponds where they often give a greenish color to the water. Ornamental lily ponds are excellent sources. *Euglena gracilis* and *Euglena viridis* are commonly studied species.

Study of Live Specimens

☞ Place a drop of 10% methylcellulose (or the commercial preparations Protoslo or Detain) on a clean slide, spread it thin, and add a drop of rich *Euglena* culture and a cover glass. Study with high power.

Members of *E. gracilis* range from 35 to 65 μm in length, are spindle shaped, and are greenish. The color is caused by the presence of **chloroplasts,** which contain **chlorophyll** and are scattered through the cytoplasm (Figure 4-7). A nucleus is located centrally, but is difficult to see in living specimens.

Locomotion. The blunt anterior end bears a little whiplike **flagellum** that you may be able to see with reduced light when a specimen has slowed down con-

siderably.* The flagellum emerges from the **reservoir,** a clear flask-shaped space in the anterior end. In some flagellates, such as *Peranema,* the flagellum extends forward, but in *Euglena* it is directed backward along the side of the body. Movement involves the generation of waves originating at the base of the flagellum and transmitted along its length to the tip (Figure 4-8). The flagellum beats at the rate of about 12 beats per second and not only moves the organism forward, but also rotates it and pushes it to one side, causing it to follow a corkscrew path, rotating about once every second as it goes.

Euglenoid Movement. When *Euglena* stops swimming, watch how it changes shape. These peculiar wormlike contractions are referred to as "euglenoid movements," and authorities believe that the movements are made possible by microtubules, tiny hollow fibrils about 20 nm in diameter, lying just beneath the pellicle (Figure 4-9).

Body Covering. The body is covered with a protective but flexible **pellicle** secreted by the clear **ectoplasm** that surrounds the **endoplasm.** The **stigma,** or "eyespot," is a reddish pigment spot that shades a swollen basal area of the flagellum that is thought to be light sensitive. Early microscopists, however, believed

---

* Dilute methyl violet or Noland's stain may help in making the flagella easier to see.

Name _____

Date _____

Section _____

Phylum _____

Subphylum _____

Volvox

1. Were the colonies you examined reproducing asexually or sexually? _____ Give reasons for your answer.

   _____

   _____

   _____

   _____

2. In what ways does *Volvox* show both animal-like and plantlike properties?
   _____

   _____

   _____

3. Why is *Volvox* considered a protist rather than a metazoan animal?
   _____

   _____

   _____

Sketch of *Volvox* Colony

Euglena

1. What is the average length of *Euglena* specimens in your culture? _____

2. Describe the different forms of movement and locomotion in *Euglena*. _____
   _____
   _____
   _____

3. How do *Euglena* satisfy their nutritional requirements? _____
   _____
   _____

4. What function might the stigma of *Euglena* serve? _____
   _____

Observations on *Euglena* Locomotion

Trypanosoma

1. Using the average diameter of the mammalian red blood cell (7.5 µm) as a guide, what do you estimate the length of *Trypanosoma* to be? _____

2. How must *Trypanosoma* organisms obtain their nutritional requirements? _____
   _____

EXERCISE 4C
Phylum Apicomplexa, Class Gregarinea, Class Coccidea—*Gregarina* and *Plasmodium*

Sporozoans are entirely endoparasitic. On the whole they lack special locomotor organelles, although some can move by gliding or by changes in body shape similar to euglenoid movement, and some species have flagellated gametes.

They are distributed via resistant sporelike stages. The process of **sporogony** (spōr-ahg′uh-nē; *spores,* seed, + *gonos,* progeny), or spore formation, follows the formation of the zygote in most sporozoans, so that sexual reproduction leads into the production of spores.* Many sporozoans also undergo multiple fission, or **schizogony** (ski-zahg′uh-nē; *schizein,* to split, + *gonos,* progeny), a means of spreading the population within the host and this too is followed by sporogony.

Sporozoans have no mouth, and nutrition is osmotrophic (nutritives are absorbed from the immediate environment).

Common sporozoans are the gregarines (class Gregarinea), parasitic in such invertebrates as annelids, echinoderms and ascidians; and the coccidians (class Coccidea), which are endoparasites in both invertebrates and vertebrates. Among the Coccidia are *Eimeria,* which causes coccidiosis in domestic rabbits and chickens, and *Plasmodium,* which causes malaria in humans.

Gregarina

Phylum Apicomplexa
    Class Gregarinea
        Order Eugregarinida
            Family Gregarinidae
                Genus *Gregarina*

Various species of *Gregarina* (L. *gregarius,* belonging to a herd or flock) can be found in the guts of cockroaches (such as *Blatta* and *Periplaneta*), mealworms (the larvae of *Tenebrio*), and grasshoppers. Some species are surprisingly large for a protozoan, and at one point were placed among the worms by nineteenth-century zoologists.

☞ Using a live cockroach or mealworm, cut the anus free from the body wall with scissors. Remove the

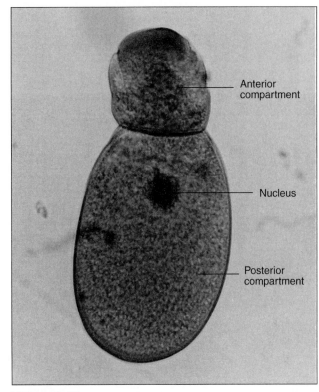

Figure 4-13
Gregarina, a parasitic sporozoan that lives in the gut of the cockroach or mealworm. The body is divided into two compartments.

complete digestive tract by holding the animal and pulling off the head and attached gut with forceps. Place the gut in a watch glass and cover with a little insect saline solution. Tease into small bits; then transfer a little of the contents in a drop of the saline to a slide for examination. Add a coverslip.

If adult feeding stages, called **trophozoites** (trōf′uh-zō′īts; *trophe,* food, + *zōon,* animal), are present, they will appear as large, distinctly shaped forms ranging from nearly round to elongate and wormlike, depending on the species (Figure 4-13). In general the body is constricted into two unequal parts: a small anterior compartment (the protomerite), which bears a holdfast device (but which usually becomes detached during transfer of the gut and its contents to the slide); and a much larger posterior compartment (the deutomerite), which contains the nucleus.

The entire life cycle takes place within the midgut of the cockroach or mealworm. Two trophozoites join together, encyst, and produce gametes, which, after fertilization, divide into **sporozoites** (spō′ruh-zō′īts; *sporos,* seed, + *zōon,* animal). These pass out with the feces. When ingested by another cockroach or mealworm, the sporozoites develop into trophozoites—the adult stage.

---

*The term "spore" is unfortunate, since protozoan spores are not homologous to the spores produced by plants, and confusing, since the term refers to a variety of propagating, infective, and resistant stages in the life cycles of many different species. In *Plasmodium,* the "spore" is the **oocyst** (Gr. *oon,* egg, + *kystis,* bladder), a cystlike stage in which numerous daughter spores, called **sporozoites,** are produced by sporogony.

Exercise 4   Protozoan Groups
       4-18
       41

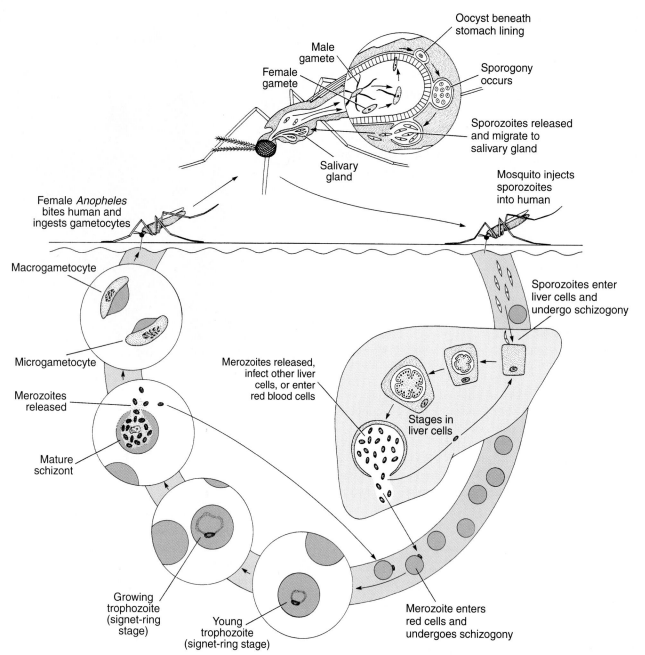

Figure 4-14

Plasmodium falciparum life cycle.

Plasmodium

Phylum Apicomplexa
  Class Coccidea
    Order Eucoccidia
      Family Plasmodiidae
        Genus *Plasmodium*

Plasmodium (*plasma,* molded, image, + *eidos,* form) is the genus of sporozoan parasites that causes malaria. *Plasmodium* is a sporozoan that requires two hosts, one a vertebrate, the other an invertebrate. It is transmitted to the human by a female *Anopheles* mosquito that has previously had a blood meal from a malaria-infected

human. The mosquito, as it feeds, also injects into the human bloodstream some of its salivary juice, which contains infective **sporozoites** (Figure 4-14).

In humans the sporozoites leave the bloodstream and penetrate into liver cells, where they undergo **schizogony** (asexual cleavage multiplication) and produce **merozoites** (mer-uh-zō′its; *meros,* part, + *zōon,* animal). These may either infect other liver cells or may be released into the blood to enter red blood cells (erythrocytes). In either case, they undergo further multiplication.

In the red blood cells they become **trophozoites** (adult stage). The cytoplasm of the trophozoite

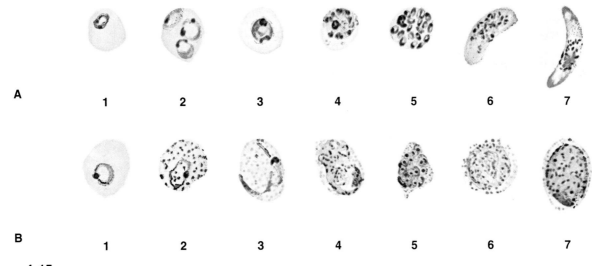

Figure 4-15

Stages in the life cycle of the malaria parasite. **A,** *Plasmodium falciparum.* 1, Young trophozoite in signet-ring form. 2, Three medium trophozoites in one red blood cell. 3, Mature trophozoite showing clumped pigment. 4, Developing schizont containing several merozoites. 5, Mature schizont filled with merozoites. 6, Mature crescent-shaped microgametocyte (length about 1.5 × diameter of red blood cell). 7, Mature macrogametocyte (about same size as microgametocyte). **B,** *Plasmodium vivax.* 1, Early ring-form trophozoite. 2, Young trophozoite with heavy chromatin dots. 3, Developing trophozoite. 4, Mature trophozoite with large mass of chromatin. 5, Schizont containing merozoites. 6, Microgametocyte almost filling red blood cell. 7, Macrogametocyte, similar to microgametocyte but with cytoplasm staining darker blue.

becomes nucleated and takes on a ringlike appearance called the "signet-ring stage" (Figure 4-14).

☞ Examine a slide containing a stained blood smear with various infective stages. Find a red blood cell containing a signet-ring stage. Slides stained with blood-differentiating dyes usually show the cytoplasm of the trophozoite blue and the nucleus reddish pink.

The signet-ring trophozoite now develops into a **schizont** (Gr. *schizō,* cleave, + *ontos,* a being) (Figure 4-14). The nucleus divides repeatedly, producing 8 to 32 merozoites. The host red blood cell ruptures, releasing the merozoites as well as metabolic wastes that are responsible in large part for the characteristic symptoms of malaria. Many merozoites enter and infect more red blood cells, with the cycle repeating until an enormous number of host cells have become parasitized.

After several such asexual generations, some of the merozoites enter red blood cells to become sexual forms called **macrogametocytes** and **microgametocytes** (Figures 4-14 and 4-15). These may be picked up by a feeding *Anopheles* female. The gametocytes give rise within the mosquito to gametes, which unite to become zygotes. The zygotes divide to produce sporozoites, which migrate to the salivary glands from which they may be injected along with the saliva into the blood of a human, perhaps to cause another malarial infection.

EXERCISE 4D

Phylum Ciliophora— *Paramecium* and Other Ciliates

Paramecium

Phylum Ciliophora
    Class Oligohymenophora
        Genus *Paramecium*

Where Found

Paramecium is an active ciliate protozoan common in most fresh water that contains vegetation and decayed organic matter. Pond scum and even cesspools are also good sources. Some of the most commonly studied forms are *P. multimicronucleatum, P. caudatum, P. aurelia,* and *P. bursaria.*

Study of Live Specimens

Locomotion and Behavior

☞ Spread out a *very few* fibers of absorbent cotton on a slide and add a drop or two of *Paramecium* culture. Cover with a coverslip. Examine first with the scanning lens to locate the paramecia, then switch to low power for study. Add water as necessary to prevent drying out.

A paramecium is slipper shaped, rather transparent and colorless, and very active. It can swim at the rate of 1 to 3 mm per second. Watch its swimming habits. Does

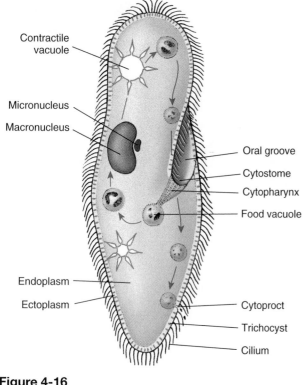

Contractile vacuole

Micronucleus

Macronucleus

Oral groove

Cytostome

Cytopharynx

Food vacuole

Endoplasm

Ectoplasm

Cytoproct

Trichocyst

Cilium

Figure 4-16
Paramecium.

it swim in a straight line, a circle, or zigzag? Does it keep one side uppermost or revolve on an axis? Does it have a definite anterior end? Can it reverse its direction? Does it seem to be contractile as *Euglena* is? Why is the term *spiral movements* used to describe its swimming habits?

What does a paramecium do when it encounters a barrier? How does it go about finding an opening? Is its body flexible enough to bend or to squeeze through tight places?

Written Report

 Describe the locomotion of *Paramecium.* Use the questions in this exercise to guide your description, answering as many as you can.

Describe the attempts of *Paramecium* to avoid or to pass under or around a cotton fiber barrier. Use diagrams and arrows if you wish.

General Structure and Function

 Locate a paramecium that has been stopped by a cotton fiber, and study its structure with both low and high power.

Note the **oral groove** that extends obliquely from the anterior end to about the middle of the body. At the posterior end of the groove is the **mouth (cytostome).** The oral groove extends from the mouth into the body as a little canal, the **gullet (cytopharynx).** The groove

and gullet are lined with strong **cilia** that are used in drawing in food (Figure 4-16).

Pellicle. The cytoplasm is made up of two zones, the clear outer **ectoplasm** and the inner **endoplasm.** Are there any granules in the endoplasm? Outside the ectoplasm is a complex living **pellicle.** A delicate **plasma membrane** lies just underneath the pellicle. The pellicle and plasma membrane are more easily seen in stained preparations.

Osmoregulation. A contractile vacuole (water-expulsion vesicle) is usually located in each end of the body. *P. multimicronucleatum* may have more than two.

 Observe the pulsations of the vacuoles caused by their alternate filling and emptying. After one empties, note the starlike **radiating canals** that appear. These radiating, or nephridial, canals collect liquid from a network of minute tubules and empty into the vacuoles, which rupture when filled, thus expelling the fluid to the outside.

Do the two vacuoles empty at the same time or alternatively? _____ How much time is there between vacuole discharges? _____ (seconds). In normal pond water a vacuole may discharge once every 6 to 10 seconds. How does this animal's rate compare (faster, slower, about the same)? _____

The pulsation period depends on the temperature and osmotic pressure of the medium. As the temperature increases, the pulsation rate increases. The pulsation rate can be slowed by adding drops of 0.25% solution of sodium chloride to the slide. Knowing this, how do you think the rate in marine ciliates would compare with that of freshwater ciliates? Why?

Written Report

Report the results of your observations with contractile vacuoles, answering the questions in this exercise.

Ciliary Action. The cilia perform very much like the oars of a boat; that is, they have an effective stroke that propels the animal forward and a recovery stroke that offers little resistance (Figure 4-17). For a discussion of ciliary action see your text.

Find a quiet specimen, cut down the light, and look closely with high power at the margin of a paramecium. Study the action of the **cilia.**

Nuclei. The nuclei may be difficult to see in living paramecia.

Add a drop of acidified methyl green to the slide at the edge of the coverslip so that the stain is drawn under by capillary action.

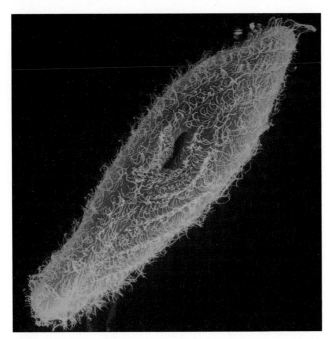

Figure 4-17

Scanning electron micrograph of *Paramecium* showing how the cilia beat in waves. The cytostome is visible at the center, ×400.

This will kill the organisms but should give the cytoplasm a bluish tinge and stain the nuclei green. Note the presence of two nuclei, a large macronucleus that regulates the metabolism of the cell (controlling cellular functions such as feeding and digestion), and a small micronucleus that is reproductive and contains the animal's genome. The micronucleus is sometimes obscured by the macronucleus.

Trichocysts. If you reduce your light properly, you can, with very careful focusing, observe the many small spindle-shaped **trichocysts** lying in the ectoplasm just under the pellicle and perpendicular to the surface. Under certain kinds of stimulation the trichocysts explode, each one releasing a liquid that hardens in water to form a long slender threadlike filament. The tangle of filaments is believed to have some protective function. In some, discharged trichocysts may be used to anchor the animal while feeding. Trichocysts are absent in some ciliates.

Perhaps the acidified methyl green you just used to demonstrate the nuclei has caused the explosion of trichocysts on your slide. Alternatively, they may be made to discharge in live paramecia by adding a drop of dilute picric acid to the slide at the side of the coverslip and drawing the acid through with filter paper touched to the opposite side of the coverslip. As the fluid reaches the paramecia, you may be able to observe the discharged trichocysts by use of subdued light and high-power lens. How does the length of the trichocysts compare with the length of cilia?

Written Report

Report the results of your observations, including a sketch of a paramecium with discharged trichocysts.

Feeding. Nutrition in ciliates is holozoic. The paramecium is a particulate feeder; that is, it feeds on small particles such as bacteria, which it moves toward the cytostome by the action of cilia in the oral groove.

On a clean slide, spread out a few cotton fibers and add a drop or two of *Paramecium* culture. Dip a toothpick into a preparation of yeast stained with Congo red, and transfer a *very small* quantity to the culture on the slide. Mix gently with the toothpick. The mixture should be light pink. If too much yeast is added, the protozoa will be obscured. Carefully apply a coverslip.

Do not allow the culture to dry out during your observation.

Find a specimen trapped by the cotton fibers. Note the currents created by the cilia in its **oral groove.** Watch the passage of the yeast particles into the groove and through the **cytostome,** or cell mouth, into a passageway called the **cytopharynx.** Watch the formation of a **food vacuole,** which is a membranous sac containing water and suspended food particles. When it reaches a certain size, the vacuole breaks away and another forms. In what direction are the food vacuoles carried by streaming endoplasm (cyclosis)? _____

Follow the course of a food vacuole. Does it vary any in size during its trip? _____ If so, how? _____ Congo red, which is red in weak acid to alkaline solutions (pH 5.0 or above), turns blue in stronger acid solutions (pH 3.0 or below). Can you observe any changes in color in the vacuoles that might indicate a change in the condition of the vacuole contents? _____ Is there any subsequent color change as the vacuoles near the anal pore? _____ How might this be explained? _____ Are the vacuoles any smaller as they complete their circuit than when first formed? _____

An **anal pore (cytoproct)** is found between the mouth and posterior end of the body. It is a temporary opening where indigestible food is discharged, and it is seen only at that time. Have you noticed such a discharge of material in one of your specimens?

Response to Stimuli. The following simple experiments are designed to determine the type of response (**taxis**) paramecia make to selected types of stimuli. A **taxis** (pl. **taxes**) is a directed reaction and orientation of the body to a specific stimulus. A movement toward the source is a positive taxis; a movement away from the source is a negative taxis. In these experiments the responses are **chemotactic.** Perhaps you can design and perform other experiments to determine responses to other types of stimuli, such as **phototaxis** (response to light rays), **thigmotaxis** (to contact), or **geotaxis** (to gravity).

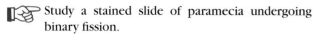 1. Place a drop of culture on a clean slide with no coverslip. Use a hand lens, scanning lens, or dissecting microscope. (Some of you will be able to see paramecia with the naked eye.) Place a drop of weak acetic acid on the slide **near** the culture but **not touching it.** Locate the specimens; while observing them, draw a line with the point of a pin or toothpick from the acid to the culture.

Describe the reaction of the animals to the approach of the acid. Are they positively or negatively chemotactic to weak acid? _____ Do they move to the area where the acid is strongest, weakest, or in between? _____

2. Place a drop of culture on a clean slide. Place a few grains of salt on the slide near the culture but not touching it. While observing the animals, draw a grain or two of salt into the side of the culture drop.

Describe the reaction of the animals. Do they choose the area nearest the salt, farthest from the salt, or in between?

Study of Stained Slides

With both low and high power, study a paramecium on a stained slide.

Focus up and down on the body and note that its entire surface is covered with **cilia.** Examine the **pellicle.** Specially prepared slides will show the peculiar pattern of hexagonal areas. Look especially for features difficult to see in the living unstained specimens, such as the **macronucleus,** one or more **micronuclei, trichocysts, oral groove, cytostome, cytopharynx,** and **contractile vacuoles.**

Binary Fission.

Study a stained slide of paramecia undergoing binary fission.

Note the constriction across the middle of the body. What is happening to the macronucleus and micronucleus? At the end of the process the halves produced by the constriction will be separate daughter animals. How does this fission process compare to that of *Trypanosoma* (Exercise 4B)?

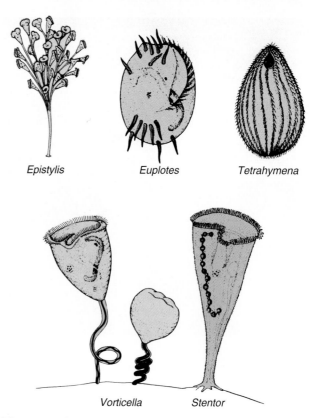

Epistylis   *Euplotes*   *Tetrahymena*

Vorticella   *Stentor*

Figure 4-18

Some representative ciliates. *Euplotes* has stiff cirri used for crawling about. Contractile myonemes in ectoplasm of *Stentor* and in stalks of *Vorticella* allow great expansion and contraction. Note the macronuclei, long and curved in *Euplotes* and *Vorticella*, shaped like a string of beads in *Stentor*.

Conjugation.

Study also a stained slide of conjugation. Look for paired individuals lying *with oral grooves attached.*

In this position they exchange micronuclear material. Consult your textbook for details of the process of conjugation.

Drawing

 Sketch a paramecium in the process of binary fission and a pair of paramecia in the process of conjugation. If you can find a live specimen dividing, watch it over a period of time and make a series of sketches to illustrate the process.

Other Ciliates

Vorticella

Vorticella (Figures 4-18 and 4-19) is a solitary sessile ciliate. It clings to aquatic vegetation of stagnant ponds and streams. The blanket algae of ponds and small lakes is a favorite place for this animal.

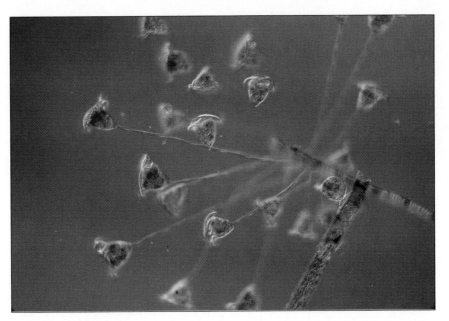

Figure 4-19
Group of living *Vorticella,* a solitary sessile ciliate.

☞ Study both living specimens and the stained slides.

What is the color of the living animal? The vorticellid is attached by a long slender **stalk** that can contract into a spiral spring shape when it is disturbed. The **body** is bell shaped with a flaring rim, the **peristome,** at its distal end. Within the peristome is a circular **oral disc.** Note the **cilia** on the edges of the peristome and the oral disc. Note the beating of the cilia. What is their function? The **cytostome** (mouth) is found between the peristome and the oral disc. From the cytostome a short tube, the **cytopharynx,** leads into the interior. Food particles are swept into the cytopharynx by the action of the cilia, and **food vacuoles** are formed as they are in other protozoa. Note that the **nucleus** is made up of an elongated U-shaped body, the **macronucleus,** and a much smaller **micronucleus.** Does the animal have a **contractile vacuole?** *Vorticella* has a surrounding **pellicle,** which helps maintain the shape of the body. Reproduction is mostly by longitudinal binary fission, but **budding** also occurs.

☞ Sketch a specimen in your notebook.

Epistylis (Figure 4-18) is a stalked ciliate similar to *Vorticella* but is colonial (*Vorticella* is solitary) and has non-contractile stalks.

Stentor

☞ Spread methylcellulose thinly on the slide before adding *Stentor* culture and coverslip.

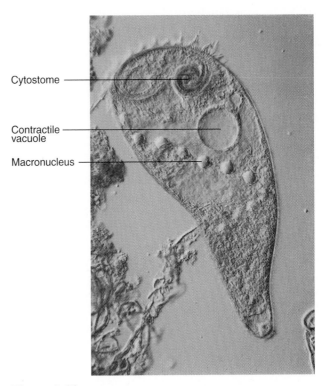

Cytostome

Contractile vacuole

Macronucleus

Figure 4-20
Living *Stentor* specimen.

Stentor (Figures 4-18 and 4-20) is a large ciliate with many of the same characteristics as *Paramecium* and *Vorticella.* How long is it when it is expanded? _____ When it is contracted? _____ How would you describe its shape? See how many of

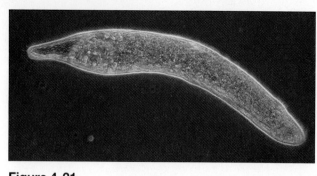

Figure 4-21

Spirostomum, an unusually large ciliate.

the structures of this ciliate you can identify from your knowledge of paramecia. Can you locate the cytostome, cytopharynx, and contractile vacuole? Is there an oral groove? Can you see food vacuoles? Are the cilia uniform in length? Observe its striped appearance caused by longitudinal bands of pigmentation. The blue pigment stentorin causes the blue-green color of *S. coeruleus.* Note the large **macronucleus,** stretched out like a string of beads. Small dots nearby are the micronuclei.

☞ In your notebook draw and label an extended specimen. Sketch its shape when the animal is swimming. Estimate its size.

Spirostomum

Spirostomum (Figure 4-21) is one of the largest common freshwater protozoa. Estimate its length. Different species range from 50 μm to 3 mm long. Can you see why it might be mistaken for a worm? Locate the large

contractile vacuole posteriorly. It is fed by a long canal. The macronucleus is long and beadlike. Locate the cytostome and long oral groove. Focus on the surface and see the arrangement of the cilia and the trichocysts. Describe its swimming or other movements. Do you think it has myonemes?

☞ In your notebook draw and label an extended specimen. Outline a contracted one.

Project

Microaquariums

A fascinating study of microorganisms and their relations to each other may be carried out in the laboratory at no expense and with only a few minutes of time at each laboratory period. Use a series of small clean empty jars such as have held baby food, jelly, or mayonnaise. Fill each jar two-thirds full with water from the tap, pond, ditch, or any other source. Add to each a teaspoonful of some source material. This material could be rich soil, plants (dry grass, leaves, hay, water plants, moss, or rotting leaf mold), pond scum, sludge from a sewage plant, or any other source that occurs to you. Label each jar. To retard evaporation, cover with a piece of glass, a plastic wrap, or the jar lid placed on loosely. Examine weekly, or more often if you like, for several weeks or months. Each jar becomes a community that provides an interesting variety and an ever-changing cycle of life. In addition to protozoa, you may find a variety of crustaceans, flatworms, rotifers, gastropods, annelids, hydras, algae, and diatoms. Keep a weekly record of what is found in each jar.

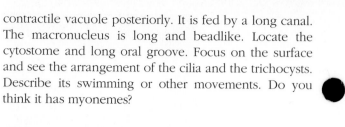

The Sponges
Phylum Porifera

The members of the phylum Porifera are among the simplest forms of metazoans (multicellular animals). Because sponges are little more than loose aggregations of cells, with little or no tissue organization, they are said to belong to the **cellular level of organization.** Thus they are more complex in organization than the single-celled protozoa, which are at the protoplasmic level of organization. There is division of labor among the cells, but there are no organs, no systems, no mouth or digestive tract, and only very primitive nervous integration. There are no germ layers, therefore sponges are neither diploblastic nor triploblastic. Adult sponges are all sessile in form. Some have no regular form or symmetry; others have a characteristic shape and radial symmetry. They may be either solitary or colonial.

Chief characteristics of sponges are their **pores** and **canal systems,** the flagellate **choanocytes,** which line their cavities and create currents of water, and their peculiar internal skeletons of **spicules** or organic fibers **(spongin).** They also have some form of internal cavity **(spongocoel)** that opens to the outside by an **osculum.** Most sponges are marine, but there are a few freshwater species. The freshwater forms are found in small slimy masses attached to sticks, leaves, or other objects in quiet ponds and streams.

Classification

Phylum Porifera

> **Class Calcarea** Cal-ca're-a (Gr. *calcis,* limy). Sponges with spicules of calcium carbonate, needle shaped or three-rayed or four-rayed; canal systems asconoid, syconoid, or leuconoid; all marine. Examples: *Sycon, Leucosolenia.*
>
> **Class Hexactinellida** (hex-ak-tin-el'i-da) (Gr. *hex,* six, + *aktis,* ray). Sponges with three-dimensional, six-rayed siliceous spicules; spicules often united to form network; body often cylindrical or funnel shaped; canal systems syconoid or leuconoid; all marine, mostly deep water. Examples: *Euplectella* (Venus's flower basket), *Hyalonema.*
>
> **Class Demospongiae** (de-mo-spun'je-e) (Gr. *demos,* people, + *spongos,* sponge). Sponges with

siliceous spicules (not six-rayed), or spongin, or both; canal systems leuconoid; one family freshwater, all others marine. Examples: *Spongilla* (freshwater sponge), *Spongia* (commercial bath sponge), *Cliona* (a boring sponge).

<table>
<tr><td>

Classification
Phylum Porifera

EXERCISE 5A

Class Calcarea—*Sycon*
Sycon, a syconoid sponge
Other types of sponge structure
Projects and demonstrations

</td></tr>
</table>

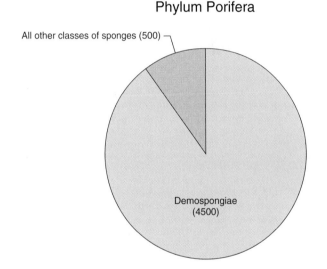

Phylum Porifera

All other classes of sponges (500)

Demospongiae (4500)

EXERCISE 5A
Class Calcarea—*Sycon*

Sycon, a Syconoid Sponge
Phylum Porifera
 Class Calcarea
  Order Heterocoela
   Genus *Sycon* (= *Scypha, Grantia*)

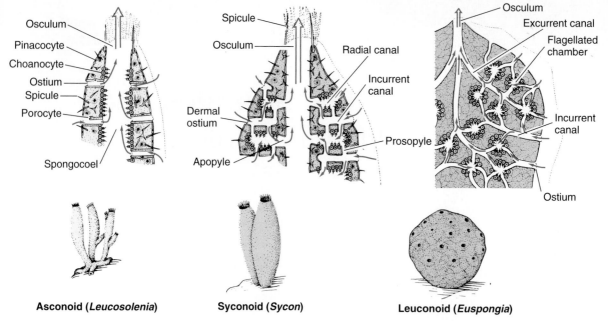

Figure 5-1

Three types of sponge structures. The degree of complexity from asconoid to complex leuconoid type involves mainly the water canal and skeletal systems, accompanied by outfolding and branching of the collar cell layer. The leuconoid type is considered the major body plan for sponges because it permits greater size and more efficient water circulation.

Where Found

Sycon is strictly a marine form, living in clusters in shallow water, usually attached to rocks, pilings, or shells. *Sycon* is chiefly a North Atlantic form. *Rhabdodermella* is a somewhat similar Pacific intertidal form, also belonging to Class Calcarea.

Gross Structure

👉 Place a preserved specimen in a watch glass and cover with water. Examine with a hand lens or a dissecting microscope.

Sycon (from the Greek meaning "like a fig") is a **syconoid** type of sponge (Figure 5-1). What is the shape of the sponge? _____ The body wall is made up of a system of tiny, interconnected dead-end canals whose flagellated cells draw in water from the outside through minute pores, take from it the necessary food particles and oxygen, and then empty it into a large central cavity for exit to the outside. What is the name of this central cavity? _____ All sponges have some variation of this general theme of canals and pores on which they depend for a constant flow of water.

External Structure. Is the base of the sponge open or closed? _____ The opening at the other end is the **osculum** (L., a little mouth), surrounded by a fringe of stiff, rodlike **spicules.** The external surface appears bristly when examined under magnification. Why? _____

Note that the body wall seems to be made up of innumerable fingerlike processes pointing outward (Figure 5-2). Inside each of these processes is a **radial canal,** which is closed at the outer end but which opens into a central cavity, the **spongocoel** (Gr. *spongos,* sponge, + *koilos,* hollow). The external spaces between these enclosed canals are **incurrent canals,** which open to the outside but end blindly at the inner end. The outside openings, or pores, are called **dermal ostia** (L., a door).

Water enters the incurrent canals and passes through minute openings called **prosopyles** (Gr. *prosō,* forward, + *pylē,* gate) into the radial canals and then to the spongocoel and out through the osculum. There is no mouth, anus, or digestive system. What kind of symmetry does this sponge have? _____

Spongocoel. To study the spongocoel, do the following:

👉 Make a longitudinal cut through the midline of the body from osculum to base with a sharp razor blade. Place the two halves in a watch glass, and cover with water.

Find the small pores, called **apopyles** (Gr. *apo,* away from, + *pylē,* gate), that open from the radial canals into the spongocoel (Figures 5-2 and 5-3). Can you distinguish the tiny canals in the cut edge of the sponge wall? _____ Which direction does the water move through these canals? _____

Study of Prepared Slide

Transverse sections of sponge are difficult to prepare for slides because the spicules prevent cutting sections thin

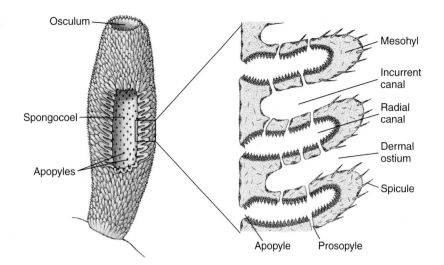

Figure 5-2

Structure of the syconoid sponge *Sycon.* The cutaway section shows the interior cavity, the spongocoel, with apopyles that lead into it from radial canals.

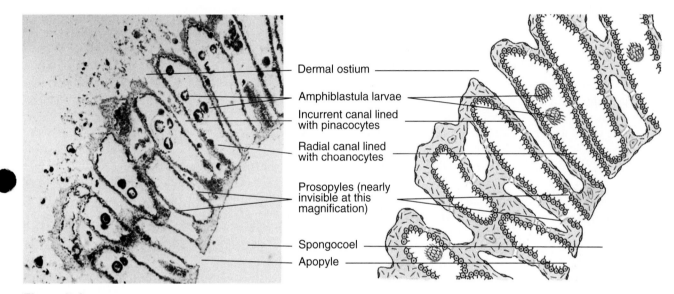

Figure 5-3

Section through wall of *Sycon.*

enough for studying the cells. Therefore the spicules have been dissolved away for slide preparation.

☞ On a prepared slide of a cross section of *Sycon,* examine the entire section with low power to get an idea of its general relations.

Note the **spongocoel** in the middle of the section (Figure 5-3). Study the canal system. Find the **radial canals,** which open into the spongocoel by way of the **apopyles.** Are apopyle openings smaller or larger in diameter than the radial canals? Some apopyle openings will be lacking in this section, and some of the radial canals will appear closed at the inner end. Follow the radial canals outward. Do they open to the outside or end blindly? The radial canals may contain young larvae, called **amphiblastula larvae.** Identify the **incurrent**

canals, which open to the exterior by the **dermal ostia.** Follow these canals inward and note that they also end blindly. Water passes from the incurrent canals into the radial canals through a number of tiny pores, or **prosopyles,** which will not be evident on the slides.

Cellular Structure

Sponge cells are loosely arranged in a gelatinous matrix called **mesohyl** (also called mesoglea or mesenchyme). The mesohyl (Gr. *mesos,* middle, + *hyle,* wood) holds together the various types of ameboid cells, skeletal elements, and fibrils that make up the sponge body.

Choanocytes.

☞ With high power, observe the "collar cells," or choanocytes (Gr. *choanē,* funnel, + *kytos,* hollow vessel) that line the radial canals (Figure 5-4).

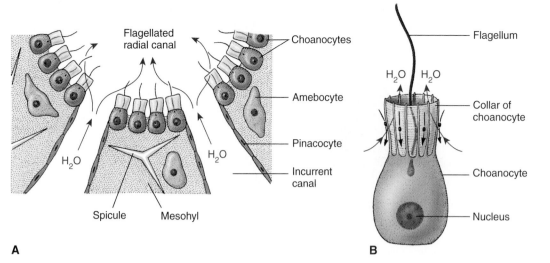

Figure 5-4
A, Small section through sponge wall showing types of sponge cells and the water current created by choanocytes in flagellated chambers.
B, Choanocyte as a food-catching cell. The "collar" is a series of cytoplasmic extensions that screen out larger food particles, letting them fall to the side of the cell for ameboid ingestion. Smaller particles flow through and are carried away in the current.

Although they are flagellated, you probably will not see the flagella. What is the function of the choanocytes? _____

Pinacocytes. Dermal amebocytes, called pinacocytes (Gr. *pinax,* tablet, + *kytos,* hollow vessel), may be seen as extremely thin (squamous) cells lining the incurrent canals and spongocoel and covering the outer surface (Figure 5-4). What is their function?

_____

Amebocytes. In the jellylike **mesohyl** that lies in the wall between the pinacocytes and choanocytes, look for large wandering amebocytes of various functions (Figure 5-4). Some are spicule-forming cells, some form sex cells, some secrete spongin or spicules, serve as contractile cells, or aid in digestion.

☞ If living sponges are available, tease a bit of tissue on a slide with a drop of seawater, and look for the various types of cells.

Reproduction

Sexual. *Sycon* sponges are monoecious, producing eggs and sperm in the mesohyl. The fertilized eggs undergo early cleavage stages in the mesohyl. The little blastula-like ciliated embryos called **amphiblastula larvae** break through into the radial canals (See Figure 5-3) and finally leave the parent by way of the osculum. They soon settle down on a substratum and grow into sessile adults.

☞ Look for the embryos in the radial canals of the cross-section slide.

What is the advantage to a sessile animal of producing free-swimming larvae? _____

Not all sponges have amphiblastula larvae. In most Demospongiae and some of the calcareous sponges, the zygote develops into a **parenchymula** (pair-en-ki′mu-la) **larva** in which the flagellated cells invaginate to form a solid internal mass.

Asexual. Many sponges, including those of the genus Sycon, also reproduce asexually by budding off new individuals from their base, thus forming sessile clusters. What would be the disadvantage if this were the sole means of reproduction? _____ Is there a bud on your specimen? _____

The freshwater sponges and some marine Demospongiae reproduce asexually by means of **gemmules,** made up of clusters of amebocytes. Gemmules (L. *gemma,* bud, + *ula,* dim.) of the freshwater sponges are enclosed in hard shells (Figure 5-5) and can withstand adverse conditions that would kill the adult sponge. In the spring they develop into young sponges. Marine gemmules give rise to flagellated larvae.

Skeleton

☞ Place a small bit of the sponge on a clean microscope slide and add a drop of commercial chlorine bleach such as Clorox (sodium hypochlorite) and set aside for a few minutes to allow the cellular matter to dissolve. Break up the piece with dissecting needles, if necessary. Add a coverslip and examine under the microscope.

Look for **short monaxons** (short and pointed at both ends), **long monaxons** (long and pointed) (Figure 5-6), **triradiates,** (Y shaped with three prongs), and **polyaxons** (T shaped). These spicules of crystalline calcium carbonate ($CaCO_3$) form a sort of network in the walls of the animal (See Figure 5-4). What is

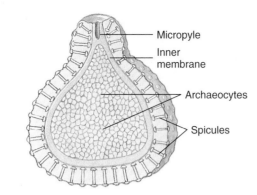

Figure 5-5
Gemmule of freshwater sponge.

Micropyle

Inner membrane

Archaeocytes

Spicules

Figure 5-6
Skeletal spicules, mostly monaxons, from the cut and dehydrated surface of *Sycon,* SEM about ×480.

the advantage of spicules to a loosely constructed animal such as *Sycon?* _____

Spicule types are used in the classification of sponges, along with the types of canal system. The Demospongiae have siliceous (mainly $H_2Si_3O_7$) spicules, or spongin fibers (composed of an insoluble scleroprotein that is resistant to protein-digesting enzymes) (Figure 5-7A), or a combination of both. Their spicules are either straight or curved monaxons or tetraxons, but never six rayed. The glass sponges (Sclerospongiae) have siliceous triaxon (six-rayed) spicules.

Drawings

On p. 55 draw:

1. External view of *Sycon,* gross structure

2. Longitudinal section showing spongocoel and internal ostia. Use arrows to show direction of water flow

3. Types of spicules you have seen

On p. 56 draw a pie-shaped segment of a transverse section through *Sycon,* showing a few canals and some of the cellular details of their structure. On the same page, sketch any other sponges you have studied.

Label all drawings fully.

Other Types of Sponge Structure

Asconoid Type of Canal System

The asconoid canal system is best seen in *Leucosolenia,* another marine sponge of the class Calcarea. *Leucosolenia* grows in a cluster, or colony (See Figure 5-1), of tubular individuals in varying stages of growth. Large individuals may carry one or more buds.

After observing the external structure of a submerged specimen, cut it in half longitudinally, place it on a slide with a little water, and cover. Study with low power. Or use a prepared slide.

The body wall is covered with pinacocytes on the outside and filled with mesohyl that contains amebocytes and spicules. Incurrent pores extend from the external surface directly to the spongocoel, which is lined with flagellated choanocytes. The choanocytes produce the water current and collect food. An osculum serves as the excurrent outlet of the spongocoel. On living specimens you may be able to see some flagellar activity in the spongocoel.

Leuconoid Type of Canal System

Most sponges are of the leuconoid type and most leuconoids belong to the class Demospongiae (See Figure 5-1). Leuconoid sponges have clusters of flagellated chambers lined with choanocytes, and water enters and leaves the chambers by systems of incurrent and excurrent canals. Water from the excurrent canals is collected into spongocoels and emptied through the oscula. In large sponges there may be many oscula. *Spongilla* and *Heteromeyenia,* which are freshwater sponges, and many marine sponges, such as *Halichondria, Microciona, Cliona,* and *Haliclona*—all belonging to Demospongiae—are of the leuconoid type.

Examine any such sponges available, in both external view and cut sections, to see this type of canal system.

Projects and Demonstrations

1. *Preparation of spicule samples.* Calcareous and siliceous sponges are identified principally from their spicules. To prepare a spicule sample, place a small piece (2 to 3 mm square block) of the sponge in a tube and add about 2 ml of Clorox (sodium

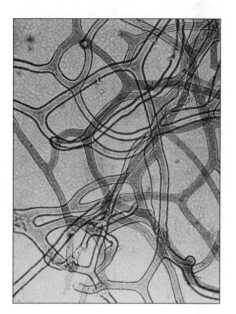

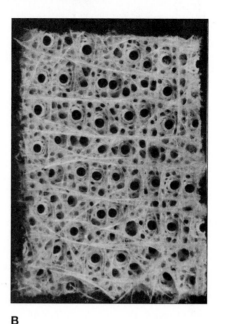

A   **B**

Figure 5-7

Skeletal elements. **A,** Spongin fibers found in Demospongiae (greatly enlarged). **B,** Portion of wall of sponge *Euplectella* (Hexactinellida), in which the spicules are arranged in a definite pattern (about natural size).

hypochlorite solution). Allow an hour or so for the organic matter to be dissolved and the spicules to settle. Pipette off the Clorox, add water to wash, allow the spicules to settle, then remove the water with a Pasteur pipette. Repeat the washing with water, then transfer the spicules by pipette to a microscope slide, add a drop of water and coverslip. Examine at about 100× or more.

Permanent spicule mounts are easily prepared by following the water washes just described with one or two washings with 95% alcohol. After the final wash, transfer the spicules with a pipette to a microscope slide and allow the alcohol to evaporate. Apply a drop of mounting medium, add a coverslip and label.

2. *Drying whole freshwater sponges.* Remove the sponges from water and place in a warm shady place to dry out completely. Dried sponges are very fragile and must be handled carefully. Pack in soft, crushed tissue paper or toilet tissue (never in cotton) for shipping or storing. **WARNING:** wash your hands and do not rub your eyes after handling a dry sponge!

3. *Demonstrations.*

   a. Different types of sponge skeletons, including some of the glass sponges.
   b. Prepared slides of spicules, spongin, gemmules, and sections of whole mounts of various sponges.
   c. Commercial (bath) sponges. The common genus of this leuconoid type is *Spongia.* Commercial sponges have a complicated organization (See Figure 5-1). When brought up alive from the sea bottom and sliced open, the inner surface looks raw and slimy, resembling liver. Dry macerated sponges of commerce (though seldom seen now) are composed only of the spongin skeleton.

Phylum _____

Class _____

Genus _____

Name _____

Date _____

Section _____

Sycon

External view

**Longitudinal section
Internal view**

Calcarerous sponge spicules

LAB REPORT

Sycon

1. Describe the pathway of water through *Sycon,* naming all canals and openings through which water passes from entrance to exit. _____

   _____

2. What drives the flow of water through the sponge? _____

   Explain how the two forms of reproduction in sponges, sexual and asexual, differ from each other. _____

   _____

3. In what way(s) does a sponge show evolutionary advancement as compared to a colonial protozoan such as *Volvox*? _____

   _____

   _____

Other Sponges

1. What is meant by the term polymorphism? _____
   _____

   Is hydra polymorphic? _____

2. Describe feeding and digestion in hydra, explaining how extracellular digestion differs from intracellular? _____
   _____

   What is the function of nutritive-muscular cells? _____

3. How does hydra respire? _____

   How does hydra excrete nitrogenous wastes?
   _____

4. Cnidarians are diploblastic. What does this term mean?
   _____

5. Describe and contrast sexual and asexual reproduction in hydra.
   _____

   _____

   Is the species of hydra you used for behavioral observations monoecious or dioecious? _____

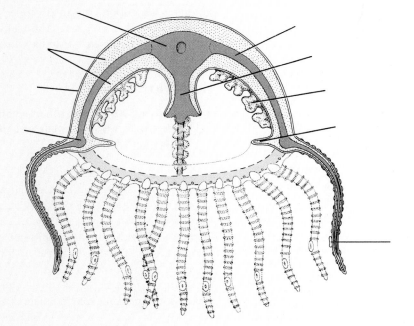

Gonionemus. Diagrammatic oral-aboral section—to be labeled.

Figure 6-6
Swimming *Aurelia aurita*.

EXERCISE 6B
Class Scyphozoa—*Aurelia*, a "True" Jellyfish

Aurelia

Phylum Cnidaria
  Class Scyphozoa
    Order Semaeostomeae
      Family Ulmaridae
        Aurelia aurita

Where Found

Aurelia aurita, the "moon jelly," is common along both coasts of North America. It is a cosmopolitan species distributed from temperate and tropical to subpolar latitudes.

    Aurelia (L. *aurum,* gold) is a scyphozoan (sy-fo-zo'an) medusa (often referred to as a scyphomedusa). Scyphomedusae are generally larger than hydrozoan medusae (hydromedusae); most of them range from 2 to 40 cm, but some reach as much as 2 m or more in diameter. The jelly layers (mesoglea) are thicker and contain cellular materials, giving medusae a firmer consistency than hydromedusae. Nevertheless, all jellyfish are largely water (94% to 96% water in marine species such as *Aurelia* and up to 99% water in some fresh-water hydromedusae) and active tissues are mostly epithelial.

    Scyphozoans are often called "true" jellyfish. Scyphomedusae are constructed along a plan similar to that of hydromedusae, but they lack a velum. Their parts are arranged symmetrically around the oral-aboral axis, usually in fours or multiples of four, so that they are said to have **tetramerous** radial symmetry. Their gastrovascular systems have more canals and more modifications than those of hydrozoans.

    The large size and fiery nematocysts of many jellyfish make them disagreeable and sometimes dangerous to swimmers. One of these is *Cyanea capillata,* the "lion's mane jellyfish" of the North Atlantic, which was central in a Sherlock Holmes mystery. Even more dangerous is the cubomedusan *Chironex fleckeri,* sea wasp of the Australian region. This jellyfish has caused numerous fatalities in Australia; deaths occur rapidly from anaphylactic shock.

Behavior

☞ If living *Aurelia* or other scyphozoan genera are available, observe their swimming movements (Figure 6-6).

How is movement achieved? _____ Are they strong swimmers? _____ How many times per minute does the medusa pulsate? _____ Does it swim horizontally or vertically? _____ Use a gentle touch with a small camel's hair brush to test reaction to touch. To test response to food chemicals, dip the brush into glucose solution, clam or oyster juice, or other food substances; or small crustaceans or other small food organisms may be placed near their tentacles.

Written Report

✎ Record your observations on separate paper.

General Structure

☞ Using a ladle (the medusa is too fragile to be handled with a forceps), transfer a preserved specimen

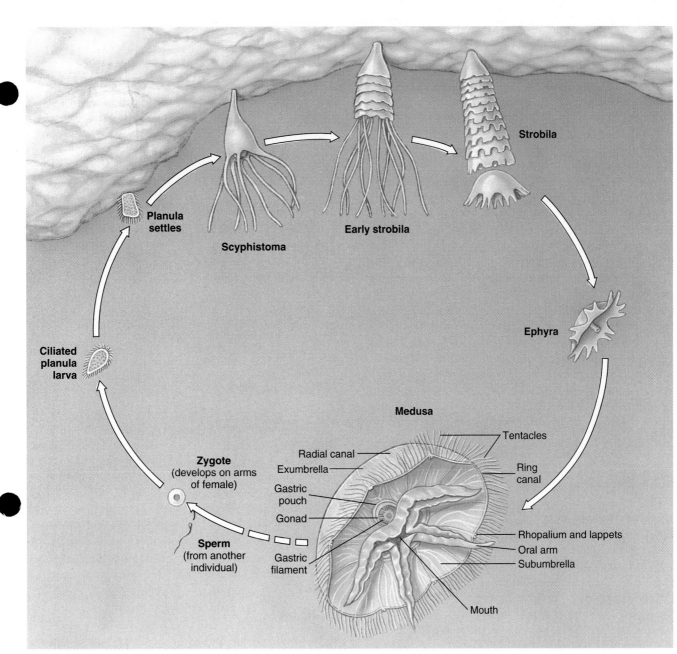

Figure 6-7
Life cycle of *Aurelia,* a marine scyphozoan medusa.

of *Aurelia* to a finger bowl of water, and spread out flat.

Note that *Aurelia* is more discoidal and less cup shaped than *Gonionemus.* When spread flat, the jelly-fish shows a circular shape broken at eight regular intervals by marginal notches (Figure 6-6). Each marginal notch contains a **rhopalium** (ro-pay′li-um; N.L. from Gr. *rhopalon,* a club), a sense organ consisting of a statocyst and an ocellus. This is flanked on each side by a marginal extension, the **lappet** (Figure 6-7). What is the function of a statocyst? _____
An ocellus? _____

☞ Snip out a rhopalium with scissors and examine under the higher power of a dissecting microscope.

In living medusae the excision of all rhopalia would interfere with swimming, either slowing down the contractions or stopping them altogether.

Note the short **tentacles** that form a fringe around the animal's margin.* Compare these tentacles with

_____

* The specific epithet *aurita* of the species *Aurelia aurita* means "eared," probably in reference to the lobelike fringes of tentacles.

those of *Gonionemus* medusae. How do they differ?

_____

Gastrovascular System. In the center of the oral side are four long troughlike **oral arms.** These are modifications of the manubrium. Note that the oral arms converge toward the center of the animal, where the square **mouth** is located. The mouth opens into a short **gullet,** which leads to the **stomach.** From the stomach four **gastric pouches** extend. They can be identified by the horseshoe-shaped **gonads** that lie within them. Near the inner edges of the gonads are numerous thin processes, the **gastric filaments,** which are provided with **nematocysts.** What would be the use of stinging cells here? _____

On the oral side of each gastric pouch is a round aperture that leads into a blind depression, the **subgenital** pit. These pits have no connection with the gonads or gastric pouches and may be respiratory in function. A complicated system of radiating canals runs from the gastric pouches to the **ring canal,** which follows the outer margin (Figure 6-7). This system of stomach and canals, resembling the hub, spokes, and rim of a wheel, forms the medusoid gut, or gastrovascular cavity. Contrast this with the simple, saclike gut of polyp individuals such as hydra or *Obelia* polyps.

Jellyfish are carnivorous, most feeding on fish and a variety of marine invertebrates. *Aurelia,* however, is a suspension feeder that feeds largely on zooplankton. Food organisms are caught in mucus secreted on the subumbrella and moved by cilia to the bell margin. Food is collected from the margin by the oral arms and is transferred by cilia to the stomach. Gastric filaments with nematocysts help pull in and subdue larger organisms and also secrete digestive enzymes. Partially digested food and seawater is then circulated by cilia through the system of canals. In this way, nutrients and oxygen are carried to all parts of the body. The canals are lined with cells that complete digestion of the food. Thus digestion is both extracellular and intracellular.

Reproduction. The sexes are separate in *Aurelia,* as they are in all scyphozoans. Sex cells are shed from the gonads into the gastrovascular cavity and are discharged through the mouth for external fertilization. Within the folds of the oral arms the young embryos develop into free-swimming **planula larvae** (Figure 6-7). These escape from the parent, attach to a substratum, and develop into tiny polyps called **scyphistomae** (Gr. *skyphos,* cup, + *stoma,* mouth). The scyphistoma later becomes a **strobila,** which begins to bud off young medusae (**ephyrae** [Gr. *Ephyra,* Greek city, in reference to castlelike appearance]) in layers resembling a stack of saucers (Figure 6-7). This budding process is called **strobilation** (Gr. *strobilos,* pinecone).

Demonstrations

1. *Slides.* Examine slides showing stages in the life cycle of *Aurelia*—scyphistoma, strobila, and ephyra.

2. *Scyphozoan jellyfish.* Examine various species of preserved scyphozoans or, if available, living specimens of the "upside-down" jellyfish *Cassiopeia.*

EXERCISE 6C
Class Anthozoa—*Metridium,* a Sea Anemone, and *Astrangia,* a Stony Coral

Metridium, a Sea Anemone

Phylum Cnidaria
    Class Anthozoa
        Subclass Zoantharia
           Order Actiniaria
               Genus *Metridium*
                  Species *Metridium senile*

Where Found

Metridium senile, a name from the Greek that means "ancient womb," is the most common sea anemone on the Atlantic coast from Delaware north to the Arctic, and on the Pacific coast from Santa Catalina Island, California, north to the polar seas. It occurs from low intertidal, where it is commonly seen on rocks and pilings, to depths of perhaps 75 m. Most sea anemones are solitary sessile animals and do not live in colonies. Members of the class Anthozoa are all polyps in form; there are no medusae. There is a great variety in size, structure, and color among the sea anemones. All are marine.

Behavior

☞ If living anemones are available, allow one to relax completely and then touch a tentacle lightly with a new (untouched) coverslip held in clean forceps.

Do the nematocysts discharge? _____ Will the nematocysts discharge if the coverslip is touched more vigorously to the tentacle? _____ Now put some saliva (which contains protein) on a dry coverslip and again touch a tentacle. Is the response stronger? Why? _____ Discharged nematocysts should stick to the glass and may be examined under the microscope.

Using another relaxed anemone, drop bits of clean filter paper on the tentacles and time the type and speed of the response. Now test with bits of filter paper soaked in shrimp, clam, or mussel juice and compare the reactions. Test again with bits of clam or other sea

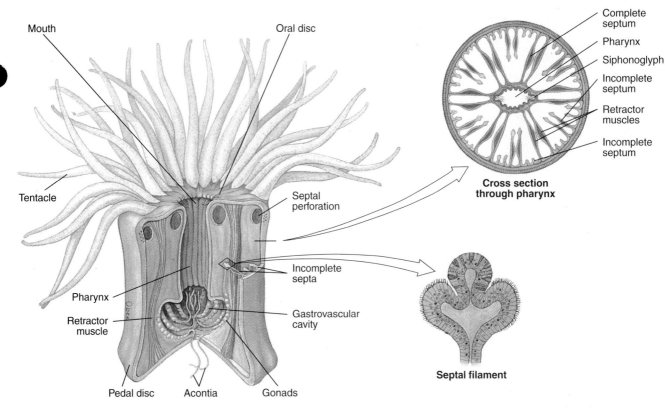

Figure 6-8
Structure of a sea anemone. The free edges of the septa and the acontia threads are equipped with nematocysts to complete the paralysis begun by the tentacles.

food. What conclusions can you draw from this simple experiment? _____

Is the response to food similar to the response to touch? _____ Is the reaction of these animals a part of the normal feeding reaction? _____ If live sea stars are available, try touching an anemone with the arm of a star. What happens? Is this a feeding response or a defense reaction? _____ Some anemones react to certain predatory stars by detaching their pedal discs and moving away from the star.

☞ Use a glass rod to probe the pedal disc. What happens? _____. If you prod the animal vigorously, it will shoot out white, threadlike **acontia** (a-con'she-a; Gr. *akontion,* dart) which are filled with nematocysts used for defense. Place an acontium thread on a slide, cover with a coverslip, and examine with a microscope. Do the acontia move? Examine the edge of an acontium with high power. What do you see that might explain acontia movement? _____

For a dramatic demonstration of nematocysts in action, draw some methyl green under the coverslip using a piece of filter paper touched to the opposite edge of the coverslip. Viewed at high power, the undischarged nematocysts in the acontia look like grains of long rice. What happens when they discharge?

Written Report

✏ Record your observations on separate paper.

External Structure

☞ Place a preserved specimen in a dissecting pan. Note the sturdy nature of its body structures compared with those of other cnidarians you have studied.

The **body** is cylindrical (Figure 6-8) but in preserved specimens it may be somewhat wrinkled. Note that the body of the animal can be divided into three main regions: (1) **oral disc,** or free end, with numerous conical **tentacles** and the **mouth;** (2) cylindrical **column,** forming the main body of the organism; and (3) **basal disc** (aboral end), by which during life the animal attaches itself to some solid object by means of its glandular secretions. Although it is called a sessile animal, the sea anemone can glide slowly on its basal disc.

Is there more than one row of tentacles? Note that the inner surface of the mouth is lined with ridges and that a smooth-surfaced ciliated groove, the **siphonoglyph** (sy'fun-o-glif; Gr. *siphōn*, tube, siphon, + *glyphē*, carving), is found at one side of the mouth. (In some specimens there may be two of these grooves.) The siphonoglyphs, aided by cilia, circulate water throughout the gastrovascular cavity. The mouth is separated from the nearest tentacles by a smooth space, the **peristome** (Gr. *peri*, around, + *stoma*, mouth).

Note the tough outer covering **(epidermis)** of the specimen. Small pores on tiny papillae are scattered over the epidermis, but they are hard to find.

Internal Structure

☞ Study of internal anatomy is best made by a comparison of two sections, one cut longitudinally through the animal and a second cut transversely. Study first one and then the other of these two sections to understand the general relations of the animal. Determine through what part of the animal the transverse section was made.

Look at the longitudinal sections and note that the **mouth** opens into a **pharynx** (fair'inks; Gr. *pharynx*, gullet) (Figure 6-8), which extends down only part way in the body to where it opens into the large **gastrovascular cavity.** Thus the upper half of the body appears as a tube within a tube; the outer tube is the **body wall,** and the inner tube is the pharynx. Look at the cross section and determine these relations. Notice that the gastrovascular cavity is not only the space aboral to the pharynx; it also extends upward to surround the pharynx.

The gastrovascular cavity is subdivided into six **radial chambers** by six pairs of **primary** (complete) **septa,** which run from the oral to the aboral end. In the gullet region these primary septa extend from the body wall to the pharynx; aboral to the pharynx their inner degree are free in the gastrovascular cavity (Figure 6-8). Are they the same width throughout their length? _____ Examine transverse sections to determine this. Note that the chambers formed by primary septa in the pharynx region communicate with each other through **septal perforations.** Find these perforations next to the pharynx near the oral end. Note the longitudinal retractor muscles on the complete septa, which, together with muscles in the body wall, enable the anemone to contract when disturbed.

Now notice that these six larger chambers are partially subdivided by smaller pairs of **incomplete septa,** which extend varying distances from the body wall into the gastrovascular cavity (Figure 6-8). Are these incomplete septa free at their inner edges? _____ Do they extend from the oral to the aboral end of the animal? _____ Each septum is composed of a double sheet of gastrodermis.

The free edges of the septa are expanded into convoluted thickenings called **septal filaments.** These bear nematocysts (to help subdue struggling prey) and glands that secrete digestive enzymes. In *Metridium* and many other sea anemones the lower edges of the septal filaments continue into the lower part of the gastrovascular cavity as long delicate threads, the **acontia** (Figure 6-8). Each acontium has stinging cells and, when the sea anemone contracts strongly, the acontia are shot out through the mouth and body pores for defense.

The **gonads** are thickened bands resembling stacks of coins, often orange-red in color, that lie in the septa just peripheral to the septal filaments. Anemones are dioecious.

Astrangia, a Stony Coral

Phylum Cnidaria
    Class Anthozoa
        Subclass Zoantharia
            Order Scleractinia
                Genus *Astrangia*
                    Species *Astrangia danae*

Stony corals resemble small anemones but are usually colonial. Each polyp secretes a protective calcareous cup into which the polyp partly withdraws when disturbed. Some corals form colonies consisting of millions of individuals, each new individual building its skeleton upon the skeletons of dead ones, thus forming, over many years, great coral reefs. Reef-building corals build only in tropical or subtropical waters, where the water temperature stays at or above 21° C.

Astrangia (Gr. *astron*, star, + *angeion*, vessel), known as the star coral, is our only shallow water northern coral. It occurs on the Atlantic and Gulf coasts locally from Florida north to Cape Cod. It does not build reefs but forms small colonies of 5 to 30 individuals encrusted upon rocks or shells. Its food consists of small organisms such as protozoa, hydroids, worms, crustaceans, and various larval forms.

Behavior

The same sort of touching and feeding experiments as suggested for *Metridium* are applicable to corals, making allowance for the smaller size.

Structure

☞ Examine living or preserved coral polyps.

Note the delicate, transparent polyps extending from the circular skeletal cups (Figure 6-9A). The polyps resemble those of anemones, with a column, oral disc, and crown of tentacles. Two dozen or more simple tentacles, supplied with nematocysts, are arranged in three rings around the mouth. Siphonoglyphs are absent. The edges of the septa can usually be seen through the transparent polyp walls. As in sea

A

Figure 6-9

Stony corals. **A,** Living polyps of *Astrangia danae.* **B,** Structure of a coral polyp, diagrammatic.

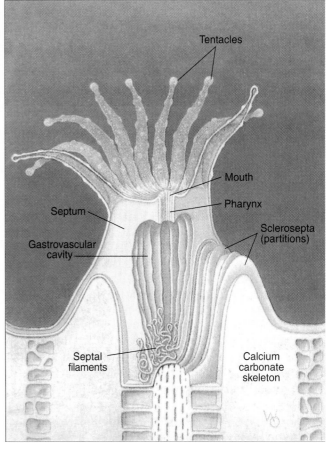

B

anemones, stony corals are built on a plan of six or multiples of six (hexamerous). Digestion in the gastrovascular cavity is similar to that of anemones.

Colonies usually arise by budding or division from a single polyp that has been sexually produced. The surface of the colony between the skeletal cups is covered by a sheet of living tissue, which is an extension of the polyp walls. This tissue connects all members of the colony.

☞ Now study a piece of skeleton from which the polyps have been removed.

Each cup (Figure 6-9B) was the home of a polyp, which secreted it. The rim of the cup is called the **theca** (L. *theca,* box), and the base is the **basal plate.** The **sclerosepta,** or radial partitions within the theca, form the same hexamerous pattern as do the septa. They are laid down by the folds of the epidermis at the base of the polyp. The theca is secreted by the epidermis of the lower part of the column. As the coral polyp grows, the cup tends to fill up with calcium carbonate, so that the theca and sclerosepta are continually extended upward. The entire skeleton is outside the body of the polyp.

Projects and Demonstrations

1. *Various hard coral skeletons.* Examine such skeletons as white corals; red coral, *Corallium;* staghorn coral, *Acropora;* organpipe coral, *Tubipora;* or brain coral, *Meandrina.*

2. *Nematocysts.* Nematocysts may be collected from various live cnidarians. Zoantharian corals (soft corals) usually have fairly large nematocysts, and the little red colonial anemone *Corynactis* is also good for this purpose. Wiping a clean coverslip across the oral disc is likely to attract both discharged and undischarged nematocysts, which can be studied by inverting the coverslip over a drop of seawater on a slide. A good quality microscope is necessary.

Collect and compare the nematocysts of various hydrozoans, scyphozoans, and anthozoans.

The Acoelomate Animals
Phylum Platyhelminthes

The Acoelomate Phyla

The acoelomates are animals that have no coelom (body cavity). They include the phylum Platyhelminthes (Gr., *platys,* flat, + *helmins,* worm), phylum Nemertea (Gr., *Nemertes,* one of the Nereids, mermaids of Greek mythology), and phylum Gnathostomulida (Gr. *gnathos,* jaw, + *stoma,* mouth). In acoelomate animals the space between the body wall and the digestive tract is not a cavity, as in coelomate animals, but is filled with muscle fibers and a loose tissue of mesenchymal origin, called **parenchyma,** both derived from the mesoderm. The presence of a well-developed mesodermal layer makes the acoelomates **triploblastic** (having three germ layers: ectoderm, endoderm, and mesoderm).

The platyhelminths, or **flatworms,** are a large and economically important group because they include not only the free-living **planarians** but also the parasitic **tapeworms** and **flukes.**

The Nemertea are the **ribbon worms,** often called nemertine or nemertean worms. Nearly all are marine and are characterized by an eversible **proboscis** that can be thrown out with great speed to capture food. The Gnathostomulida is a small phylum of delicate wormlike marine animals called **jaw worms,** that live in sandy coastal sediments. Because nemerteans and jaw worms are generally unavailable for study in the classroom, these phyla will not be included in the laboratory exercises.

Acoelomates are advanced over the radiate animals in several ways. (1) **Bilateral symmetry.** What advantages does bilateral symmetry have as compared to radial symmetry? _____ (2) Tissues defined and organized into functional **organs.** (3) Highly organized nervous system with concentration of nervous tissue and sense organs in the anterior end **(cephalization).** What advantages does cephalization offer for a bilateral animal? _____ (4) **Excretory system** of specialized flame cells and tubules for elimination of nitrogenous wastes. How does a radiate animal, such as a hydra or sea anemone, rid itself of wastes? _____ (5) The platyhelminths have a gastrovascular system, but the ribbon worms have separated the two functions and have a complete mouth-to-anus digestive tract and a circulatory system.

The Acoelomate Phyla
Classification
Phylum Platyhelminthes

EXERCISE 7A
Class Turbellaria—The Planarians
Dugesia

EXERCISE 7B
Class Trematoda—The Digenetic Flukes
Clonorchis, the liver fluke of humans
Schistosoma, the human blood fluke

EXERCISE 7C
Class Cestoda—The Tapeworms
Taenia or *Dipylidium*
Projects and demonstrations

EXPERIMENTING IN ZOOLOGY
Planaria regeneration experiment

Flatworms have a **tissue-organ level of organization.**

Classification

Phylum Platyhelminthes

Class Turbellaria (tur′bel-lar′e-a) (L. *turbellae* [pl.], stir, bustle, + *aria,* like or connected with). Turbellarians. Mostly free-living, with a ciliated epidermis. Example: *Dugesia tigrina,* the brown planarian.

Class Monogenea (mon′o-gen′e-a) (Gr. *mono,* single, + *gene,* origin, birth). Monogenetic flukes. Adult body covered with syncytial tegument without cilia; leaflike to cylindrical in shape; posterior attachment organ with hooks, suckers, or clamps, usually in combination; all parasitic, mostly on skin or gills of fishes; single host; monoecious; usually free-swimming ciliated larva. Examples: *Polystoma, Gyrodactylus.*

Class Trematoda (trem′a-to′da) (Gr. *trematodes,* with holes, + *eidos,* form). Digenetic flukes. Adult body covered with nonciliated syncytial tegument; leaflike or cylindrical in shape; usually with oral

Phylum Platyhelminthes

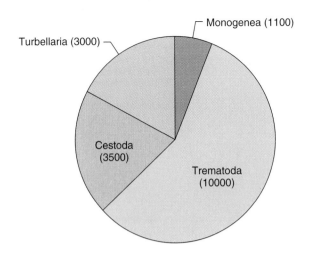

Turbellaria (3000)
Monogenea (1100)
Cestoda (3500)
Trematoda (10000)

and ventral suckers, no hooks; development indirect, first host a mollusc, final host usually a vertebrate; parasitic in all classes of vertebrates. Examples: *Fasciola, Clonorchis, Schistosoma.*

Class Cestoda (ses-to′da) (Gr. *kestos,* girdle, + *eidos,* form). Tapeworms. Adult body covered with nonciliated, syncytial tegument; scolex with suckers or hooks, sometimes both, for attachment; long ribbonlike body usually divided into series of proglottids; no digestive organs; parasitic in digestive tract of all classes of vertebrates; first host may be invertebrate or vertebrate. Examples: *Taenia, Diphyllobothrium.*

EXERCISE 7A
Class Turbellaria— The Planarians

Dugesia
Phylum Platyhelminthes
   Class Turbellaria
      Order Tricladida
         Family Planariidae
            Genus *Dugesia*

Where Found

Freshwater triclads,* or planarians, are found on the underside of stones or submerged leaves or sticks in freshwater springs, ponds, and streams. They are often confused with leeches, which they resemble in color and somewhat in shape. There are nine genera and more than 30 species of planarians in North America.

*Triclads, members of the order Tricladida (Gr. *treis,* three, + *klados,* branched), are so named for the three-branched intestine characteristic of all planarians.

Common and widely distributed species are the brown planarians, *Dugesia*† tigrina and *Dugesia dorotocephala,* adapted to warm and standing (or slowly moving) water. The black planarian *Dendrocoelopsis*‡ *vaginata* is found west of the Continental Divide.

Observation of Live Planarians

☞ Using a small camel's hair brush, place a live planarian on a ringed slide, depression slide, or Syracuse watch glass in a drop of culture water. Replace the water as it evaporates. Keep the surrounding glass dry to prevent the animal from wandering out of range. By holding the slide above a mirror you can also observe its ventral side.

External Structure.
☞ Observe the animal and decide which are its **anterior, posterior, dorsal,** and **ventral** aspects.

Note the triangular **head.** Its earlike **auricles** (Figure 7-1A) bear many sensory cells, but they are tactile and olfactory, not auditory, in function. Are the **eyes** movable? _____ Do they have lenses? _____ Note the pigmented **skin.** Is it the same color on the underside? _____ Is its coloring protective? _____ Are the length and breadth of the worm constant? _____ Holding the slide for a moment over a bright light, can you locate the muscular **pharynx** along the midline? When the animal feeds, the pharynx can be protruded through a ventral **mouth** opening. Can you verify this by use of the mirror?

☞ Use a hand lens, dissecting microscope, or low power of a compound microscope for further examination of the eyes and body surface.

Locomotion. Observe the animal's gliding movement. Glands in its ciliated epidermis secrete a path of mucus on which the planarian propels itself by means of its cilia. Do you think cilia alone are responsible for its movement? _____ What do you think causes the waves of contractions along its body? _____ Does the animal ever leave the drop of water and travel on the dry glass? _____ Why? _____ How does it use the head and auricles? _____ Does it ever move backward? _____

Reactions to Stimuli.
☞ Observe the responses of planarians to touch (**thigmotaxis**), food (**chemotaxis**), and light (**phototaxis**) by doing some of these simple experiments.

† The genus *Dugesia* was named to honor Antoine-Louis Dugès, a nineteenth-century French zoologist and physician.

‡ *Dendrocoelopsis* derives from the Greek *dendron,* stick, + *koilos,* hollow.

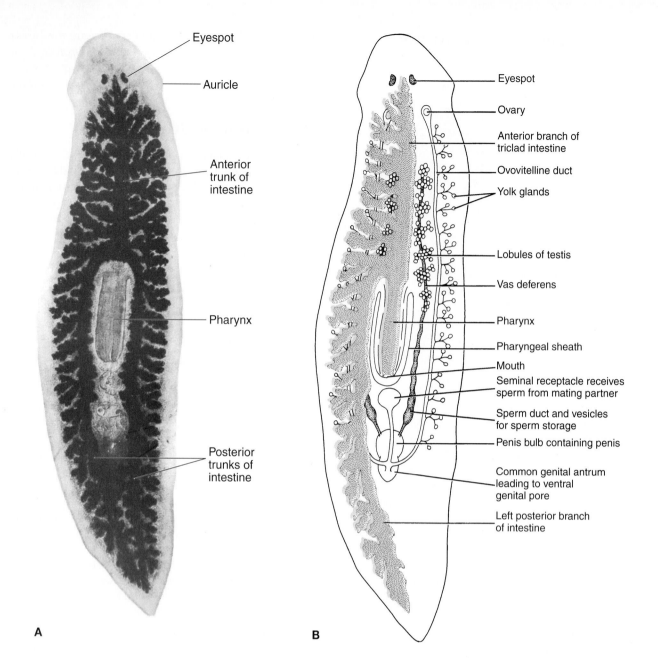

Figure 7-1

Planaria, a freshwater turbellarian. **A,** Stained whole mount. **B,** Internal anatomy; half the anterior trunk of the intestine is removed to reveal the testes.

1. *Response to touch. Very gently* touch the outer edges of the worm with a piece of lens paper or a soft brush. What parts of its body are most sensitive to touch? Are its reactions more localized or less localized than those of the hydra?

2. *Response to food.* To observe the pharynx, stick a *very small* bit of fresh beef liver (or cut-up mealworms) on the center of a coverslip. Invert the coverslip over a deep depression slide containing one or two planarians in pond water. Examine under a dissecting microscope. If the planarians have not been fed for several days, they will soon find the meat by gliding upside down across the coverslip; you will then be able to watch them extend the pharynx to feed.

3. *Response to directional illumination.* Direct a strong beam of light at the planarians from one side of the dish and observe their movements. Move the light 90 degrees around the dish and direct it again at the planarians. Do you think their movements indicate a "trial and error" response or a directed response? _____

4. *Response to light and dark backgrounds in nondirectional illumination.* Prepare the lower half of a Petri dish by painting its sides and one half of its bottom on the outside with black

Paint one-half of Petri dish on outside with flat black paint.

After introducing planarians, cover with black paper with a hole cut in the center.

Preparing painted dishes for the nondirectional illumination study.

paint (or covering with black tape or black paper). Set the dish on a white surface so that half of the bottom is black and the other half is white. Cut a circle of black paper a little larger in diameter than the top of the Petri dish. Cut out the center of the circle, leaving a ring of paper about 3 cm wide, or wide enough to extend inward from the edge of the dish about 1.5 cm, to shade the sides of the dish from overhead illumination. Clean and rinse the dish thoroughly, then place it in a dark room or box, with a light source several feet above the dish. Place a few planarians in the center of the dish and leave for a while. When they have ceased moving about, count the animals on the dark surface and those on the white surface and compare the numbers. Now remove the animals from the dish, add a suspension of carmine to the water, and rotate the dish gently. Drain off the carmine suspension and rinse the dish *gently* with water. The movements of the planarians can be seen where the carmine particles adhere to the mucous trails left on the bottom of the dish. Do you note any difference in the length of the trails on the black and white surfaces? _____ Is there a directional response? _____

Written Report

On pp. 77 report your observations and the conclusions you drew from observing the live planarians.

Observation of Stained Whole Mounts

The stained whole mount of a planarian shows an animal that was fed food mixed with India ink, carmine, or some other suitable stain before killing and fixing, resulting in a darkly stained gastrovascular tract.

☞ Using the dissecting microscope or the low power of a compound microscope, study stained whole mounts of the freshwater planarian and the marine *Bdelloura,* identifying the structures listed here.

The Digestive System. As in cnidarians the digestive tract of a turbellarian is a **gastrovascular cavity,** the branches of which fill most of the body (Figure 7-1B).

Because there is no anus, undigested food is ejected through the mouth.

The muscular **pharynx** is enclosed in a **pharyngeal sheath,** but its free end can be extended through the ventral **mouth.** Ingestion occurs through the muscular sucking action of the pharynx. The pharynx opens into the intestine, which has one **anterior trunk** and two **posterior trunks,** one on each side of the pharynx. What might be an advantage to the digestive process of the branching diverticula that extend laterally from the intestinal trunks? _____ Some digestion may occur within the lumen of the digestive cavity by means of enzymes secreted by intestinal gland cells **(extracellular digestion).** As in the cnidarians, digestion is completed within the phagocytic cells of the gastrodermis **(intracellular digestion).**

Reproduction. Flatworms are monoecious, and their reproductive system is complex (Figure 7-1B). Because of the large digestive tract, most of the reproductive organs of turbellarians are obscured on the stained whole mounts. However, the penis and genital pore may be seen on the *Bdelloura* slides. See your text for further description of the reproductive system. The reproductive system will be more easily studied in the liver flukes and tapeworms.

Planarians and other freshwater turbellarians also reproduce asexually by transverse fission; in some species chains of zooids are formed asexually.

Excretion and Osmoregulation. The excretory system, consisting of **excretory canals** and **protonephridia (flagellated flame cells),** cannot be seen in the whole mounts; see your text for a discussion of this system. The instructor may prepare a demonstration of living flame cells from planarians. The main function of the protonephridial system may be the regulation of the internal fluid content of the animal (osmoregulation). The system is often absent in marine turbellarians.

Nervous System. *Bdelloura* is a marine turbellarian that lives as a commensal on the external surface of the horseshoe crab, *Limulus.* Properly stained slides show the **ladder type of nervous system,** as well as the triclad digestive tract.

On a stained whole mount of *Bdelloura* find the **cerebral ganglia** at the anterior ends of the two lateral **nerve cords. Transverse nerves** connecting the cords, and **lateral nerves** extending outward from the cords form the "rungs" of the ladder.

Sense Organs. A pair of **eyespots,** or **ocelli,** are light-sensitive pigment cups (Figure 7-1A). Chemoreceptive and tactile cells are abundant over the body surface, especially on the auricles.

Drawings

✎ On separate paper, sketch the external morphology of a planarian and label fully.

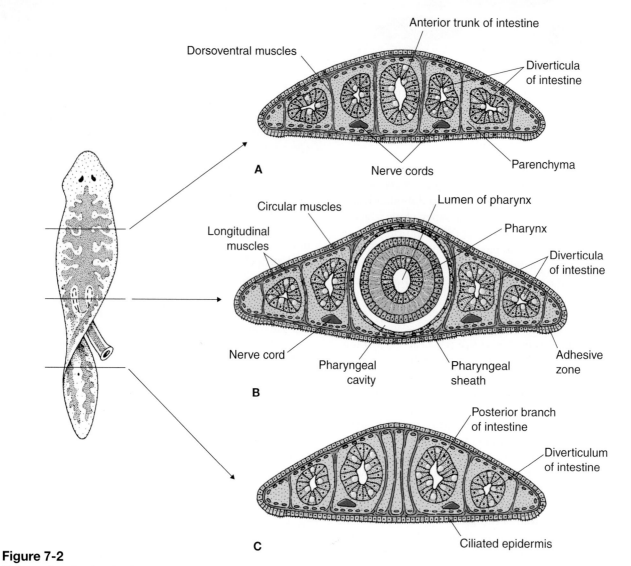

Figure 7-2

Cross section of a planarian (diagrammatic). **A,** Anterior section. **B,** Section through the region of the pharynx. **C,** Posterior section.

Transverse Sections of Planarian

The appearance of a cross section will depend on whether it is cut from the anterior, middle, or posterior part of the planarian. Sections cut anterior to the pharynx will contain a centrally located section of the anterior trunk of the intestine; those cut posterior to the pharynx will contain laterally located sections of the two posterior trunks of the intestine; those cut through the pharynx will show the conspicuous round muscular pharynx, with branches of intestine on each side (Figure 7-2).

The **epidermis** of cuboidal epithelial cells (derived from ectoderm) contains many dark rodlike **rhabdites** (Gr. *rhabdos*, rod), which, when discharged in water, swell and form a protective mucous sheath around the body. The epithelial cells are ciliated only on the animal's ventral surface. What function do they serve? _____

Inside the epidermis is a layer of **circular muscles** and then a layer of **longitudinal muscles** (cut trans- versely and appearing as dark dots). **Dorsoventral muscle fibers** are also visible, particularly at the sides of the animal. Do these muscles explain the waves of contractions you saw in the living animal? _____ Note that there is *no body cavity*. The flatworms are called **acoelomate** animals. **Parenchyma** (pair-en'ka-ma; Gr., anything poured in beside), largely of mesodermal origin, is the loose tissue filling up space between the organs.

Several hollow sections of the intestine and its diverticula (derived from endoderm) may be seen, depending on the location of the section. What kind of epithelial cells makes up the intestinal walls? _____ In a middle section, note the thick circular **pharynx,** covered and lined with epithelium and containing layers of circular and longitudinal muscle similar to those in the body wall. The **pharyngeal chamber** is also lined with epithelium.

Nerve cords, reproductive and excretory ducts, testes, and ovaries are found in the parenchyma of adult animals, but they are difficult to identify.

Planarian Behavior

Name_____

Date_____

Section_____

Locomotion

Response to touch (thigmotaxis)

Response to food (chemotaxis)

Response to directional illumination (phototaxis)

Response to light and dark background in nondirectional illumination (phototaxis)

EXERCISE 7B

Class Trematoda— The Digenetic Flukes

Clonorchis, the Liver Fluke of Humans

Phylum Platyhelminthes
  Class Trematoda
    Order Digenea
      Species: *Clonorchis sinensis* (human liver fluke)

Where Found

All trematodes are parasitic, harbored in or on a great variety of animals. Many of them have three different hosts in their life cycle.

The adult, or sexual, stage of *Clonorchis* lives in the human bile duct. It is widely distributed in Japan, Korea, China, Taiwan, and Vietnam, where it causes widespread suffering and economic loss. Prevalence ranges from 14% in cities such as Hong Kong to 80% in some rural areas. The asexual, or larval, stages are found in aquatic snails and fishes.

Study of a Stained Whole Mount

☞ Study a stained slide of an adult fluke, first with a hand lens, then with the low power of the microscope.

How long is the specimen? _____ Compare it with the planarian in size and shape. Note the **oral sucker** at the anterior end and the **ventral sucker (acetabulum)** on the ventral surface (Figure 7-3). What is their function? _____

Body Wall. The body covering of both flukes and tapeworms was formerly believed to be a nonliving cuticle, but the electron microscope reveals it to be a living syncytial **tegument** (L. *tegumentum,* to cover) consisting of cytoplasmic processes of cells that dip down into **parenchyma,** a meshwork of cells derived from mesoderm.

Circular and longitudinal **muscle layers** of the body wall are similar to those of turbellarians, and body spaces are filled with parenchyma.

Digestive System. The **mouth** lies anteriorly in the oral sucker. It leads to a muscular **pharynx** and short **esophagus** that divides into two lateral branches of the digestive tract. The sucker serves for attachment and for the abrasion of the bile duct in which the fluke lives. Food—blood cells and lacerated cells from the inflamed bile duct—is aspirated by the muscular pharynx. There is no anus.

Protonephridial System. The **excretory pore** at the posterior end is the outlet of the **bladder.** Follow the bladder forward to see where it divides into two long tubules. The tubules collect from flame cells, which you will not be able to see.

Reproduction and Life Cycle. Flukes are monoecious; each animal has both male and female reproductive systems (Figure 7-3). The male **testes** are conspicuous branched organs located one in front of the other in the posterior half of the body. From each testis a slender duct (**vas efferens**) extends forward and then unites with the duct on the other side to form a single sperm duct (**vas deferens**). This duct enlarges anteriorly to form a swollen **seminal vesicle** that empties through a small **genital pore** just anterior to the ventral sucker. A penis and cirrus sac are found in many trematodes but not in this species.

After cross-fertilization with another worm, sperm pass through the uterus to the **seminal receptacle** where they are stored, probably for the life of the worm, which may be several years. Both seminal receptacle and a trilobed **ovary** are connected to an oviduct, which gives off a tube (Laurer's canal, probably a vestigial vagina) that curves posteriad around the seminal receptacle and opens by a tiny pore to the dorsal surface. The oviduct also receives two **yolk ducts** from **yolk glands** located laterally in the worm's body. The oviduct connects to an **ootype** surrounded by a loosely organized **Mehlis' gland** (both ootype and Mehlis' gland may be difficult to see). In this region eggs are combined with yolk and enclosed in a shell. Shelled eggs then pass to a conspicuous **uterus,** usually filled with eggs, and finally to the outside through the genital pore.

Eggs containing ciliated larvae are shed into the bile, carried to the host's intestine, and voided with the feces. If feces are passed into water, the larvae hatch and are eaten by a snail where they multiply asexually through additional larval stages before leaving the snail as tadpole-like **cercariae.** If successful in finding a fish host, the cercariae bore into the muscle where they encyst. Humans (and other mammals) are infected when they eat raw or undercooked fish bearing the cysts. Young flukes then emerge, travel to the bile ducts, and mature into adults. Consult your textbook for details of this animal's life cycle. Why do you think this parasite is prevalent in Southeast Asia? _____

Schistosoma, the Human Blood Fluke

Phylum Platyhelminthes
  Class Trematoda
    Subclass Digenea
      Order Strigeata
        Family Schistosomatidae
          Genus: Schistosoma
            Species: *Schistosoma mansoni*
            (or *Schistosoma haematobium*)

Where Found

Schistosomes of the genus *Schistosoma* are blood flukes of humans that affect an estimated 200 million people in Asia, Africa, the Caribbean (including Puerto Rico), and northeastern South America. As a major global

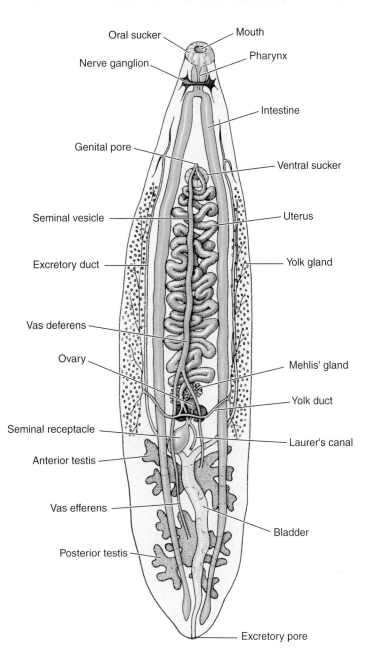

Figure 7-3

Clonorchis, the liver fluke of humans. Is this species monoecious or dioecious? _____ By what pathway do sperm from the testes reach the genital pore? _____ After cross-fertilization, how do sperm reach the seminal receptacle? _____ What is the function of the seminal receptacle? _____

health problem, **schistosomiasis** is exceeded only by malaria. In many areas, the disease is commonly known as "bilharzia" rather than the more correct schistosomiasis.

Three species of schistosomes are of enormous medical significance. *S. mansoni* lives primarily in the venules draining the large intestine. *S. haematobium* lives in the venules of the urinary bladder; *S. japonicum* inhabits the venules of the small intestine. All three species cling to the venule walls with their suckers and feed on blood.

The life cycle of *Schistosoma* is similar in all species (Figure 7-4). If discharged eggs from human feces or urine get in water, they hatch as ciliated **miracidia.** If

they find the right snail, within a few hours they burrow in and transform into a saclike form, the **sporocyst.** These multiply asexually to produce another generation of sporocysts, within each of which develop numerous **cercariae.** Successive sexual generations of the sporocysts and the cercariae they contain over a period of several months ensure an enormous increase in numbers: each miracidium may give rise to more than 200,000 cercariae.

Cercariae escape from the snail and swim about until they contact bare skin of a human. They penetrate the skin, enter the circulatory system, and make their way to the hepatic portal system, where they develop

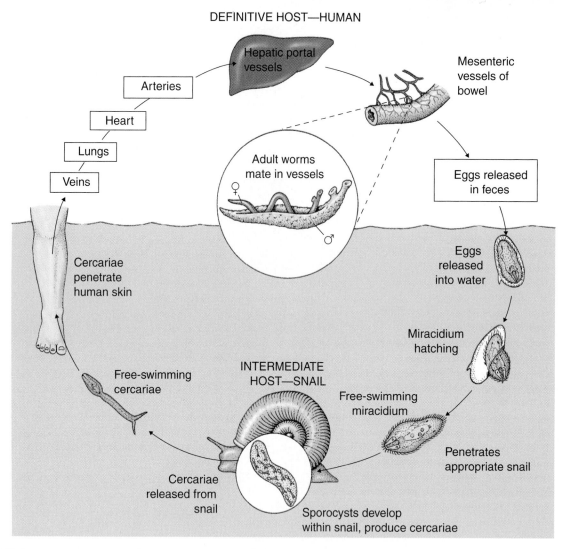

Figure 7-4

Life cycle of *Schistosoma mansoni*.

before migrating to their characteristic sites. Here they mate with another worm and the female begins producing eggs.*

Study of Prepared Slides

☞ Examine a slide of *Schistosoma mansoni* adults in copula.

An odd feature of this genus is the strong sexual dimorphism (Figure 7-4). Males are stouter than females and have a ventral longitudinal groove, the **gynecophoric canal** (gi′ne-ka-fore′ik; Gr. *gyne,* woman, + *pherein,* to carry). (The name *Schistosoma,* meaning "split body," refers to this canal.) Even more

oddly, the thinner female normally resides there, permanently embraced by the male. Note the strong oral sucker and a secondary sucker near the anterior end called the **acetabulum.**

☞ Examine a slide of schistosome eggs.

Note the elliptical shape of the eggs, each of which bears a sharp spine (the spine is terminal in *S. haematobium* and lateral in *S. mansoni;* the eggs of *S. japonicum* lack a spine). The main ill effects of schistosomiasis result from the eggs, which, as they work their way out of the venules where the adults live, cause ulceration, abscesses, bloody diarrhea, and abdominal pain.

Calcified eggs of *S. haematobium* have been found in Egyptian mummies dating from 1200 B.C. There is a well-reasoned hypothesis that the curse Joshua placed on Jericho (after destroying the city and putting all of its inhabitants to the sword! [Joshua 6:26]) was the introduction of *S. haematobium* into the communal

---

* Although the life cycle is similar in all species of *Schistosoma,* it differs from that of other digeneans in that the cercariae penetrate the definitive host directly. There is no second intermediate host (unlike *Clonorchis,* for example, in which there are two intermediate hosts, a snail and a fish).

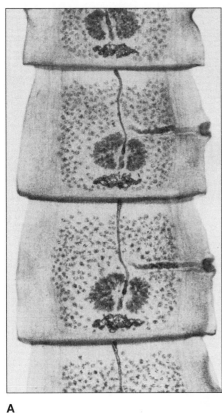

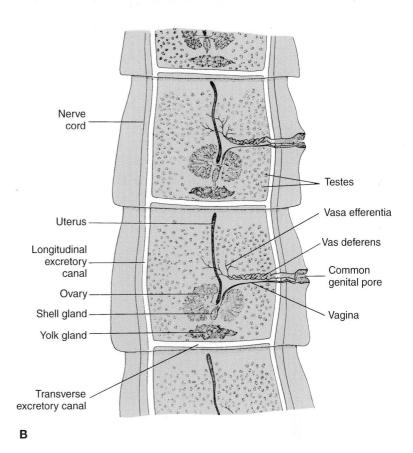

Figure 7-5

Photomicrograph and interpretive drawing of *Taenia pisiformis,* the dog tapeworm.

well. The curse was removed after Jericho was abandoned and subsequent droughts killed the snail host. Today the people of Jericho (Ariha, in Jordan) are free of schistosomiasis.

Of Egypt's population 30% to 40% suffered from the disease at the time of Napolean's invasion (1799-1801) and the disease became prevalent in his troops. It is still a problem today in Egypt and in fact has become much worse since construction of the High Aswan Dam because conditions for the intermediate snail host were unintentionally vastly improved. It is estimated that half the population of Egypt now have the infection.

EXERCISE 7C
Class Cestoda—The Tapeworms

Taenia or *Dipylidium*

Phylum Platyhelminthes
    Class Cestoda
        Order Cyclophyllidea
            Species *Taenia pisiformis* (dog tapeworm) or
            Dipylidium caninum (small dog tapeworm)

Tapeworms are all endoparasitic and show extreme adaptations for their parasitic existence. For example, they lack a digestive system and there is great emphasis on reproduction. Most require two hosts of different species, with the adult tapeworm characteristically living in the digestive tract of a vertebrate.

As their name implies, tapeworms are ribbonlike and their long bodies (the **strobila**) are usually made up of segments called **proglottids** (Gr. *proglottis,* tongue tip). (Proglottids are not to be confused with metamerism in higher forms; they are formed by a continuous process of **budding** from the anterior end or **scolex** [Gr. *skōlēx,* worm, grub].) As new proglottids are formed anteriorly, the older ones are pushed backward so that the oldest, or most mature, proglottids are always at the posterior end. The scolex, which serves as a holdfast, is usually equipped with **suckers, hooks,** or both. The body covering, or **tegument,** is similar to that of the flukes.

The forms described here are *Taenia pisiformis* (Gr. *tainia,* ribbon; L. *pisum,* pea, + *forma,* shape), a dog or cat tapeworm whose larval stage is found in the liver of rabbits (Figure 7-5), and *Dipylidium caninum* (Gr. *di,* two, + *pyle,* entrance, + *idion,* dim. suffix; L. *caninus,* belonging to a dog), a small tapeworm of dogs or cats (Figure 7-6), with fleas as the alternate host. Other forms may be substituted in the laboratory. Among those that parasitize humans are *Taenia solium,* the larval stage of which encysts in muscle of pigs and which, like *Taenia pisiformis,* possesses both hooks and suckers;

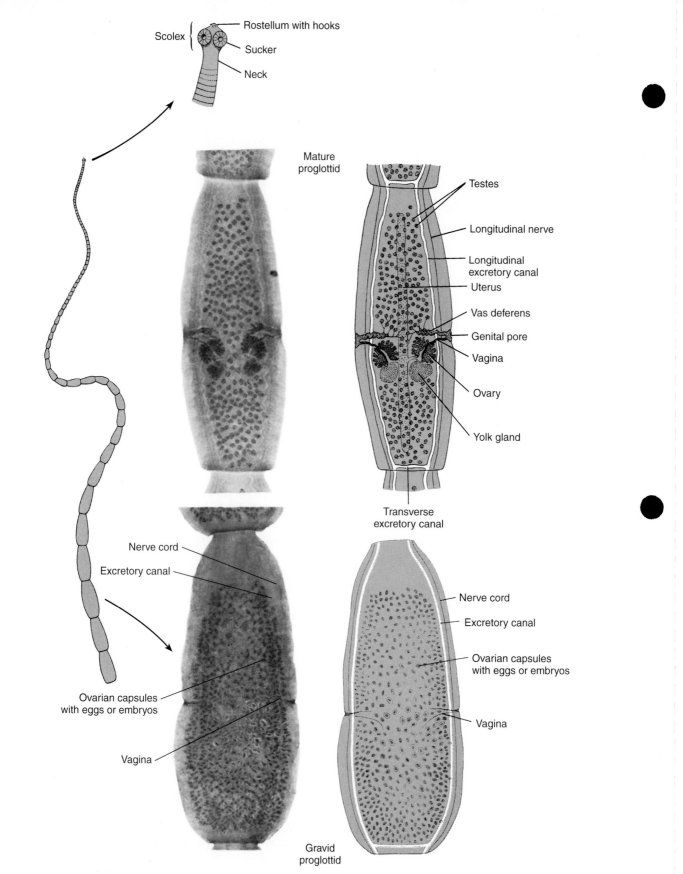

Rostellum with hooks

Scolex

Sucker

Neck

Mature proglottid

Testes

Longitudinal nerve

Longitudinal excretory canal

Uterus

Vas deferens

Genital pore

Vagina

Ovary

Yolk gland

Transverse excretory canal

Nerve cord

Excretory canal

Ovarian capsules with eggs or embryos

Vagina

Gravid proglottid

Nerve cord

Excretory canal

Ovarian capsules with eggs or embryos

Vagina

Figure 7-6

Dipylidium caninum, the small tapeworm of dogs and cats.

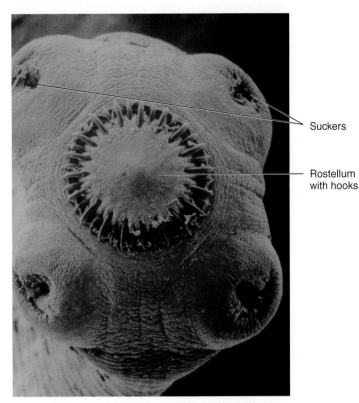

Figure 7-7
Scolex of *Taenia* sp., showing four suckers and a circlet of hooks on the rostellum. SEM ×125.

Taenia saginata, which has cattle as the alternate host and has no hooks; and *Diphyllobothrium latum,* whose larval stage is found in crustaceans and fishes. *D. latum* is the largest of all human cestodes, some reaching a length of 20 m (65 ft)! Humans become infected by eating uncooked freshwater fish or uncooked salmon.

General Structure

 Study a preserved whole specimen or one embedded in plastic.

How long is the specimen? _____ Examine the **scolex** under a dissecting microscope and identify the **hooks** and **suckers.** Note the **neck,** from which new proglottids are budded off. As older proglottids are pushed back by addition of new ones, they mature and become filled with reproductive organs. Can you distinguish maturing from very young proglottids by the presence of a **genital pore** at one side? As eggs develop and become fertilized, the proglottid becomes distended (gravid), with the uterus filled with embryos. Where do you find **gravid proglottids** (Figure 7-6)? _____ Such proglottids soon break off and are shed in the feces of the host. Ingested by a suitable alternate host, young embryos of *T. pisiformis* migrate to the liver and encyst there as cysticerci (bladder worms). Eggs of *D. caninum* are eaten by larval

fleas which, when mature, are nipped or licked out of the fur of a dog or cat.

Microscopic Study

Examine with low power a prepared slide of a tapeworm whole mount containing (1) scolex with neck and a few immature proglottids, (2) mature proglottid, and (3) a gravid proglottid.

The Scolex. Note the **suckers** (how many? _____) and a **rostellum** bearing a circle of **hooks** (Figure 7-7). What is the function of hooks and suckers? _____ Is there a mouth? _____ Is the **neck** segmented? _____ Two lightly stained tubes, one on each side, are **excretory canals,** which extend the entire length of the animal. Some of the **immature proglottids** have developed **genital pores.**

The Mature Proglottid. The mature proglottid of *T. pisiformis* is a little narrower at its anterior end (See Figure 7-5). From the lateral **genital pore** two tubes extend medially. The more anterior of these, the **sperm duct,** is convoluted and branches from the middle of the proglottid into many small efferent ducts that collect sperm from many small **testes** scattered through the proglottid. The more posterior duct is the slender

vagina, which curves posteriorly between the branched **ovaries** to connect with short **oviducts.** The oviducts also connect with the long central **uterus. Mehlis' gland** (function unknown) and a larger **vitelline gland** (yolk gland) are found posterior to the uterus.

Each mature proglottid of *Dipylidium caninum* bears a set of male and a set of female reproductive systems (See Figure 7-6). Are tapeworms monoecious or dioecious? _____ Locate the convoluted **vas deferens,** which branches to numerous testes lobules. The more slender **vagina,** sharing a **common genital pore** with the vas deferens, connects by an **oviduct** to branched **ovaries** and also to **yolk glands** and the **uterus.**

Note the **excretory canals** on each side of the proglottids of either tapeworm species. They are part of the protonephridial system and empty to the outside at the posterior end of the worm. They are connected by a **transverse canal** across the posterior margin of each segment.

The **nerve cords** run just lateral to the excretory ducts. The tapeworm nervous system resembles that of the planarian and fluke, although it is less well developed.

Do you find any digestive organs in the tapeworm? _____ Why? _____ What is a tapeworm's food? _____ Where is it digested? _____ How does the worm obtain it? _____ How is the structure of tapeworms adaptive for their environment? _____ What control methods can be used against pork and beef tapeworms? _____

The Gravid Proglottid.

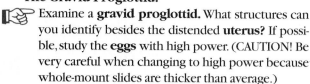 Examine a **gravid proglottid.** What structures can you identify besides the distended **uterus?** If possible, study the **eggs** with high power. (CAUTION! Be very careful when changing to high power because whole-mount slides are thicker than average.)

Can you distinguish in any of the eggs a six-hooked larval form called an **onchosphere** (Gr. *onkinos,* hook, + *sphaira,* globe)? _____

Drawings

On separate paper, draw and label the scolex and neck, a mature proglottid, and a gravid proglottid. Include the scale of the drawings and be sure to identify the species you are drawing.

Projects and Demonstrations

1. *Slides of cross sections of tapeworm.* Identify the tegument, parenchyma, excretory ducts, nerve cords, uterus, and other reproductive organs that appear in section.

2. *Preserved specimens or prepared slides.* Examine (a) preserved specimens or prepared slides of various species of tapeworms and (b) prepared slides of ova and cysts.

EXPERIMENTING IN ZOOLOGY
Planaria Regeneration Experiment

Planarians are easy to work with and have remarkable powers of regeneration.

Before starting, assemble your materials: sharp razor blade or scalpel blade, snap-cap vial or screw-cap specimen jar for each *planarian,* clean culture water (pond water or dechlorinated tap water), pipette, camel's hair brush, lens paper, and an ice cube. Decide how you want to cut the worms (Figure 7-8).

Fold a lens tissue over an ice cube and, with a camel's hair brush, transfer a planarian to the top surface of the ice cube; it will become quiescent almost immediately. Using a dissecting microscope, make the desired cut or cuts with the razor blade. If the worm is being partially split longitudinally, make certain the cut is clean and has completely separated the cut surfaces. Make a sketch in your notebook or on p. 86. With the pipette, rinse the pieces off the lens paper into the vial (previously half-filled with culture water). Label each vial with your name, date, and a sketch of the cut made. Put in a cool place.

The culture water *must* be changed every 3 to 4 days to keep the planarians healthy. *Do not* feed the animals. Remove any dead pieces, which will appear grayish or fuzzy.

Examine the worms about every 3 days for 2 weeks. To obtain two-headed or two-tailed planarians from the partial cuts shown in the second, third, and fifth sketches in Figure 7-8, you will need to recut them after 12 to 18 hours and probably again at 2 days to prevent the parts from fusing back together. Soon a blastema

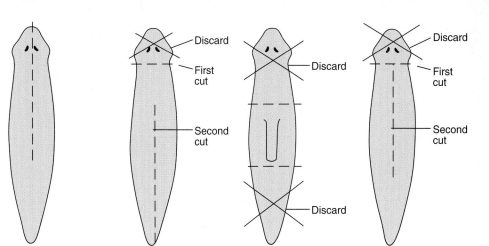

Figure 7-8

Some suggestions for regeneration experiments.

will form on the cut edge. Is the growth faster at the posterior end of a cephalic piece or the anterior end of a caudal piece? _____ On decapitated pieces, when do new eyes appear? _____ When can you distinguish new auricles? _____ Which is regenerated first, eyes or pharynx? _____ Do you find any evidence of polarity in the manner of growth of these pieces? _____

Drawings

On the record sheet, p. 86, make a sketch of the shape of each piece. Examine the specimens twice a week if possible, each time recording the date and sketching the appearance of each regenerating piece. Summarize the results on p. 87. In your report, describe (1) your technique, (2) any problems you encountered, (3) potential or real sources of error in the experimental approach, and (4) your results. If you were to repeat the experiment, what would you do differently?

Name_____

Date_____

Section_____

Regeneration Experiment with *Dugesia*

Sketches of original cuts of specimens:

Dates	1	2	3	4	Control

Record of growth:

Dates	1	2	3	4	Control

Methods used

Observations and conclusions

LAB REPORT

The Pseudocoelomate Animals

Phylum Nematoda
Phylum Rotifera
Phylum Gastrotricha
Phylum Nematomorpha
Phylum Acanthocephala

All bilateral animal phyla except the acoelomates possess a **body cavity** belonging to one of two types: (1) **true coelom,** in which a peritoneum (an epithelium of mesodermal origin) covers both the inner surface of the body wall and the outer surface of the visceral organs in the cavity; or (2) **pseudocoel,** a body cavity not entirely lined with peritoneum.

A body cavity of either type is an advantage because it provides room for organ development and storage and allows some freedom of movement within the body. Because the cavity is often fluid filled, it also provides for a hydrostatic skeleton in those forms lacking a true skeleton.

There are seven pseudocoelomate phyla, of which phylum Nematoda is by far the largest. All of the pseudocoelomate phyla are at the **organ-system level of organization.**

In general the pseudocoelomates tend to be cylindrical in body form, unsegmented, and have a complete (mouth-to-anus) digestive tract (this is absent in acanthocephalans). The epidermis is usually covered with a cuticle. There are both aquatic and terrestrial members, and parasitism is fairly common.

EXERCISE 8A

Phylum Nematoda—
Ascaris and Others

The nematodes are an extensive group with worldwide distribution. They include terrestrial, freshwater, marine, and parasitic forms. They are elongated roundworms covered with a flexible, nonliving cuticle. Circular muscles are lacking in the body wall and, in *Ascaris,* the longitudi-

---

> **EXERCISE 8A**
>
> **Phylum Nematoda—*Ascaris* and Others**
> *Ascaris,* the intestinal roundworm
> Some free-living nematodes
> Some parasitic nematodes
> Projects and demonstrations
>
> **EXERCISE 8B**
>
> **A Brief Look at Some Other Pseudocoelomates**
> Phylum Rotifera
> Phylum Gastrotricha
> Phylum Nematomorpha
> Phylum Acanthocephala

nal muscles are arranged in four groups separated by epidermal cords (some nematodes have six or eight groups of longitudinal muscles). Cilia are completely lacking. Are cilia present in any acoelomates? _____ In cnidarians? _____ Nematodes—both parasitic and free-living—are incredibly abundant. A handful of good garden soil contains thousands of nematodes. Some 50 species of nematodes occur in humans, most of them nonpathogenic.

Ascaris, the Intestinal Roundworm

Phylum Nematoda
  Class Rhabditea
    Order Ascaridida
      Genus *Ascaris*
        Species *Ascaris suum*

Where Found

Ascaris lumbricoides (Gr. *askaris,* intestinal worm) is a common intestinal parasite of humans. *A. suum,* which parasitizes pigs, is so similar to the human parasite that

The Pseudocoelomates

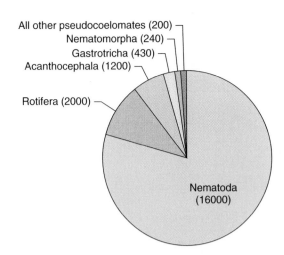

All other pseudocoelomates (200)
Nematomorpha (240)
Gastrotricha (430)
Acanthocephala (1200)

Rotifera (2000)

Nematoda
(16000)

it was long considered merely a different strain of *Ascaris lumbricoides. Ascaris megalocephala* is common in horses.

Infections with *Ascaris* are extremely common. Surveys in the southeastern United States showed prevalences of 20% to 60%. Worldwide nearly one quarter of the human population is infected. Careless defecation near habitation by individuals harboring the worms seeds the soil with eggs that may remain infective for years. The eggs are extraordinarily resistant to chemicals, remaining viable in 2% formalin and in 50% solutions of the common laboratory acids. Moderate to heavy infections cause malnutrition and underdevelopment in children, and the worms' metabolites may cause immune reactions. Heavy infections can lead to fatal intestinal blockage or to wandering of overcrowded worms; the psychological trauma caused by the latter can only be imagined.

General Features

☞ Place a preserved *Ascaris* in a dissecting pan and cover with water. A word of caution: Although the chance of infection from eggs from formalin-preserved female *Ascaris* is remote, be sure to wash your hands after dissecting the worm.

Females, which run 20 to 40 cm in length, are more numerous and are larger than males, which average 15 to 31 cm in length. Males have a curved posterior end and two chitinous **spicules** projecting from the anal region. The spicules are used to hold the female's vulva open during copulation. How long is your specimen? _____ Is the body segmented? _____ Compare your specimen with one of the other sex.

With a hand lens, find the **mouth** with three **lips,** one dorsal and two ventral (Figure 8-1). Find the ventral **anus** at the posterior end. The anus in a male not only discharges feces from the rectum but also serves as a genital opening. The female genital opening **(vulva)** is located

on the ventral side about one-third the length of the body from the anterior end. It may be hard to distinguish from scars. Use the hand lens or dissecting microscope.

Note the shiny **cuticle** that covers the body wall. It is nonliving and consists primarily of **collagen,** which also is found in vertebrate connective tissue.

Four **longitudinal lines** run almost the entire length of the body—the **dorsal** and **ventral median lines** and two **lateral lines.** The dorsal and ventral lines, which indicate location of bundles of nerve fibers, are very difficult to see on preserved specimens. However, along the lateral lines the body wall is thinner, and the lines usually appear somewhat transparent. Excretory canals are located inside the lateral lines.

Internal Structure

☞ Select a female specimen, place the worm in a dissecting pan, and cover it with water. Locate the lateral lines, where the body wall seems somewhat thinner. Now find the anus and vulva on the ventral side. This should help you identify the opposite, or middorsal line. Now, with a razor blade, slit open the body wall *along the middorsal line,* being careful to avoid injuring the internal structures. Pin back the body wall to expose the viscera, *slanting the pins outward* to allow room for dissection.

Body Wall and Pseudocoel. Note the body cavity. Why is it called a **pseudocoel?** _____ How does it differ from a true coelom? _____ _____ Note fluffy masses lining the body wall. These are large nucleated cell bodies of the **longitudinal muscle cells,** whose fibers extend longitudinally in the body wall. With your teasing needle, tease out some of the fibers from the cut edge of the wall. Examine fibers and cells under the microscope. Absence of circular muscles accounts for the thrashing movements of these animals. Note the absence of muscle cells along the **lateral lines.**

Excretory System. Excretory canals located in the lateral lines unite just back of the mouth to empty ventrally through an **excretory pore.** The canals are largely osmoregulatory in function. Excretion also occurs through the cuticle. Flame cells are lacking in *Ascaris* and other nematodes, although they are found in some other pseudocoelomate phyla.

Digestive System. The mouth empties into a short muscular **pharynx,** which sucks food into the ribbon-like **intestine** (Figure 8-1A). The intestine is thin walled for absorption of digested food products into the pseudocoel. Trace it to the **anus.** What is meant by "tube within a tube" construction? _____ _____ Does *Ascaris* fit this description? _____ Does the planarian? _____

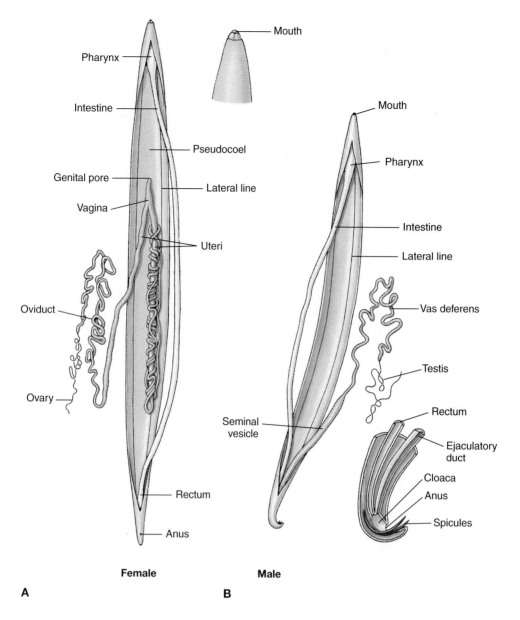

Figure 8-1

Ascaris. **A,** Internal structure of female, with insert showing detail of anterior end. **B,** Internal structure of male, with insert showing detail of cloacal region.

Digestion is begun extracellularly in the lumen of the intestine and is completed intracellularly in the cells of the intestinal wall.

There are no respiratory or circulatory organs. Oxygen is obtained mainly from the breakdown of glycogen within the body, and distribution is handled by the pseudocoelomic fluid.

Reproductive System. The female reproductive system fills most of the pseudocoel. The system is a Y-shaped set of long, convoluted tubes. Unravel them carefully with a probe. The short base of the Y, the **vagina,** opens to the outside at the **vulva.** The long arms of the inverted Y are the **uteri.** These extend posteriorly and then double back as slender, much-coiled **oviducts,** which connect the uteri with the threadlike terminal **ovaries.** Eggs pass from the ovaries through

the oviducts to the uteri, where fertilization occurs and shells are secreted. Then they pass through the vagina and vulva to the outside. The uteri of an *Ascaris* may contain up to 27 million eggs at a time, with as many as 200,000 eggs being laid per day. Study the life history of *Ascaris* from your text. Is there an intermediate host in the life cycle? _____

The male reproductive system is essentially a single, long tube made up of a threadlike **testis,** which continues as a thicker **vas deferens.** Both are much coiled. The vas deferens connects with the wider **seminal vesicle,** which empties by a short, muscular **ejaculatory duct** into the anus. Thus the male anus serves as an outlet for both the digestive system and the reproductive system and is often called a **cloaca** (L. sewer). **Spicules** secreted by and contained in spicule pouches

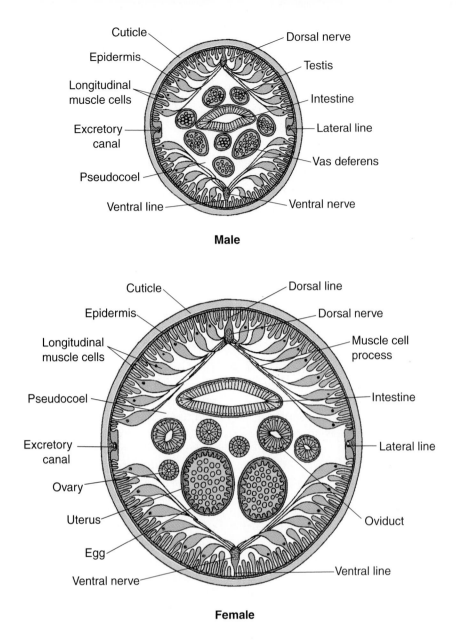

Figure 8-2

Transverse sections through male and female *Ascaris* worms.

may be extended through the anus. In copulation a male inserts the copulatory spicules into the vulva of a female and discharges spermatozoa through the ejaculatory duct into the vagina.

What is meant by "sexual dimorphism"? _____ _____ How does *Ascaris* illustrate this? _____ How is *Ascaris* transmitted from host to host? _____ How can infestation be prevented? _____ In what ways is *Ascaris* structurally and functionally adapted to life as a parasite in the intestine? _____ _____

Transverse Sections of Ascaris

☞ Study a prepared stained slide, at first under low power. If both female and male cross sections

are present, examine first the larger female cross section.

Note the thick noncellular **cuticle** on the outside of the body wall (Figure 8-2). Below the cuticle is the thinner syncytial **epidermis,** which contains nuclei but few cell walls. The **longitudinal muscles** making up most of the body wall appear as fluffy, irregular masses dipping into the **pseudocoel,** with the tips of the cells directed toward the nearest nerve cord. Muscle continuity is interrupted by the **longitudinal lines.** Look for **excretory canals** in the lateral lines and look for the **dorsal** and **ventral nerve cords** in the dorsal and ventral lines. The lateral lines appear free of muscle cells. In the pseudocoel of the female the large **uteri** (Figure 8-2) are filled with eggs enclosed in shells and in cleavage stages. The thin-walled **oviducts** also contain eggs, whereas the wheel-shaped **ovaries** are composed of tall epithelial cells and have small lumens. The **intestine** is composed of a single layer of tall columnar cells (endodermal). The pharynx and the rectal region of the intestine are lined with cuticle. Why? _____ Why is *Ascaris* not digested in the human intestine? _____

Examine the male cross section. It is similar to the female in all respects except for the reproductive system. You should see several rounded sections of **testes** packed with spermatogonia (precursors to male reproductive cells). There may also be several sections of **vas deferens** containing numerous spermatocytes, and possibly a section of a large **seminal vesicle** filled with mature spermatozoa. Note that male reproductive structures visible in the cross section of the roundworm will depend on the body region from which the cross section is taken (refer to Figure 8-1B).

Table of Comparison

On pp.96 and 97 is a table for comparing representatives of three of the metazoan phyla that you have studied so far. Filling in this table affords a survey of the development of these phyla and is an excellent form of review.

Some Free-Living Nematodes

Soil Nematodes

Nematodes are present in almost every imaginable habitat. N. A. Cobb emphasized their abundance in this quotation from a 1914 U.S. Department of Agriculture yearbook. "If all the matter in the universe except the nematodes were swept away, our world would still be dimly recognizable, and if, as disembodied spirits, we could then investigate it, we should find its mountains, hills, vales, rivers, lakes, and oceans represented by a thin film of nematodes. The location of towns would be decipherable, since for every massing of human beings there would be a corresponding massing of certain nematodes. Trees would still stand in ghostly rows representing our streets and highways. The location of the various plants and animals would still be decipherable, and, had we sufficient knowledge, in many cases even their species could be determined by an examination of their erstwhile nematode parasites."

The main limiting factor is the presence of water because nematodes are aquatic animals in the strictest sense. They are capable of activity only when immersed in fluid, even if it is only the microscopically thin film of water that normally covers soil particles. If no water film is present, the nematodes either die or pass into a quiescent resting stage.

Soil nematodes are more abundant among the roots of plants than in open soil, so the best collecting source would probably be the top few centimeters of a long-established meadow turf.

Some nematodes are herbivorous, feeding on algae, fungi, or higher plants; many feed on plant roots. Examples are species of *Tylenchus, Heterodera, Dorylaimus,* and *Monhystera.* Carnivorous nematodes feed on other nematodes, rotifers, small oligochaetes, and so on. These include species of *Dorylaimus, Diplogaster,* and *Monochus.* Some nematodes are **saprophagous** (suhprof'uh-gus; Gr. *sapros,* rotten, + *phagos,* to eat), such as *Rhabditis, Cephalobus,* and *Plectus,* which probably live on bacteria or other microorganisms. Some species are omnivorous. Some soil nematodes are parasitic on plant or animal life.

Some Parasitic Nematodes

Hookworm

Hookworms, *Necator americanus* ("American killer"), live in the intestines of their vertebrate hosts. They attach themselves to the mucosa and suck up the blood and tissue fluids from it. The species most important to humans are *N. americanus* and *Ancylostoma duodenale* (an-ke-los'ta-muh, Gr. *ankylos,* crooked, + *stoma,* mouth). Hookworms infect about 4% of the population in the southern United States, where 95% of the cases are *Necator* infections. Caucasians are 10 times more susceptible to hookworm than are African Americans. *A. caninum* is the common hookworm of domestic dogs and cats (Figure 8-3).

Hookworms mature and mate in the small intestine of the host. Developing eggs are passed out in the feces. On the ground they require warmth (preferably 20° to 30° C), shade, and moisture for continued development. They hatch in 24 to 48 hours into young juveniles, which feed on the fecal matter, molt their cuticles twice, and in a week or so are ready to infect a new host.

If the ground surface is dry, they migrate into the soil, but after a rain or morning dew they move to the surface, extend their bodies in a snakelike fashion, and wave back and forth. Thousands may group together waving rhythmically in unison. Under ideal conditions they may live for several weeks.

Infection occurs when the juveniles contact the skin of the host and burrow into it. Those that reach blood vessels are carried to the heart and then to the lungs. Here they are carried by ciliary action up the respiratory passages to the glottis and swallowed. In the small intestine they grow, molt, mature, and mate. In 5 weeks after entry they are producing eggs.

Whether hookworm disease results from infection depends on number of worms present and nutritional condition of the infected person. Massive infections in the lungs may cause coughing, sore throat, and lung

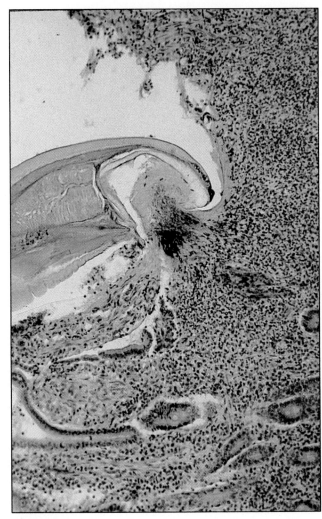

Figure 8-3
Section through anterior end of *Ancylostoma caninum,* the dog hookworm, in the intestine. Note the lacerating cutting plates, used for attachment to the intestinal mucosa.

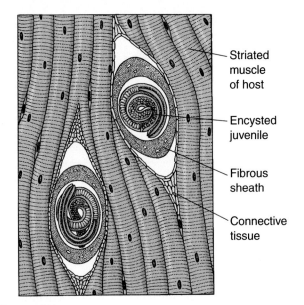

Figure 8-4
Trichinella spiralis, the trichina worm. Juveniles are shown encysted in skeletal muscle.

present, which in *Necator* are fused at the distal ends to form a characteristic hook. In the female the vulva is located in about the middle of the body.

Trichina Worm

The trichina worm, *Trichinella spiralis* (Gr. *trichinos,* of hair, + *ella,* dim. suffix), is a nematode parasite in humans, hogs, rats, and other omnivorous or carnivorous mammals.

☞ **Study a slide of larvae encysted in pork muscle (Figure 8-4).**

How many cysts does your slide show? _____ Would you be able to see the cysts in meat with the naked eye? _____ The cyst wall is made of fibrous tissue that gradually becomes calcified from the host's immune reactions. How many worms are coiled in a cyst? _____ Study the life cycle of *Trichinella* in your text. Are they oviparous or ovoviviparous? _____ How can you prevent trichinosis? _____ Study a slide showing male and female adults. Where would the adults live in the human host? _____

Pinworm

Pinworms, *Enterobius vermicularis* (en-te-robe′ee-us, Gr. *enteron,* intestine, + *bios,* life) (Figure 8-5), are the most common nematode parasite of humans in the United States. They live in the large intestine and cecum. Females, up to 12 mm in length, lay their eggs at night around the anal region of their host. A single female may lay 4600 to 16,000 eggs. Scratching contaminates the hands and bedding of the host. The eggs,

infection. In the intestinal phase moderate infections cause an iron-deficiency anemia. Severe infections may result in severe protein deficiency. When accompanied by chronic malnutrition, as in many tropical countries, there may be irreversible damage, resulting in stunted growth and below-average intelligence.

☞ **Examine prepared slides of hookworms.**

Adult males of *Necator* are typically 7 to 9 mm long, and females 9 to 11 mm long. Specimens of *Ancylostoma* are slightly longer. The anterior end curves dorsally, giving the worm a hooklike appearance. Note the large buccal capsule, which bears a pair of dorsal and a pair of ventral cutting plates surrounding its margin (Figure 8-3). A stout muscular esophagus serves as a powerful pump.

Note on the male a conspicuous copulatory bursa consisting of two lateral lobes and a smaller dorsal lobe, all supported by fleshy rays. Needlelike spicules are

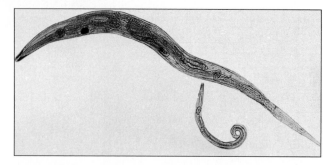

Figure 8-5

Pinworms, *Enterobius vermicularis*. The male is smaller and has the curled posterior end.

when swallowed, hatch in the duodenum and mature in the intestine.

Although at least 400 million people are infected with pinworms, their presence is mainly an irritation and an embarrassment, since pinworms cause no obvious debilitating effects. But unmeasurable is the mental stress suffered by families in their efforts to rid their households of the worms.

Projects and Demonstrations

1. *Prepared slide of* Wuchereria bancrofti (wu-ka-rir′ee-a, after Otto Wucherer, nineteenth-century German physician). Examine prepared slides of the filarial worm, *Wuchereria bancrofti.* What disease is caused by this nematode? _____ In what climate is infestation common? _____ What is the alternate host? _____ What control methods might be used? _____

2. *Prepared slides of* Dracunculus medinensis (dra-kunk′u-les, L. dim. of *draco,* dragon). Examine prepared slides of the guinea worm, *Dracunculus medinensis.* In what part of the world is this nematode common? _____ Where do the larvae develop? _____ How is this parasite acquired? _____ What control methods would you use to prevent infestation? _____

3. *Prepared slides of other nematode parasites.* Examine prepared slides of eggs, larvae, or cysts of any of the nematode parasites.

4. *Prepared slide of* Dirofilaria immitis *(dog heartworm).* Examine a prepared slide of microfilariae in a smear of dog blood. Heartworm is especially prevalent along the Atlantic and Gulf Coast states and northward along the Mississippi River drainage where dogs are infected by mosquitoes which ingest and transmit the microfilariae with their blood meals. The worms mature in the right heart and pulmonary artery. Heavy infestations cause cardiopulmonary failure.

EXERCISE 8B
A Brief Look at Some Other Pseudocoelomates

Phylum Rotifera

Philodina or Others

☞ Place a drop of rotifer culture on a depression slide, cover, and examine with subdued light under low power.

How does the animal attach itself? _____ Is it free-swimming? _____ Does it have a definite head end? _____ Note the anterior discs of cilia **(corona)** that give the impression of wheels turning. Are they retractile? _____ The cilia function both in swimming and in feeding.

The tail end (or **foot**) bears slender toes. How many? _____ The foot contains a pedal gland that secretes a cement used for clinging to objects.

Locate the pharynx **(mastax),** which is fitted with jaws for grinding up food particles. The mastax is conspicuous in living rotifers because of its rhythmic contractions (Figure 8-6). Rotifers feed on small plankton organisms swept in by cilia. Can you identify the digestive tract?

In *Philodina* (fill-uh-dine′uh, Gr. *philos,* fond of, loving, + *dinos,* whirling) (Figure 8-6), the **cuticle** is ringed (annulated) so that it appears segmented. From watching its movements would you say it had circular muscles? Longitudinal? Oblique? _____ Estimate the size of the rotifers. Is there more than one variety in the culture? _____

In many rotifers, the cuticle is thickened and rigid and is called a **lorica.** *Monostyla* and *Platyias* are examples. Some rotifers, such as *Floscularia* (Figure 8-6), live in a secreted tube. Most rotifers live in fresh water.

Phylum Gastrotricha

Chaetonotus or Others

Gastrotrichs include both freshwater and marine organisms. They are similar in size and general habits to the rotifers and are often found in the same cultures.

☞ Place a drop of culture on a slide, cover, and study first with low and then with high power.

Observe the manner of locomotion of the gastrotrichs. They glide along on a substratum by means of ventral cilia. *Chaetonotus* (NL. *chaeta,* bristle, + Gr. *nōtos,* back), a common genus, is covered with short curved dorsal spines (Figure 8-6). The rounded head bears cilia and little tufts of sensory bristles. The tail end is forked and contains cement glands similar to those of

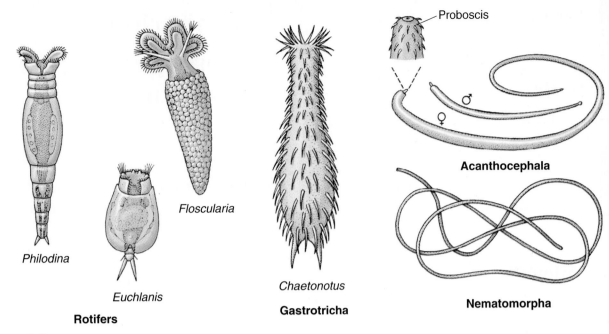

Figure 8-6
Some representative pseudocoelomates.

rotifers. Do gastrotrichs use the forked tail in the same manner as rotifers use the toes? _____

Gastrotrichs have a syncytial epidermis covered with a cuticle. In feeding they use the head cilia to sweep algae, detritus, and protozoans into the mouth. How long are the specimens? _____

Most marine gastrotrichs are hermaphroditic, but in freshwater species only parthenogenetic females are known.

Phylum Nematomorpha

The Threadworms, or "Horsehair" Worms

Threadworms, or "horsehair" worms (Figure 8-6), such as *Paragordius* and *Gordius,** are long cylindrical, hairlike worms often found wriggling in watering troughs, puddles, ponds, and quiet streams. Most of them range between 0.5 and 3 mm in diameter and from 10 to 300 mm in length, but some may reach a length of 1 m.

How long are your specimens? _____ Is the diameter uniform throughout? _____ If you have live specimens, what would you conclude about their muscular makeup, judging from their movements? _____

---

*L. *Gordius,* after Gordius, king of Phrygia, who tied an intricate knot that, according to legend, could be untied only by one destined to rule Asia. Many generic names have been taken from mythology, and in many cases their application is quite obscure. Because of the habit of horsehair worms becoming inextricably entangled in knots comprising several worms, the mythological application in this case, however, is apt.

Nematomorphs have no lateral lines or excretory system, and in the adults the digestive system is degenerate. They differ from nematodes in having a cloaca in both sexes. Females lay long gelatinous strings of eggs on water plants. Larvae, which are encysted on plants, are sometimes eaten by arthropods, in whom the larvae are parasitic for a while.

Phylum Acanthocephala

The Spiny-Headed Worms, Macracanthorhynchus

The adult *Macracanthorhynchus hirudinaceus* (mak'ruh-kan-thuh-rink'us, Gr. *makros,* long, + *akantha,* thorn, + *rhynchos,* snout) (Figure 8-6) is parasitic in the small intestine of pigs, where it attaches to the intestinal lining and absorbs digested food of its host. Like tapeworms it has no digestive system at all.

The body is cylindrical and widest near the anterior end. A small spiny proboscis on the anterior end bears six rows of recurved hooks for attachment to the intestinal wall. The proboscis is hollow and can be partially retracted into a proboscis sheath.

The worms are dioecious. The male is much smaller than the female and has a genital bursa at the posterior end that may be partly evaginated through the genital pore and is used in copulation. Eggs discharged by the female into the host feces may be eaten by white grubs (larvae of the beetle family *Scarabeidae)* in whom they develop. Pigs are infected by eating the grubs or the adult beetles.

Name _____

Date _____

Section _____

Comparing Representatives of Three Phyla

Features	Cnidaria Hydra	Platyhelminthes Planaria	Nematoda *Ascaris*
Symmetry			
Shape			
Germ layers			
Body covering			
Cephalization (present or absent)			
Coelomic cavity (If present, state what type)			

Features	Cnidaria Hydra	Platyhelminthes Planaria	Nematoda *Ascaris*
Musculature (layers present and how arranged)			
Digestive tract and digestion			
Excretion			
Nervous system			
Sense organs			
Reproduction, sexual			
Reproduction, asexual			

The Molluscs
Phylum Mollusca
A Protostome
Eucoelomate Group

The molluscs, with nearly 50,000 species, rank next to the arthropods in number of named species. They include chitons, snails, slugs, clams, oysters, squids, octopuses, cuttlefish, and some others. They have retained the basic features introduced by the preceding phyla, such as triploblastic structure, bilateral symmetry, cephalization, and a body cavity. The body cavity, though small, is now a **true coelom,** a characteristic shared by all remaining phyla. All organ systems are present.

Molluscs have a specialized muscular **foot,** generally used in locomotion. A fold of the dorsal wall, called the **mantle,** or **pallium,** encloses a **mantle cavity,** which usually contains the **gills** and secretes an **exoskeleton,** or shell. There is an open **circulatory system** with a pumping **heart** and a complete mouth-to-anus digestive system. Is a circulatory system present in any of the phyla already studied? _____ Most molluscs, with bivalves being a conspicuous exception, have within the mouth a unique rasping organ, the **radula,** used for scraping off food materials. Most molluscs have a well-developed head—again, a feature absent in the bivalves.

Molluscs have left an extensive fossil record, indicating that their evolution has been a long one. They occupy numerous ecological niches and are found in the sea, in fresh water, and on land. Behavior ranges from those that are sedentary herbivores to other species that are fast-swimming predators.

Classification

Phylum Mollusca

Class Monoplacophora (mon′o-pla-kof′o-ra) (Gr. *monos,* one, + *plax,* plate, + *phora,* bearing). Body bilaterally symmetrical, with broad flat foot; a single dome-shaped shell; five or six pairs of gills in shallow mantle cavity; radula present; separate sexes. Example: *Neopilina.*

Classification
Phylum Mollusca

EXERCISE 9A

Class Bivalvia
(= Pelecypoda)—The Freshwater Clam
Freshwater clam
Projects and demonstrations

EXERCISE 9B

Class Gastropoda—The Pulmonate Land Snail
Land snail
Projects and demonstrations

EXERCISE 9C

Class Cephalopoda—*Loligo,* the Squid
Loligo
Demonstrations

Class Polyplacophora (pol′y-pla-kof′o-ra) (Gr. *polys,* many, + *plax,* plate + *phora,* bearing). Chitons. Elongated, dorsally flattened body with reduced head; bilaterally symmetrical; radula present; shell of eight dorsal plates; foot broad and flat; gills multiple, along sides of body between foot and mantle edge; sexes usually separate. Examples: *Katharina, Mopalia, Chaetopleura.*

Class Caudofoveata (kaw′do-fo-ve-at′a) (L. *cauda,* tail, + *fovea,* small pit). Wormlike; shell, head, and excretory organs absent; radula usually present; mantle with chitinous cuticle and calcareous scales; oral pedal shield near anterior mouth; mantle cavity at posterior end with pair of gills; sexes separate; formerly united with solenogasters in class Aplacophora. Examples: *Chaetoderma, Limifossor.*

Class Solenogastres (so-len′o-gas′trez) (Gr. *solen,* pipe, + *gaster,* stomach). Solenogasters. Wormlike; shell, head, and excretory organs absent; radula usually absent; mantle usually covered with scales or spicules; mantle cavity posterior, without true

Phylum Mollusca

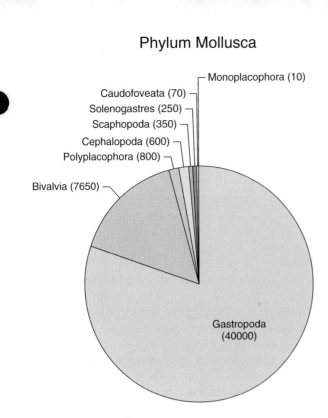

Monoplacophora (10)
Caudofoveata (70)
Solenogastres (250)
Scaphopoda (350)
Cephalopoda (600)
Polyplacophora (800)
Bivalvia (7650)
Gastropoda (40000)

gills but sometimes with secondary respiratory structures; foot represented by long, narrow, ventral pedal groove; hermaphroditic. Example: *Neomenia.*

Class Scaphopoda (ska-fop′o-da) (Gr. *skaphe,* boat, + *pous, podos,* foot). Tusk shells. Body enclosed in a one-piece, tubular shell open at both ends; conical foot; mouth with radula and tentacles; head absent; mantle for respiration; sexes separate. Example: *Dentalium.*

Class Gastropoda (gas-trop′o-da) (Gr. *gaster,* belly, + *pous, podos,* foot). Snails, slugs, conchs, whelks, and others. Body asymmetrical, usually in a coiled shell (shell uncoiled or absent in some); head well developed, with radula; foot large and flat; one or two gills, or with mantle modified into secondary gills or lung; dioecious or monoecious. Examples: *Busycon, Physa, Helix, Aplysia.*

Class Bivalvia (bi-val′vi-a) (L. *bi,* two, + *valva,* valve). Bivalves. Body enclosed in a two-lobed mantle; shell of two lateral valves of variable size and form, with dorsal hinge; cephalization much reduced; no radula; foot usually wedge-shaped; gills platelike; sexes usually separate. Examples: *Anodonta, Venus, Tagelus, Teredo.*

Class Cephalopoda (sef′a-lop′a-da) (Gr. *kephalē,* head, + *pous, podos,* foot). Squids, nautiloids, and octopuses. Shell often reduced or absent; head well developed with eyes and radula; foot modified into arms or tentacles; siphon present; sexes separate. Examples: *Loligo, Octopus, Sepia.*

EXERCISE 9A

Class Bivalvia (= Pelecypoda)— The Freshwater Clam

Freshwater Clam

Phylum Mollusca
  Class Bivalvia
    Subclass Palaeoheterodonta
      Order Unionoida
        Genus *Anodonta,* others

Where Found

Bivalves are found in both fresh water and salt water. Many of them spend most of their existence partly or wholly buried in mud or sand. Freshwater clams (also called mussels) are found in rivers, lakes, and streams and were once particularly abundant in the Mississippi River watershed before stream pollution and increasing acidity from acid rain greatly depleted their populations.* Native bivalves are also the biggest losers in the recent invasion of zebra mussels, unintentionally introduced from Eastern Europe into the Great Lakes in the late 1980s and now running amok throughout the central United States and Canada. Freshwater clams are overwhelmed by zebra mussels, which attach to their shells in enormous numbers. Some common freshwater genera are *Anodonta, Lampsilis, Elliptio,* and *Quadrula.*

Traditionally freshwater clams are used in general laboratories because of their availability, but the sea clam, *Spisula* is quite similar to the freshwater clams and makes a good substitute. Marine mussels, *Mytilus,* or quahogs, *Mercenaria,* can also be substituted.

Behavior and General Features

☞ Observe living bivalves in an aquarium if they are available.

Freshwater clams lie half buried in the sand. They are sluggish, and their reactions are slow.

What is the natural position of the clam at rest? _____ When moving? _____ Note that it leaves a furrow in the sand when it moves. The soft body is protected by a hard **exoskeleton** composed of a pair of **valves,** or shells, hinged on the dorsal side. When the animal is at rest, the valves are slightly agape ventrally, and you can see at the posterior end the fringed edges of the **mantle,** which lines the valves. The posterior edges of the mantle are shaped so as to form two openings **(apertures)** to the inside of the mantle cavity (Figure 9-1).

---

* Clams are especially vulnerable to aquatic pollution. Approximately 10% of all freshwater clams have become extinct in the last century and most of the rest are endangered.

Figure 9-1
Freshwater clam showing apertures in mantle. The dorsal, or upper, aperture is excurrent; the larger, lower aperture is incurrent.

☞ With a Pasteur pipette or hypodermic syringe and needle, carefully introduce a small amount of carmine dye into the water near the apertures and see what happens to it.

Which of the apertures has an incurrent flow, and which has an excurrent flow? _____ A steady flow of water through these apertures is necessary to bring oxygen and food to the animal and to carry away wastes. Most bivalves are filter feeders that filter minute food particles from the water, trap them in mucus, and carry them by ciliary action to the mouth.

Some marine clams have the mantle drawn out into long muscular **siphons.** When the animal burrows deeply into the mud or sand, the siphons extend up to the surface to bring clear water into the mantle cavity.

☞ Gently touch the mantle edge with a glass rod.

What happens? The mantle around the apertures is highly sensitive, not only to touch but also to chemical stimuli, a necessity if the animal is to close its valves to exclude water containing unpleasant or harmful substances.

If the supply of live clams is plentiful, the instructor may want to remove one of the valves from a clam and sprinkle a few carmine granules on the gills, the labial palps, and the inside of the mantle to let you observe the ciliary action that maintains and controls water flow.

☞ Lift a clam out of the sand to observe the hasty withdrawal of the foot. Then lay it on its side on the sand to see if it will right itself.

The foot is as soft, as flexible, and as sensitive as the human tongue. Mucous glands keep the foot well protected with mucus.

☞ If there is a marine tank containing scallops, compare the method of locomotion of the scallops with that of the clams.

Written Report

🖎 Report your observations on clam behavior on separate paper.

External Structure

The two valves of the clam are attached by a **hinge ligament** on the dorsal side; the ventral side is free for the protrusion of the foot. A swollen hump, the **umbo** (pl. **umbones**), near the anterior end of the hinge is the oldest and thickest part of the shell, and the part most resistant to boring gastropod predators. Concentric **lines of growth** around the umbo indicate growth periods. Where is the youngest part of the shell? _____

The outer horny layer of the valve is the *periostracum,* which is secreted by a fold at the edge of the mantle. Has any of this layer been eroded away?

Now determine the correct orientation of the clam. Note that the umbones are dorsal and located toward the anterior end. Identify the right and left valves.

Internal Structure

These directions apply to living clams but may be easily adapted to preserved specimens.

☞ Obtain a living clam that has been heated to about 40° C (104° F), causing the valves to gap slightly. *Caution! Do not overheat.* The instructor or teaching assistant will open the clam by inserting a strong, short-bladed knife (*not* a scalpel) between the *left* valve and the *left* mantle at the posterior end and cutting the posterior adductor muscle as close as possible to the left valve (Figure 9-2A). Cut the anterior adductor muscle in the same manner. Hold the clam against a firm surface when cutting, and keep your hand clear of the knife blade. The valve will gap open because of the action of the hinge ligament. Separate the mantle completely from the left valve and lift the loosened left valve. Place the clam in a dissecting pan and flood the body with pond or dechlorinated water, allowing the right valve to serve as a container.

The Shell

The bivalve shell protects the animal from predators, serves as a skeleton for muscle attachment, and, in burrowing forms, helps to keep mud and sand out of the mantle cavity.

☞ Examine the inner surface of the left valve.

The inner, iridescent mother-of-pearl surface is the **nacreous layer,** which lies next to the mantle and is secreted continuously by the mantle surface (refer to Figure 9-4). Between the inner nacreous layer and the outer periostracum is the **prismatic layer,** made up of

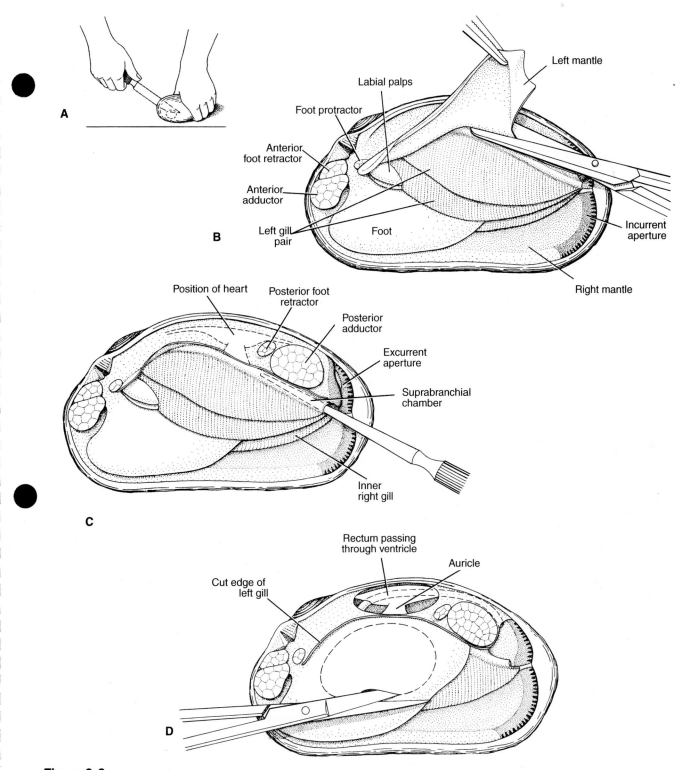

Figure 9-2

Dissection of a freshwater clam. **A,** Cutting the posterior adductor. **B,** Trimming off the left mantle. **C,** Probing the suprabranchial chamber. **D,** Cutting to expose the visceral mass.

crystalline calcium carbonate. It is secreted by glands in the edge of the mantle.

With the aid of Figure 9-3, locate on the valve the scars of the **anterior** and **posterior adductor muscles,** which close and hold the valve together; the **anterior** and **posterior foot retractor muscles,** which pull in the foot; and the **foot protractor muscle,** which loops around the visceral mass and helps extend the foot by squeezing the viscera. The shell is opened by the elastic **hinge ligament,** which acts like a spring to force the shells apart when the adductors relax. Are there also muscles to open the shell? _____

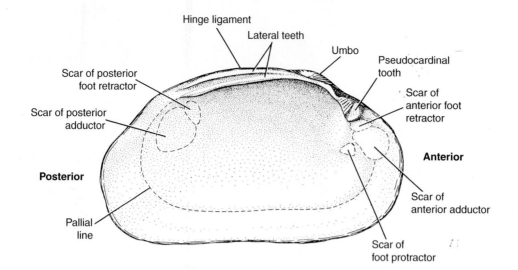

Figure 9-3

Left valve of a freshwater clam, showing the muscle scars.

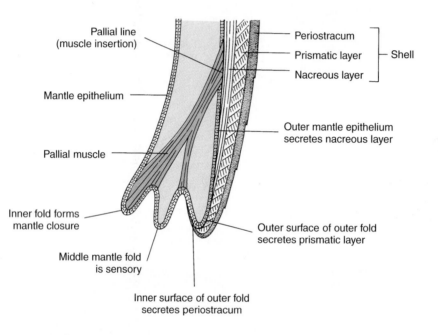

Figure 9-4

Section through margin of freshwater clam shell and mantle.

Find the **pallial line,** where the pallial muscle of the mantle was attached to the valve.

Identify near the dorsal edge of the valves the **teeth,** which fit tightly together when the valves are closed, forming an effective locking mechanism. The **pseudocardinal teeth** are pointed projections and the **lateral teeth** are long ridges along the dorsal edge. Shape and prominence of the teeth vary greatly among different genera.

The Mantle

☞ With the left valve removed, now examine the thin **mantle** covering all the soft tissues of the clam.

Posteriorly the edges of the two mantles are thickened, darkly pigmented, and fused together dorsally to form the ventral **incurrent aperture** and dorsal **excurrent aperture.** The apertures permit a continuous flow of water through the mantle cavity. In many burrowing clams, the apertures are extended into siphons.

☞ Examine the edge of the mantle, which forms three parallel folds: an outer, a middle, and an inner fold (Figure 9-4).

The outer fold (closest to the shell) secretes the horn-like periostracum as well as the prismatic layer of the shell. The middle lobe is sensory in function; in some

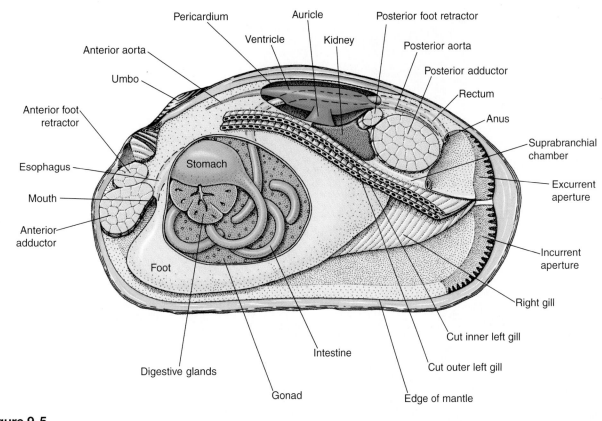

Figure 9-5
Anatomy of a unionid clam.

bivalves (scallops, for example) it is drawn out into specialized sensory structures such as tentacles and eyes. The ciliated surface of the inner fold assists in water circulation within the mantle cavity and sweeps out debris; the inner fold also seals the mantle cavity when the clam closes its shell. It is this inner fold that, on the posterior side, forms the clam's incurrent and excurrent apertures.

The mantle is attached to the shell by a muscle along a semicircular line (the **pallial line**) just inside the shell edge (See Figure 9-3). This muscle retracts the mantle folds when valves are closed (if you are dissecting a living clam, you will probably find the mantle edge strongly retracted by this muscle).

Sometimes a foreign object, such as a sand grain or a parasite, becomes lodged between the mantle and shell. The outer epithelium of the mantle then secretes nacre around the object, forming a pearl. Often, however, the formative pearl becomes fused with the nacreous layer of the shell. Do you find any pearls or other evidence of such irritation in the valve or mantle?

Muscles. Locate the large **adductor muscles,** which close the valves (Figure 9-5). Slightly dorsal to them are the **foot retractor muscles.** The **foot protractor muscle** is small and will be found lying in the visceral mass just posterior to the anterior adductor.

Pericardium. On the dorsal side of the animal locate the delicate, almost transparent **pericardial**

membrane surrounding the heart. In living clams the beating ventricle is visible through the pericardium. Do not open the pericardium until instructed to do so later.

Mantle Cavity

Lift up the mantle to expose the outer pair of **gills** and body mass beneath. The entire space between the right and left lobes of the mantle is **mantle cavity.** Cilia on both mantle and gills keep water flowing through the mantle cavity. If the outer gill is much thicker than the inner gill, the animal is probably a female in which the gill is serving as a **brood chamber** for developing embryos. Are clams monoecious or dioecious?

_____

The soft portion of the body is the **visceral mass.** The muscular **foot** lies ventral to the visceral mass; it will be retracted but will contract even further if you touch it with a probe. The foot operates by a combination of muscular movement and hydraulic mechanisms. The clam can extend or enlarge the foot hydraulically by engorgement with blood and uses the extended foot for anchorage or for drawing the body forward.

Attached to the anterior end of the visceral mass are two pairs of **labial palps,** a pair on each side of the body (See Figure 9-2B). The left and right outer palps join anteriorly to form a protective lip for the slitlike **mouth.** The palps secrete a great amount of mucus

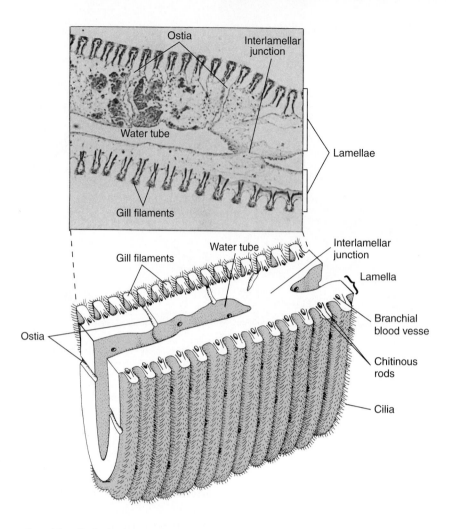

Figure 9-6

Cross section through a portion of the gill of a freshwater clam.

and are ciliated to guide food particles trapped in the mucus toward the mouth.

Respiration and the Gills

☞ Carefully trim off the mantle where it is attached dorsally to the gills (See Figure 9-2B).

With the mantle removed, the gills are conspicuous. Gaseous exchange occurs in both the mantle and the gills. Both mantle and gills are ciliated to promote water flow. Note that only the dorsal margins of the gills are attached; the ventral edge hangs freely in the mantle cavity. Note also that there are *two* gills on each side of the visceral mass, an outer and a somewhat larger inner one.

Each gill is a double fold (Figure 9-6). The walls **(lamellae)** are ridged in appearance, consisting of numerous parallel **filaments** that are supported by chitinous rods. The two lamellae of each gill are connected to each other by a series of thin partitions, which divide the gill into many vertical spaces, the **water tubes.** Water enters the tubes through innumerable small holes **(ostia)** in the lamellae.

☞ Cut a transverse section about 1 mm wide from the lower part of the gill. Lay the section on a slide and examine with the dissecting microscope. Identify the lamellae, filaments, and water tubes, and the partitions **(interlamellar junctions)** that hold the two lamellae apart (Figure 9-6).

The water tubes connect dorsally with the **suprabranchial chamber,** which in turn empties to the outside through the excurrent aperture.

☞ Sprinkle a few grains of powdered carmine on the outer gill to see the current produced by the cilia.

The cilia beat constantly, pulling water in the incurrent aperture, through the ostia and into the water tubes, and then into the suprabranchial chamber. From here it is carried out through the excurrent aperture.

☞ From the excurrent aperture, run a probe into the suprabranchial chamber (See Figure 9-2C). With scissors, slit the suprabranchial chamber to see the tops of the water tubes of the outer gill.

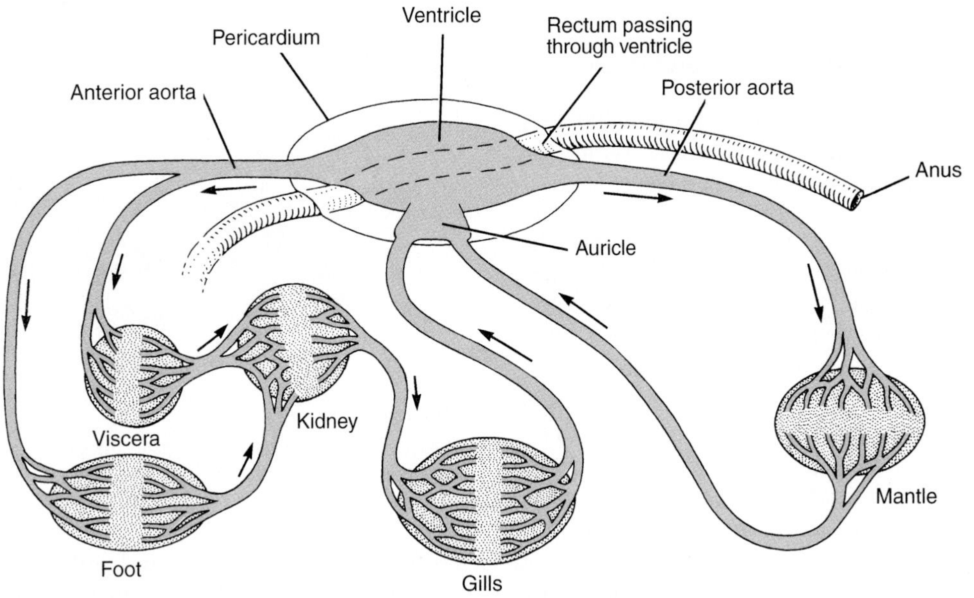

Figure 9-7

Scheme of circulation of a freshwater clam.

A sedentary animal such as a clam, which lives half buried in mud or sand, must have some means of clearing the water that goes through the gills of sediment, detritus, and fecal matter. This is accomplished partially by the small size of the ostia and partially by mucus secreted by glands in the roof of the mantle cavity, which traps particles too large for the ostia. Larger debris drops off the gills by gravitational pull and smaller food particles are carried on toward the mouth.

The water tubes in the female serve as brood pouches for eggs or larvae during breeding season.

Circulatory System

Near the dorsal midline, just below the hinge, is the thin-walled **pericardial sac,** within which lies the **heart.**

☞ Carefully slit open the pericardium, using fine-tipped surgical scissors. Be cautious not to injure the delicate heart inside.

The three-chambered heart is composed of a single **ventricle** and a pair of **auricles** (See Figure 9-5).

☞ If the ventricle is still beating, count the rate (beats/minute).

Note that the ventricle surrounds the intestine. Two aortae leave the ventricle; the **anterior aorta** passes to the visceral mass and intestine, and the **posterior aorta** runs along the ventral side of the rectum to the mantle.

The paired auricles are fan shaped and very thin walled. Pass a probe under the uppermost (left) auricle and lift carefully to see its connection to the ventricle.

At the instructor's option, you may wish to inject a small amount of carmine solution into the ventricle

using a tuberculin syringe and a small (26 or 27 gauge) hypodermic needle. This will flow into and reveal the two aortae.

The pericardial space around the heart is a part of the **coelomic cavity,** which is greatly reduced in molluscs.

Figure 9-7 shows the general plan of the **open system of circulation** found in the freshwater clam. From the ventricle, the aortae carry blood to sinuses in the body tissues. From the visceral organs, blood is carried first to the kidney for removal of wastes, then to the gills for gaseous exchange, and then back to the auricles and ventricle. Blood from the mantle, rich in oxygen, returns directly to the auricles. Although the blood of molluscs is colorless, it contains either hemoglobin or hemocyanin for oxygen transport. It contains nucleated, ameboid corpuscles.

Coelom

Although molluscs have a true coelom, it is small. The pericardial cavity is part of the coelom, as is the small space around the gonads. A true coelom, you recall, is distinguished from other cavities by being lined with epithelium that arises from the mesoderm. What is the name of the lining of the coelom? _____

Excretory System

A pair of dark kidneys lie under the floor of the pericardial sinus. They are roughly U-shaped tubes. The kidneys pick up waste from blood vessels, with which they are richly supplied, and from the pericardial sinus, with which they connect. Waste is discharged into the suprabranchial chamber and carried away with the exhalant current.

Digestive System

Locate again the **labial palps** and the **mouth.**

☞ To reveal the alimentary canal, cut through the surface tissue on one side of the visceral mass and foot and strip away the epithelium (See Figure 9-2D).

The mouth leads into a short **esophagus** that widens into the **stomach,** surrounded by greenish brown **digestive glands.** The stomach narrows into a tubular **intestine,** which can be seen looping back and forth through the visceral mass. The intestine connects to the rectum, which was seen earlier passing through the ventricle. Trace the rectum as it passes dorsal to the posterior adductor muscle to its end, the **anus,** which empties feces into the exhalant current. Surrounding the intestine is light brown tissue of the **gonad** (Figure 9-5).

In freshly collected clams you may find a solid gelatinous rod, the **crystalline style,** projecting into the stomach. It is composed of mucoproteins and digestive enzymes (chiefly amylase), which are released into the food. The crystalline style disappears within a few days after clams are collected and usually is absent from specimens purchased from biological supply houses.

Clams are suspension feeders (also called filter feeding) that depend on respiratory currents to bring in a food supply, which is mostly phytoplankton and organic debris. Food particles passed anteriorly along the edge of the gills become entangled in strings of mucus, which are carried into the mouth. Larger particles are dropped to the edge of the mantle and discarded. Digestion is mostly intracellular.

Reproductive System

The sexes are separate but are difficult to distinguish, except by the swollen gills of the pregnant female. The **gonads (ovaries** or **testes)** are a brownish mass of minute tubes filling the space between the coils of the intestine.

☞ Make a wet mount of gonadal tissue and determine whether there are eggs or sperm in it.

The gonads discharge their products into the suprabranchial chamber. Spermatozoa pass into the surrounding water. They enter a female with the inhalant current and fertilize eggs in the suprabranchial chamber. The **zygotes** settle into the water tubes of the outer gill (brood pouch), where each zygote develops into a tiny bivalved larval form known as a **glochidium** (Gr. *glochis,* point, + *idion,* dim.) (found only in freshwater clams) (Figure 9-8). Glochidia, about the size of dust particles, escape through the excurrent siphon. Glochidia have valves bearing hooks, by which they fasten themselves to the gills, fins, or skin of a passing fish. Here they encyst and live as parasites for several weeks. After a growth period the

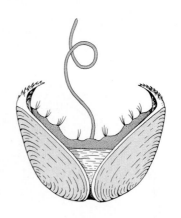

Figure 9-8
Glochidium, the larval form of freshwater bivalve molluscs. Each glochidium is about 0.3 mm in diameter.

young clams break loose and sink to the bottom sand, where they develop as free-living adults. What advantages do the young clams derive from this relationship with the fishes? _____

☞ If a female with swollen brood pouch is available, make a wet mount of some of the gill contents and examine with the microscope to determine whether eggs or glochidia are present.

Nervous System

The nervous system of the clam is not very highly centralized. Dissection of the nervous system is difficult and often impractical. Three pairs of **ganglia** (small groups of nerve cells) are connected to each other by nerves. **Cerebropleural ganglia** are found one on each side of the esophagus on the posterior surface of the anterior adductor muscle. **Pedal ganglia** are fused and are found in the anterior part of the foot. **Visceral ganglia** are fused into a star-shaped body just ventral to the posterior adductor muscle. They are covered by a yellowish membrane and are connected to the cerebropleural ganglia by nerves.

How is the placement of the ganglia and their connectives related to the other organs of the body and to the important receptors and effectors?

Sense organs are poorly developed in the clam. They are involved with touch, chemical sensitivity, balance, and light sensitivity. They are most numerous on the edge of the mantle, particularly around the incurrent aperture, but you will not be able to see them. However, in the scallop, *Pecten,* the ocelli are large and numerous, forming a distinctive row of steel-blue "eyes" along the edges of each mantle.

Oral Report

🗨 Be prepared to demonstrate your dissection and explain the clam's structures and their functions.

Projects and Demonstrations

1. *Cross section of entire clam after removal from the shell.* This study is made from prepared slides. What you will see in it depends to some extent on the region through which the body was cut. Identify **mantle, mantle cavity, gills, lamellae, water tubes, intestine, foot, suprabranchial space, gonad,** and other structures revealed in the cross section.

2. *Oysters, scallops, and sea clams.* Examine the shells of oysters, scallops, and sea clams.

3. *Shipworm,* Teredo. Examine a piece of wood into which the wormlike molluscs, *Teredo,* have bored. Examine a preserved specimen of the shipworm. Note the small valves, small body, and the prolonged siphon that make up the bulk of the shipworm.

EXERCISE 9B

Class Gastropoda— The Pulmonate Land Snail

Land Snail

Phylum Mollusca
    Class Gastropoda
        Subclass Pulmonata
            Superorder Stylommatophora
                Genus *Helix, Polygyra,* others

Where Found

Most land snails (Figure 9-9) prefer fairly moist habitats. They are common in wooded areas, where they spend their days in the damp leaf mold on the ground, coming out to feed on the vegetation at night. *Helix* (Gr. *helix,* twisted) (Figure 9-9, *top*) is a large snail from southern Europe; *Polygyra* (Gr. *poly,* many, + *gyros,* round) is an American land snail.

The subclass Pulmonata constitutes one of three major groups of gastropods. Two other subclasses— Prosobranchia and Opisthobranchia—contain mainly marine snails and marine nudibranchs and tectibranchs.

Behavior

☞ Place a living snail in a finger bowl and, using these suggestions, observe its behavior.

Studying a snail will take patience, for snails are unhurried. Can you detect muscular waves along its **foot** as it moves? Does it leave a trail on the glass as it travels? What causes this? _____ Place a piece of lettuce leaf near a snail and see whether it will eat. How long does it take the snail to find the food? Note the **head** with its **tentacles.** How do the two pairs of tentacles differ? _____ Where are the **eyes** located? Touch an anterior tentacle gently and describe what

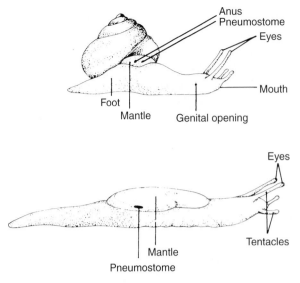

Figure 9-9
Common pulmonates. **Top,** *Helix,* a common land snail. **Bottom,** *Limax,* a common garden slug.

happens. Touch some other parts of the body and note results. Pick up a snail very carefully from the glass on which it is moving and observe exactly what it does.

☞ Place a land snail on a moistened glass plate and invert the plate over a dish so that you can watch the ventral side of the foot under a dissecting microscope.

Do you see the waves of motion? Can you see evidence of mucus? Of ciliary action? _____ If you are fortunate you may see the action of the **radula** (L., scraper) in the mouth. The radula is a series of tiny teeth attached to a ribbonlike organ that moves rapidly back and forth with an action like that of a rasp or file (Figure 9-10).

☞ Prop the plate up in a vertical position. Which direction does the snail move? Now rotate the plate 90° so that the snail is at right angles to its former position. When it resumes its travels, in which direction does it move? Now rotate the plate again and observe.

Do you think the snail is influenced in its movements by the forces of gravity (geotaxis; refer to pp. 46 for an explanation of taxes, and the difference between a positive and a negative taxis)? Try several snails. Do they respond in a similar manner?

The Shell

Examine the shell of a preserved specimen. Note the nature of the spiral shell. Is it symmetrical? _____ To what part of the clam shell does the **apex** correspond? _____ The body of the snail extends through the **aperture.** A **whorl** is

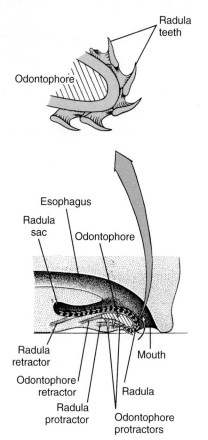

Figure 9-10
Diagrammatic longitudinal section of a gastropod head showing the radula and radula sac. The radula moves back and forth over the odontophore cartilage. As the animal grazes, the mouth opens, the odontophore is thrust forward against the substratum, the teeth scrape food into the pharynx, the odontophore retracts, and the mouth opens. The sequence is repeated rhythmically.

one complete spiral turn of the shell. On the whorls are fine **lines of growth** running parallel to the edge of the aperture. Holding the apex of the shell upward and the aperture toward you, note whether the aperture is at the right or left. **Dextral,** or right-handed, shells will have the aperture toward the right, and **sinistral,** or left-handed, shells toward the left (Figure 9-11). Is your snail dextral or sinistral? _____ Examine a piece of broken shell for the characteristic three layers—the outer **periostracum,** the middle **prismatic layer,** and the shiny inner **nacre.**

Surface Anatomy

☞ If the snail has not contracted enough during killing and preservation* to allow you to slip it carefully from the shell, you may use scissors to carefully cut around the spiral between the whorls, removing pieces of shell as you go and leaving only the central parts, or **columella** (L., pillar). Try not to damage the coiled part of the visceral mass.

The body is made up of **head,** muscular **foot,** and coiled **visceral hump** (Figure 9-12). Identify the **tentacles, eyes,** ventral **mouth** with three lips, and the **genital aperture** just above and behind the right side of the mouth. The foot bears a **mucous gland** just below the mouth. Mucous secretions aid in locomotion.

Note the thin **mantle** that covers the visceral hump and forms the roof of the **mantle cavity.** It is thickened anteriorly to form the **collar** that secretes the shell.

Find a small opening, the **pneumostome** (Gr. *pneuma,* air, + *stoma,* mouth), under the edge of the collar (See Figure 9-9). It opens into a highly vascular portion of the mantle cavity, located in the first half-turn of the spiral, that serves as a respiratory chamber **(lung)** in pulmonates. Here diffusion of gases occurs between the air and the blood. Oxygen is carried by the pigment hemocyanin. Most aquatic gastropods possess gills. The mantle cavity in the second half-turn contains the heart and a large kidney.

The rest of the coiled visceral mass contains the dark lobes of the digestive gland, the intestine, the lighter-colored albumen gland (part of the reproductive system), and the ovotestis (Figure 9-12).

Projects and Demonstrations

1. *Example of a pond snail.* Watch a pond snail attached to the glass side of an aquarium. Note the broad foot by which it clings to the glass. Can you see the motion of the **radula** as the animal eats algae that has settled on the glass?

2. *Ciliary action in the intestine of snail.* Obtain a live aquatic snail (preferably *Lymnaea stagnalis*). Cut open the shell and remove the viscera. Carefully slit open the body and remove the intestine. Slit the intestine and place a small portion, intestinal surface up, on a slide in a drop of saline solution. Examine with high power of your microscope. Note the progressive undulations of cilia over the surface. Ciliary action is best seen some time after you have made the preparation. Are the cilia independent of nerve action? Place a drop of warm (45°C) saline solution on the preparation and note what happens. What is the function of cilia in this region? _____

3. *Isolation of the radula of a snail.* Cut off the head of a snail and soak it in a 10% solution of KOH for 2 or 3 days or until the soft tissues are destroyed and only the radula remains. Transfer the radula to water and wash for 1 or 2 hours in running water. Pieces of attached tissues may be removed by gentle

---

*Specimens from biological supply houses are usually narcotized before killing, which leaves the animal relaxed and expanded. This is best done by sealing snails in a jar of water, capped so as to exclude all air. Boiling (and then cooling) the water beforehand to drive out oxygen will shorten the asphyxiation time. Animals thus treated will be fully relaxed with antennae and foot extended.

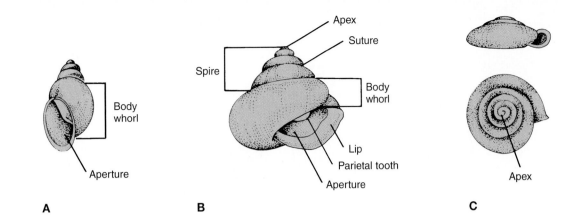

Figure 9-11

Structure of a snail shell. **A,** *Physa,* a sinistral, or left-handed, freshwater snail with lymnaeiform shell (height exceeds width). **B,** *Mesodon* (= *Polygyra*), a dextral (right-handed) snail with heliciform shell (width exceeds height). **C,** *Helicodiscus,* a land snail with planaspiral, or flattened, spire (two views).

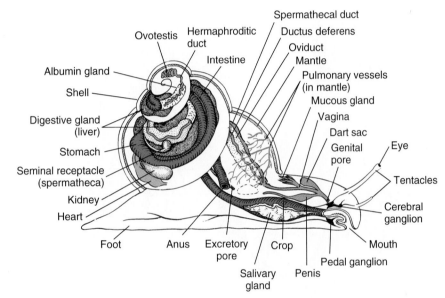

Figure 9-12

Anatomy of a pulmonate snail (diagrammatic).

teasing with a needle. With a piece of paper on each side, the radula should be placed between two glass slides bound together by strong rubber bands. In this position, dehydrate it in 50%, 95%, and 100% alcohols, clear in xylol, and mount in balsam or other mounting medium.

EXERCISE 9C
Class Cephalopoda— *Loligo,* the Squid

Loligo

Phylum Mollusca
    Class Cephalopoda
        Subclass Coleoidea
            Order Teuthoidea
                Genus *Loligo*

Where Found

Squids and octopuses are all marine animals. The active squids are free-swimming and are found in offshore waters at various depths, whereas octopuses are often found in shallow tidal waters.

Squids range in size from 2 cm up. The giant squid *Architeuthis* (Gr. *archaios,* ancient, + *teuthis,* squid) may measure 15 m from tentacle tip to posterior end. *Loligo* (L., a cuttlefish) averages about 30 cm.

Behavior

Because squids ordinarily do not survive long in aquaria, they may not be available for observation. If they are, notice the swimming movements. A squid can move swiftly, either forward or backward, by using its fins and ejecting water through its **funnel.** When water is ejected forward, the squid moves backward; when

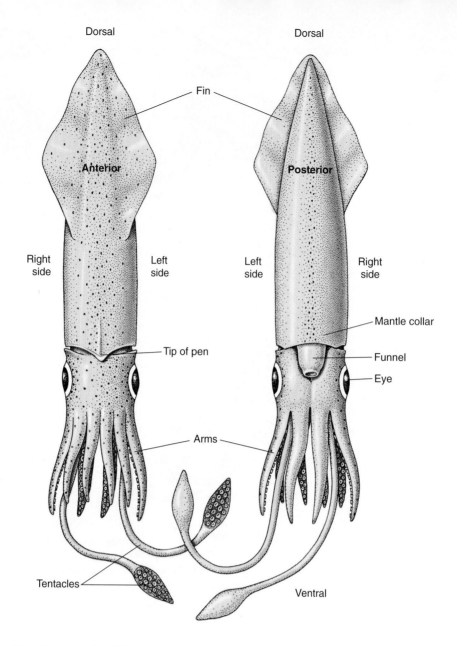

Figure 9-13

External structure of a squid. **Left,** anterior view. **Right,** posterior view.

water is ejected backward, the animal moves forward— movement by jet propulsion! Even when it is resting there is a gentle rhythmical movement of the fins, a muscular movement that also aids in bringing water to the gills in the mantle cavity.

Notice the color changes in the integument caused by contraction and expansion of many **chromatophores** (pigment cells) in the skin. Each chromatophore is a minute organ filled with pigment granules and surrounded by a series of radial muscles. Contraction of the radial muscles stretches the entire organ to form a sheet of pigmented cytoplasm. When the muscles relax, the organ with its pigment shrinks to a small sphere. The pigments, called **ommochromes,** may be black, red, brown, yellow, or orange. Colors appear suddenly when the chromatophores expand and may disappear just as quickly

when the chromatophores shrink. Thus, the colors of squids (and those of most other cephalopod molluscs, such as octopuses) can change rapidly. Elaborate color changes that may sweep quickly across the squid's body serve as a complicated and highly developed means of communication. The chromatophores of cephalopod molluscs are quite different from those of vertebrates (frogs, for example), which are irregularly shaped cells containing pigment granules that can be concentrated into a small area or dispersed throughout the cell. No muscles are involved and pigment changes are much slower than those of the cephalopods.

When the animal is attacked, it can emit a cloud of ink from its ink sac through its funnel.

Food is caught by a pair of retractile tentacles extended with lightning speed, then is held and manipulated by the eight arms and killed by a poison injection. Small fishes, shrimps, and crabs are favorite foods.

External Structure

Squids are called "head-footed" because the end of the head, which bears tentacles, is really homologous to the ventral side of the clam, and the tentacles represent a modification of the foot. To orient the animal morphologically, hold it with its pointed end uppermost; this is the dorsal "side" of the animal. The head, with the arms, tentacles, and funnel, is the ventral surface. The mouth is on the anterior surface, and the funnel is posterior. Functionally, however, the animal swims horizontally by moving forward and backward. Thus, the *morphological* anterior surface is the *functional* dorsal side, and the *morphological* ventral surface is the *functional* anterior side. We will use the morphological orientation in the directions shown in Figure 9-13.

Notice the streamlined body with its **head** and **arms** at one end and a pair of **lateral fins** at the other (Figure 9-13). The visceral mass, dorsal to the head, is covered with a thick **mantle,** the free end of which forms a loosely fitting **collar** about the neck. The head bears a pair of complex, highly advanced **eyes,** each with pupil, iris, cornea, lens, and retina, that can form clear images. The remarkable similarity of cephalopod and vertebrate eyes is an example of **convergent evolution,** in which two groups of organisms of completely different ancestry develop similar structures.

The head is drawn out into 10 appendages: 4 pairs of **arms,** each with 2 rows of stalked suckers, and 1 pair of long retractile **tentacles,** with 2 rows of stalked suckers at the ends. The arms of males are longer and thicker than those of females, and in the mature male the left fourth arm becomes slightly modified for the transfer of spermatophores to the female. This transition is called **hectocotyly** (Gr. *hekaton,* hundred, + *kotylē,* cup). On the hectocotylized arm some of the suckers are smaller and form an adhesion area for carrying the spermatophore. The long tentacles can be shot out quickly to catch prey.

The **mouth** lies within the circle of arms. It is surrounded by a **peristomial membrane,** around which is a **buccal membrane** with seven projections, each with suckers on the inner surface. Probe in the mouth to find two horny beaklike **jaws.**

A muscular **funnel (siphon)** usually projects under the collar on the posterior side, but it may be partially withdrawn. Water forced through the funnel by muscular contraction of the mantle furnishes the power for "jet propulsion" locomotion. Wastes, sexual products, and ink are carried out by the current of water that enters through the collar and leaves through the funnel. The siphon, or funnel, of the squid is not homologous to the siphon of the clam; the clam siphon is a modification of the mantle, whereas the squid siphon, along with the arms and tentacles, is a modification of the foot.

The mottled appearance of the skin is caused by **chromatophores,** the pigment cells.

Mantle Cavity

☞ Beginning near the funnel, make a longitudinal incision through the mantle from the collar to the tip. Pin out the mantle and cover with water.

The space between the mantle and the visceral mass is the **mantle cavity.** The mantle itself is made up largely of circular muscles covered with integument. The funnel contains both circular and longitudinal muscles.

Locate the interlocking **pallial cartilages.** They help support the funnel and close the space between the neck and the mantle so that water inhaled around the collar can be expelled only by way of the funnel. Lateral to the funnel, find large saclike valves that prevent outflow of water by way of the collar.

☞ Slit open the funnel to see the muscular tongue-like valve that prevents inflow of water through the funnel.

This valve allows a buildup of hydrostatic pressure in the mantle cavity before a jet stream of water is ejected through the funnel.

Locate a large pair of **funnel retractor muscles** and beneath them the even larger **head retractor muscles.** Locate the free end of the **rectum** with its **anus** near the inner opening of the funnel. Between it and the visceral mass is the **ink sac.** Do not puncture it. When it is endangered, the squid can send out a cloud of black ink through the funnel as it darts off in another direction.

A pair of long **gills** is attached at one end to the visceral mass and at the other to the mantle. They are located so that water entering the mantle cavity passes directly over them. The gills are not ciliated, as in the bivalves. The mantle cavity is ventilated by action of the mantle itself. Contraction of the radial muscles in the body wall causes the wall to become thinner and the capacity of the mantle cavity to become greater, so that water flows in around the collar. Water is then expelled through the funnel when mantle muscles contract. This movement, also used in locomotion, permits very efficient ventilation of the gills.

In most introductory laboratories observation of the squid will end with the study of the mantle cavity.

Demonstrations

1. *Microslides showing spermatophores of* Loligo.
2. *Preserved octopuses and cuttlefish* (Sepia).
3. *Shells of* Nautilus.
4. *Dried cuttlebone of* Sepia.
5. *Dissection of an injected cephalopod to show circulatory system.*
6. *Dissection of a cephalopod brain.*
7. *Living cephalopod, if available.*

The Annelids
Phylum Annelida
A Protostome
Eucoelomate Group

The annelids include a variety of earthworms, leeches, and marine polychaetes. Their various adaptations fit them for freshwater, marine, terrestrial, and parasitic living. They are typically elongate wormlike animals, circular in cross section, and have muscular body walls. The most distinguishing characteristic that sets them apart from other wormlike creatures is their **segmentation.** They are often referred to collectively as the "segmented worms." This repetition of body parts, called **metamerism** (me-ta′me-ri′sum; Gr. *meta,* between, + *meros,* part), is not only external but is also seen internally in the serial repetition of body organs. The development of metamerism is of significance in the general evolutionary trend toward specialization, for along with segmentation comes the opportunity for the various segments, or metameres, to become specialized for certain functions.

The evolutionary potential of a metameric body plan is amply demonstrated by the large and diverse phylum Arthropoda, where it may have arisen independently. Metamerism also arose independently in the deuterostome line, which includes the numerous and adaptively diverse vertebrates.

The division of the coelomic cavity into fluid-filled compartments also has increased the usefulness of hydrostatic pressure in the locomotion of annelids. By shifting coelomic fluid from one compartment to another through perforations in the dividing septa, differential turgor can be effected, permitting a preciseness of body movement not possible in the pseudocoelomates. The coordination between their well-developed neuromuscular system and more efficient hydrostatic skeleton makes the annelids proficient in swimming, creeping, and burrowing.

Annelids have a complete mouth-to-anus digestive tract with muscular walls so that its movements are independent of body movements. There is a well-developed closed circulatory system with pumping vessels, a high degree of cephalization, and an excretory system of nephridia. Some annelids have respiratory organs.

Classification
Phylum Annelida

EXERCISE 10A
Class Polychaeta—The Clamworm
Nereis
Other Polychaetes

EXERCISE 10B
Class Oligochaeta—The Earthworms
Lumbricus, the common earthworm

EXERCISE 10C
Class Hirudinea—The Leech
Hirudo, the medicinal leech

EXPERIMENTING IN ZOOLOGY
Behavior of the Medicinal Leech, *Hirudo medicinalis*

Classification

The three groups of annelids are classified chiefly on the basis of presence or absence of a clitellum, parapodia, setae, annuli, and other features.

Phylum Annelida

Class Polychaeta (pol′e-ke′ta) (Gr. *polys,* many, + *chaitē,* long hair). Segmented inside and out; parapodia with many setae; distinct head with eyes, palps, and tentacles; no clitellum; separate sexes, trochophore larva usually; mostly marine. Example: clamworm (*Nereis*).

Class Oligochaeta (ol′i-go-ke′ta) (Gr. *oligos,* few, + *chaitē,* long hair). Body segmented inside and out; number of segments variable; clitellum present; few setae; no parapodia, head poorly developed; coelom spacious and usually divided by intersegmental septa; direct development; chiefly terrestrial and freshwater. Examples: *Lumbricus, Tubifex.*

Phylum Annelida

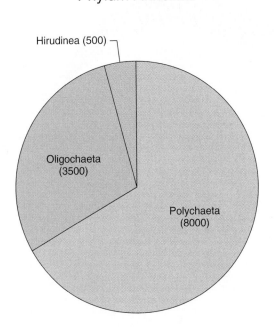

Class Hirudinae (hir′u-din′ea) (L. *hirudo*, leech, + *ea*, characterized by). Segments 33 or 34 in number, with many annuli; clitellum present; anterior and posterior suckers; setae absent (except *Acanthobdella*); parapodia absent; coelom closely packed with connective tissue and muscle; terrestrial, freshwater, and marine. Examples: *Hirudo*, *Placobdella*.

EXERCISE 10A
Class Polychaeta—
The Clamworm

Nereis

Phylum Annelida
  Class Polychaeta
    Order Phyllodocida
      Family Nereididae
        Genus *Nereis*

Where Found

Clamworms (also called sandworms or ragworms) are strictly marine. They live in the mud and debris of shallow coastal waters, often in burrows lined with mucus. Largely nocturnal in habit, they usually are concealed by day under stones, in coral crevices, or in their burrows. The common clamworm *Nereis virens* (Gr. *Nēreis*, a sea nymph) may reach a length of one half meter.

Behavior

☞ If living nereid worms are available, study their patterns of locomotion.

When the animal is quiescent or moving slowly, note that the **parapodia** (Gr. *para*, beside, + *pous, podos*, foot) (lateral appendages) undergo a circular motion that involves an effective stroke and a recovery stroke, each parapodium describing an ellipse during each two-stroke cycle. In the effective stroke the parapodium makes contact with the substratum, lifting the body slightly off the ground. The two parapodia of each segment act alternately, and successive waves of parapodial activity pass along the worm.

For more rapid locomotion, the worm uses undulatory movements of the body produced by muscular contraction and relaxation in addition to parapodial action. As the parapodia on one side move forward in the recovery stroke, the longitudinal muscles on that side contract; as the parapodia sweep backward in their effective stroke, the muscles relax. Watch a worm in action and note these waves of undulatory movements. Can the worm move swiftly? Does it seek cover? Does it maintain contact with substratum, or does it swim freely?

Place near the worm a glass tube with an opening a little wider than the worm. Does the worm enter it? Why? Place a bit of fresh mollusc or fish meat near the entrance of the tube. You may be able to observe feeding reactions.

Written Report (Optional)

✍ Record your observations on separate paper.

External Features

☞ Place a preserved clamworm in a dissecting pan and cover with water.

Note the body with its specialized **head,** variable number of segments bearing **parapodia,** and the caudal segment bearing the **anus** and a pair of feelers, or **cirri** (sing. **cirrus;** sir′us; L., curl). Compare the length and number of segments of your specimen with those of other specimens at your table. Do they vary? The posterior segments are the smallest because they are the youngest. As the animal grows, new segments are added just anterior to the caudal segment.

Examine the **head.** It comprises a **prostomium** (Gr. *pro*, before, + *stoma*, mouth), which is a small protuberance bearing two small median **tentacles,** a pair of fleshy **palps,** and four small dark **eyes;** and a **peristomium** (Gr. *peri*, around, + *stoma*, mouth), or first segment, bearing four pairs of **peristomial tentacles** (Figure 10-1). The palps and tentacles are sensory organs of touch and taste and the eyes are photoreceptors. The **mouth** is a slit on the ventral side of the peristomium through which the **pharynx** can be protruded. If the pharynx is everted on your specimen, do not confuse

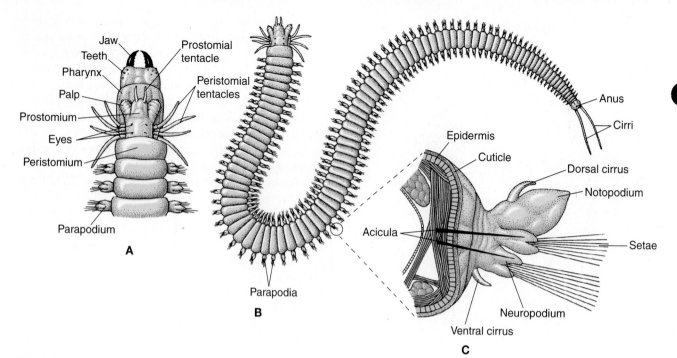

Figure 10-1
Structure of the clamworm *Nereis*. **A,** Anterior view with pharynx extended. **B,** Dorsal view of the entire worm with pharynx withdrawn. **C,** Structure of parapodium.

it with the head structures. The pharynx is large and muscular, bearing a number of small horny teeth and a pair of dark, pincerlike chitinous **jaws.** If the pharynx is fully everted, the jaws will be exposed. If not, you may be able to probe into the pharynx to find them.

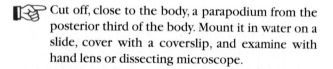 Cut off, close to the body, a parapodium from the posterior third of the body. Mount it in water on a slide, cover with a coverslip, and examine with hand lens or dissecting microscope.

The **parapodia** are used for respiration as well as for locomotion. Are they all identical? Each parapodium has a dorsal lobe called the **notopodium** (Gr. *nōtos,* back, + *pous, podos,* foot) and a ventral lobe called the **neuropodium** (Gr. *neuron,* nerve, + *pous, podos,* foot) (Figure 10-1). Each lobe bears a bundle of bristles called **setae** (see'tee, sing. **seta;** L. bristle). A small process called the **dorsal cirrus** projects from the dorsal base of the notopodium, and a **ventral cirrus** is borne by the neuropodium. Each lobe has a long chitinous, deeply embedded spine called an **aciculum** (L. *acicula,* dim. of *acus,* a point). The acicula are the supporting structures of the parapodium (they are more conspicuous in the posterior parapodia). Each aciculum is attached by muscles that can protrude it as the parapodium goes into its effective stroke and retract it during the recovery stroke. How might the spines help in locomotion? _____

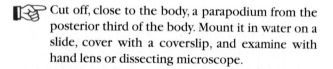 Peel off a piece of the thin cuticle that covers the animal and study it in a wet mount under the microscope.

It is fibrous, and its iridescence is caused by its cross striations. It is full of small pores through which the gland cells of the underlying epidermis discharge their products.

Internal Structure

Because the internal structure of the polychaetes bears many general resemblances to that of the oligochaetes, the study of the internal anatomy will be limited to that of the earthworm.

Other Polychaetes

There are over 10,000 species of polychaetes, most of them marine. They include some unusual and fascinating animals. Besides the errant, or free-moving, worms such as *Nereis* and its relatives (the Errantia), there are many sedentary species (the Sedentaria), including the burrowing and tube-building forms (Figure 10-2).

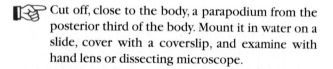 Examine the marine aquarium for living polychaetes and also examine the preserved material on the demonstration table.

A

B

Figure 10-2
Annelid tubeworms secrete the tubes in which they live. They are sessile suspension feeders that have feathered crowns well adapted for the capture of food. **A,** Giant featherduster worm, *Eudistyllia polymorpha,* **B,** Christmas tree worms, *Spirobranchus giganteus.*

EXERCISE 10B

Class Oligochaeta— The Earthworms

Lumbricus, the Common Earthworm

Phylum Annelida
    Class Oligochaeta
        Order Haplotaxina
            Family Lumbricidae
                Genus *Lumbricus*
                    Species *Lumbricus terrestris*

Where Found

Earthworms prefer moist rich soil that is not too dry or sandy. They are found all over the earth. They are chiefly nocturnal and come out of their burrows at night to forage. A good way to find them is to search with a flashlight around the rich soil of lawn shrubbery. The large "night crawler" is easily found this way, especially during warm moist nights of spring and early summer. *Lumbricus terrestris* (L. *lumbricum,* earthworm), named by Linnaeus, is one of the most common earthworms in Europe, Asia, and North America and has been introduced all over the world.

Behavior

☞ Place a sheet of paper toweling on your work area and wet the center of the paper with pond or dechlorinated water, leaving the rest of the paper dry. Place a live earthworm on the moist area. Using these suggestions, observe its behavior.

Is the skin of the worm dry or moist? _____
Do you find any obvious respiratory organs? _____ Where do you think exchange of gases occurs? _____ Would this necessitate a dry or damp environment? _____ _____ Does the worm respond positively or negatively to moisture? _____ _____

Notice the mechanics of crawling. Its body wall contains well-developed layers of circular and longitudinal muscles. As it crawls, notice the progressing peristaltic waves of contraction. These are produced by alternate contraction and relaxation of longitudinal and circular muscles in the body wall acting against noncompressible coelomic fluid. The earthworm's body is divided internally into segments by septa. When the longitudinal muscles of a segment contract, the segment will become shorter and thicker because the volume within each body segment remains constant. Conversely, when the circular muscles contract, the segment elongates. Do the waves of circular muscle contraction move anteriorly or posteriorly when the worm as a whole is moving forward? How far apart (what proportion of total body length) are the waves of contraction? Watch the anterior end as the worm advances. Do the short and thick or the long and thin regions advance the head end forward?

Run a finger along the side of the worm. Do you detect the presence of small setae (bristles)? How might these setae be used? _____

How does the animal respond when you gently touch its anterior end? Its posterior end? Draw the towel to the edge of the desk and see what happens when the worm's head projects over the edge of the table. Is it positively or negatively thigmotactic (responsive to touch)? Turn the worm over and see if and how it can right itself.

Can you devise a means of determining whether the earthworm is positively or negatively geotactic (responsive to gravity)?

☞ Place the earthworm on a large plate of wet glass. Does this difference in substratum affect its locomotion? Is friction important in earthworm locomotion?

Does the earthworm have eyes or other obvious sensory organs? Can you devise a means of determining whether it responds positively or negatively to light?

Written Report

✍ On p. 122, record the responses of the earthworm. Comment on hydrotaxis, locomotion, thigmotaxis, phototaxis, and the importance of friction.

External Structure

☞ Anesthetize an earthworm by immersing it for 30 to 40 minutes in 7% ethanol.* When the worm is completely limp when picked up, transfer it to a dissecting tray that has been dampened with water. Examine the worm with a dissecting microscope or hand lens as necessary.

What are the most obvious differences between the earthworm and the clamworm? List two or three here. _____

The first four segments make up the head region. The first segment is the **peristomium.** It bears the **mouth,** which is overhung by a lobe, the **prostomium.** The head of the earthworm, lacking in specialized sense organs, is considered degenerate and is not a truly typical annelid head.

Find the **anus** in the last segment. Observe the saddlelike **clitellum** (L. *clitellae,* packsaddle), which in mature worms secretes the egg capsules into which the eggs are laid. In what segments does it occur? _____

How many pairs of setae are on each segment, and where are they located? _____ Use the hand lens or dissecting microscope to determine this. What does the name "Oligochaeta" mean? "Polychaeta"? Are these names well chosen?

There are many external openings other than the mouth and anus. Earthworms are monoecious. Note the **male pores** on the ventral surface of somite 15. These are conspicuous openings of the sperm ducts from which spermatozoa are discharged. Note the two long

_____
* Prepared by diluting 74 ml of 95% ethanol with 1 liter of water.

seminal grooves extending between the male pores and the clitellum. These guide the flow of spermatozoa during copulation. The small **female pores** on the ventral side of segment 14 will require a hand lens. Here the oviducts discharge eggs. You may not be able to see the openings of 2 pairs of **seminal receptacles** in the grooves between segments 9 and 10 and 10 and 11, or the paired excretory openings, **nephridiopores,** located on the lateroventral surface of each segment (except the first 3 and the last).

A **dorsal pore** from the coelomic cavity is located at the anterior edge of the middorsal line on each segment from 8 or 9 to the last. Many earthworms eject a malodorous coelomic fluid through the dorsal pores in response to mechanical or chemical irritation or when subjected to extremes of heat or cold. The dorsal pores may also help regulate the turgidity of the animal. How would the loss of coelomic fluid affect the animal's escape mechanism (quick withdrawal into its burrow)? How does the lack of dorsal pores in its anterior segments protect its burrowing ability?

Drawings

✏ Complete the external ventral view of the earthworm on p. 121. Draw in and label prostomium, peristomium, mouth, setae, male pores, female pores, seminal grooves, clitellum, and anus.

Internal Structure and Function

☞ Reanesthetize the earthworm in 7% ethanol if necessary. Place the anesthetized worm dorsal side up in a dissecting pan and straighten it by passing one pin through the fourth or fifth segment (just behind the peristomium) and another pin through any segment near the posterior end of the worm. With a razor blade or new scalpel blade, and beginning at about the fortieth segment (just behind the clitellum), cut through the body wall at a point just to one side of the dark middorsal line (the **dorsal blood vessel**). Use fine-tipped scissors to complete the middorsal cut all the way to the head, pulling up on the scissors as you proceed to avoid damaging internal organs. Keep your incision slightly to one side of the dorsal blood vessel. With a pipette, squirt some isotonic salt solution on the internal organs to keep them moist. Now, starting at the posterior end of the incision, pin the animal open. You will need to break the septa (partitions between the metameres) with a needle as you proceed anteriorly. When you have finished, remove all the pins except those anchoring the worm at the anterior end. Stretch out the worm by pulling gently on the posterior end and repin, placing the pins at an oblique angle. If you are using a dissecting microscope rather than a hand lens, you may need

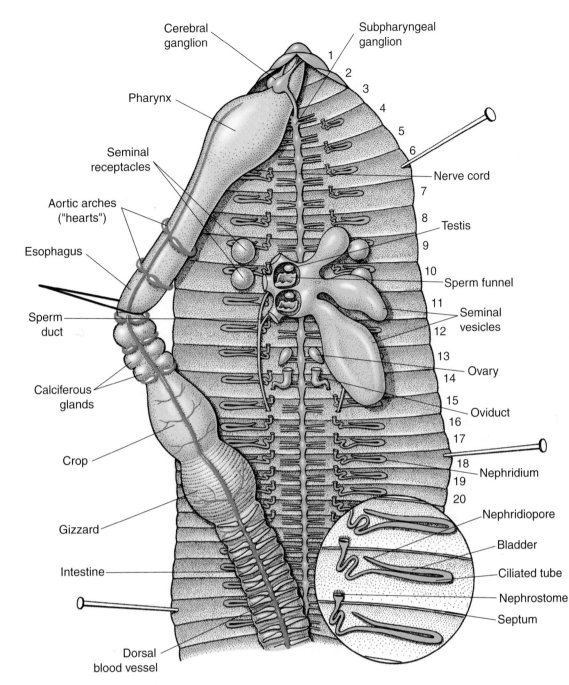

Figure 10-3

Internal structure of *Lumbricus,* dorsal view.

to position the worm to one side of the dissecting tray for viewing. Now flood the tray with enough isotonic saline (0.6% NaCl) to completely cover the earthworm.

Note the peristaltic movements of the **digestive tract,** which serve to propel food posteriorly. Find 3 pairs of cream-colored **seminal vesicles** in somites 9 to 12 (Figure 10-3), 2 pairs of glistening white **seminal receptacles** in somites 9 and 10, and a pair of delicate, almost transparent tubular **nephridia** in the coelomic cavity of each segment. Note the **dorsal vessel** riding

on the digestive tract. In which direction is the blood flowing in this vessel? A total of 5 pairs of pulsating **aortic arches,** sometimes called "hearts," surround the **esophagus** in somites 7 to 11 (some of these arches are covered by the seminal vesicles).

Digestive System. Identify the **mouth;** the muscular **pharynx** attached to the body by **dilator muscles** for sucking action (the muscles, torn by the dissection, give the pharynx a hairy appearance); the slender **esophagus** in somites 6 to 13, which is hidden by the aortic arches and seminal vesicles; the large thin-walled

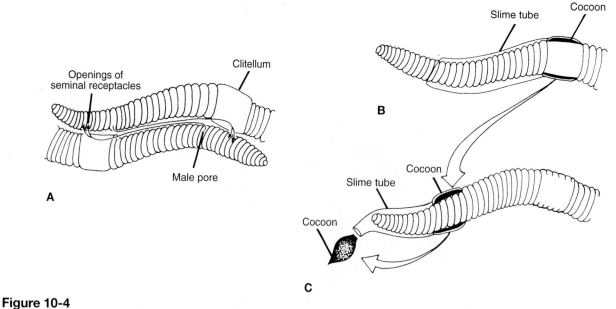

Figure 10-4
A. Earthworm copulation. **B** and **C,** Formation and deposit of the cocoon.

crop (15, 16) for food storage; the muscular **gizzard** (17, 18) for food grinding; the **intestine** for digestion and absorption; and the **anus.** Two or three pairs of yellowish to brownish **calciferous glands** lie on either side of the esophagus (usually partly concealed by the seminal vesicles). They are believed to remove excess calcium and carbonates taken in with the soil; these ions are accumulated as calcite crystals and passed out with the feces. Bright yellow or green **chloragogue** cells often cover the intestine and much of the dorsal vessel. They are known to store glycogen and lipids, but probably have other functions as well, similar to those of the vertebrate liver. Make an off-center longitudinal cut into the intestine in the region of the clitellum to expose the **typhlosole** (Gr. *typlos,* blind, + *sōlēn,* channel), a ridgelike structure projecting into the lumen of the intestine (See Figure 10-5). The typhlosole increases the surface available for digestive enzyme production and absorption.

Circulatory System. The earthworm has a **closed** circulatory system. Note that both the **dorsal vessel** and the **aortic arches** (identified earlier) are contractile, with the dorsal vessel being the chief pumping organ and the arches maintaining a steady flow of blood into the **ventral vessel** beneath the digestive tract.

☞ Retract the digestive tract. Lift up the white nerve cord in the ventral wall. Note the **subneural vessel** clinging to its lower surface and a pair of **lateroneural vessels,** with one located on each side of the nerve cord. Be able to trace blood flow from the dorsal vessel to the intestinal wall and back, to the epidermis and back, and to the nerve cord and back.

Reproductive System. The earthworm is monoecious; it has both male and female organs in the same individual, but cross fertilization occurs during copulation. First, consider the **male organs** (Figure 10-3). The 3 pairs of **seminal vesicles** (sperm sacs in which spermatozoa mature and are stored before copulation) are attached in somites 9, 11, and 12; they lie close to the esophagus. The 2 pairs of small, branched **testes** are housed in special reservoirs in the seminal vesicles, and the 2 small sperm ducts connect the testes with the **male pores** in somite 15; but both testes and ducts are too small to be found easily. The **female organs** are also small. The two pairs of small, round **seminal receptacles,** easily seen in somites 9 and 10, store spermatozoa after copulation. You should be able to find the paired **ovaries** that lie ventral to the third pair of seminal vesicles. The paired **oviducts** with ciliated funnels that carry eggs to the female pores in the next segment will probably not be seen.

Earthworm Copulation. When mating, two earthworms, attracted to each other by glandular secretions, extend their anterior ends from their burrows and, with their heads pointing in opposite directions, join their ventral surfaces in such a way that the seminal receptacle openings of one worm lie in opposition to the clitellum of the other (Figure 10-4). Each worm secretes quantities of mucus so that each is enveloped in a slime tube extending from segment 9 to the posterior end of the clitellum. Seminal fluid discharged from the sperm ducts of each worm is carried along the seminal grooves by contraction of longitudinal muscles and enters the seminal receptacles of the mate. After copulation the worms separate, and each clitellum produces a secretion that finally hardens over its outer surface. The

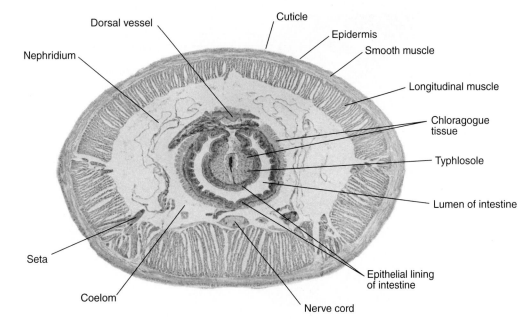

Dorsal vessel — Cuticle — Epidermis — Smooth muscle — Longitudinal muscle — Chloragogue tissue — Typhlosole — Lumen of intestine — Epithelial lining of intestine — Nerve cord — Coelom — Seta — Nephridium

Figure 10-5
Cross section of an earthworm through the intestinal region.

worm moves backward, drawing the hardened tube over its head (Figure 10-4C). As it is moved forward, the tube receives eggs from the oviducts, sperm from the seminal receptacles, and a nutritive albuminous fluid from skin glands. Fertilization occurs in the cocoon. As the worm withdraws, the cocoon closes and is deposited on the ground. Young worms hatch in 2 to 3 weeks.

Excretory System. A pair of tubular **nephridia** lie in each somite except the first three and the last one. Each nephridium begins with a ciliated, funnel-shaped **nephrostome,** which projects through the anterior septum of the segment and opens into the next anterior segment.

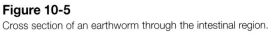 Use a dissecting microscope to examine a nephridium. They are largest in the region just posterior to the clitellum. With fine-tipped scissors carefully remove a nephridium along with a small portion of the septum through which the nephrostome projects. Mount on a slide with a drop or two of saline solution, cover with a coverslip, and examine with a compound microscope.

Note the slender tubule passing from the nephrostome to the looped nephridium. Note the ciliary activity in one narrow portion of the tubule. You may also see parasitic nematodes in the large bladder segment of the tubule. Coelomic fluid is drawn by ciliary activity into the nephrostome and then flows through the narrow tubule where ions, especially sodium and chloride, are reabsorbed. The urine, containing wastes, collects in the bladder, which empties to the outside through a **nephridiopore.**

Nervous System

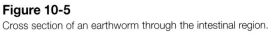 If you have not already done so, extend the dorsal incision to the first somite.

Find the small pair of white **cerebral ganglia** (the brain), lying on the anterior end of the pharynx and partially hidden by dilator muscles; the small white **nerves** from the ganglia to the prostomium; a pair of **circumpharyngeal connectives,** extending from the ganglia and encircling the pharynx to reach the **subpharyngeal ganglia** under the pharynx; and a **ventral nerve cord,** extending posteriorly from the subpharyngeal ganglia for the entire length of the animal. Remove or lay aside the digestive tract and examine the nerve cord with a hand lens to see in each body segment a slightly enlarged **ganglion** and **lateral nerves.**

Oral Report

Be prepared to (1) demonstrate your dissection to the instructor, (2) point out both the external and internal structures you have studied, and (3) explain their functions.

Histology of Cross Section (Figure 10-5)

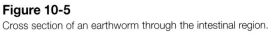 Examine a stained slide with low power. Note the tube-within-a-tube arrangement of intestine and body wall. Identify:

Cuticle. Thin, noncellular, and secreted by the epidermis.

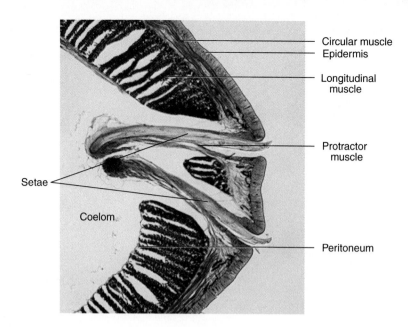

Figure 10-6
Portion of a cross section of an earthworm showing one set of setae with their protractor muscles.

Epidermis. (Ectodermal.) Columnar epithelium containing mucous gland cells. Mucus prevents the skin from drying out.

Circular Muscle Layer. Smooth muscle fibers running around the circumference of the body. How does their contraction affect the body shape?

Longitudinal Muscle Layer. Thick layer of feather-like fibers that run longitudinally. The muscle layers may be interrupted by the setae and dorsal pore.

Peritoneum. (Mesodermal.) The peritoneum (Gr. *peritonaios,* stretched around), a thin epithelial layer lining the body wall and covering the visceral organs (Figure 10-6). Peritoneum lining the body wall is called parietal (L., *paries,* wall) peritoneum. Peritoneum covering the digestive tract and other visceral organs is called visceral (L. *viscera,* bowels) peritoneum.

Setae. If present, they are brownish spines in a sheath secreted by epidermis. They are moved by tiny muscles (Figure 10-6).

Coelom. Space between the **parietal peritoneum,** which lines the body wall, and the **visceral peritoneum,** which covers the intestine and other organs.

Alimentary Canal. The intestine, surrounded by chloragogue tissue, which plays a role in intermediary metabolism similar to that of the liver in vertebrates. Inside the chloragogue layer is a layer of **longitudinal muscle.** Why does it appear as a circle of dots? Next is a **circular muscle layer,** followed by a layer of ciliated columnar epithelium (endodermal), which lines the intestine. Intestinal contents are moved along by peristaltic movement. Is such movement possible without longitudinal and circular muscles? _____ Is peristalsis possible in the intestine of the flatworm or *Ascaris?* _____

Ventral Nerve Cord. Use high power to identify the three **giant fibers** in the dorsal side of the nerve cord, and the nerve cells and fibers in the rest of the cord.

Blood Vessels. Identify the **dorsal** vessel above the typhlosole, **ventral** vessel below the intestine, **subneural** vessel below the nerve cord, and **lateral neurals** beside the nerve cord.

Some slides may also reveal parts of **nephridia, septa, mesenteries,** and other structures.

Drawing

Sketch and label a cross section of the earthworm as it appears on your slide. Use separate paper.

Phylum _____

Genus _____

Name _____

Date _____

Section _____

Earthworm

External structure of the earthworm, ventral view

Transverse section through the body of an earthworm

Observations on the Behavioral Responses of the Earthworm

Hydrotaxis _____

_____

_____

_____

_____

_____

_____

Locomotion _____

_____

_____

_____

_____

_____

_____

Thigmotaxis (response to touch) _____

_____

_____

_____

_____

_____

_____

_____

Phototaxis _____

_____

_____

_____

_____

_____

_____

Importance of friction _____

_____

_____

_____

_____

_____

Demonstrations

1. Living or preserved specimens of tarantula; trap-door spiders; black widows, or other spiders as available. (Living tarantulas are easily maintained in the laboratory and a source of fascination to all.) Captive garden spiders will spin webs in most any container, which need be no more sophisticated than a large glass jar covered with cheese cloth or other netting.

2. Various mites and ticks.

3. Scorpions, alive or preserved.

4. Fossil trilobites and/or eurypterids.

5. Mounted stained preparations of the "trilobite larva."

The Crustacean Arthropods

EXERCISE 12

Subphylum Crustacea—The Crayfish (or Lobster) and Other Crustaceans

EXERCISE 12

Subphylum Crustacea—The Crayfish (or Lobster) and Other Crustaceans
Crayfish or lobster
Other crustaceans
Crustacean development, exemplified by the brine shrimp
Projects and demonstrations

EXPERIMENTING IN ZOOLOGY
The Phototactic Behavior of *Daphnia*

The large lineage of crustaceans, insects, and myriapods that possess **mandibles,** or jawlike appendages, instead of chelicerae are often called the mandibulate arthropods. All of them have, in addition to the mandibles, at least one pair of **antennae** and a pair of **maxillae** on the head.

Until recently, the crustaceans, the insects, and the myriopods (millipedes and centipedes) were all placed in a single subphylum Mandibulata of the Phylum Arthropoda because all possess mandibles. However, a growing body of opinion among zoologists contends that the mandibles of the crustaceans and the mandibles of insects and myriopods are not homologous, but are the result of convergent evolution from separate origins. Consequently the crustaceans are now placed in a separate subphylum Crustacea and the insects and myriopods are grouped in the subphylum Uniramia.

The crustaceans are **gill-breathing arthropods,** with **two pairs of antennae** and **two pairs of maxillae** on the head, and usually a pair of appendages on each body segment. Some of the appendages of present-day adult crustaceans are biramous (two-branched).

Crayfish or Lobster

Phylum Arthropoda
    Subphylum Crustacea
        Class Malacostraca
            Order Decapoda
                Genus *Cambarus*

This description will apply to either the crayfishes (*Cambarus, Procambarus, Pacifastacus,* and *Orconectes*) or the lobster (*Homarus* and others), for crayfishes and lobsters are very similar except in size.

Where Found

The crayfish, called locally and variously crawfish, crawdad, or mudbug, is found in freshwater streams and ponds all over the world. *Pacifastacus* (Pacific, fr. L. *pacificus,* peace, + Gr. *astakos,* crayfish) is found mainly west of the Rocky Mountains, *Procambarus* in the southern states, and *Cambarus* (Gr. *kammaros,* lobster) and *Orconectes* are widely distributed east of the Rocky Mountains. There are about 350 species of crayfish in the United States, more species of crayfish than in all the rest of the world. The southern United States, especially the Gulf Coast, has the largest number of crayfish species. This area also produces most of the edible crayfish, and millions are raised each year on watery crayfish "farms" in Louisiana.

Crayfish are omnivorous, feeding on all kinds of succulent aquatic vegetation and on animal foods such as snails, worms, and small vertebrates. Crayfish are hearty animals, able to survive in almost any type of freshwater habitat. They are frequently sold as bait to fishermen and may either escape from bait buckets or be intentionally released. When crayfish are introduced into new habitats they quickly become established and can harm the biological community by feeding on native freshwater invertebrates and amphibians. Crayfish should not be introduced into habitats where they are not naturally found.

The lobster is a marine form. *Homarus americanus* (F. *homard,* lobster) is found along the eastern North American coast from Labrador to North Carolina. Both lobsters and crayfish are widely used for food in various parts of the world. The spiny, or rock lobster, *Panulirus* (anagram of *Palinurus,* Gr. *palin,* backwards, + *oura,* tail), from the West Coast and southern Atlantic coast, has no pincers and differs in several other respects.

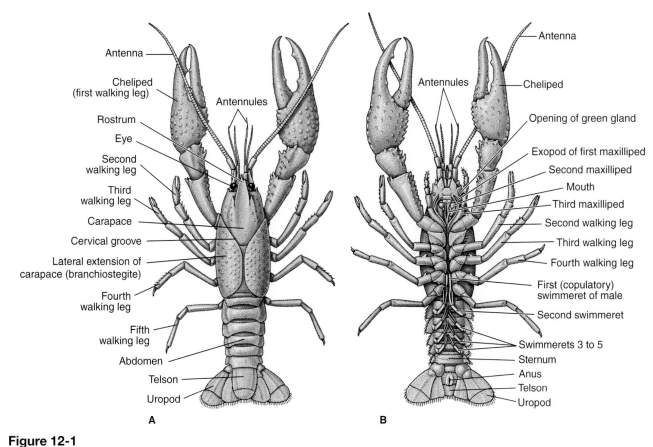

Figure 12-1
External structure of the crayfish. **A,** Dorsal view. **B,** Ventral view.

Behavior

👉 Place a live crayfish or lobster in a bowl or small aquarium partly filled with water. Provide some shells or stones for shelter.

How are the antennae used? What is the response when you stroke the antennae? Notice the compound eyes on the ends of stalks. Are the eyestalks movable? Of what advantage would such eyes be to the animal? What evidence of segmentation do you see on the dorsal side of the animal?

Note the five pairs of legs. The first pair are called chelipeds because they bear the large claws (chelae). How are the chelipeds used? When the animal is startled, what does it do? Is the escape movement forward or backward? How is it accomplished? Is the tail fan involved?

Lift up a crayfish and notice the swimmerets on the abdomen. Release it near the surface of the water. Can it swim? How? The females use the swimmerets to carry and aerate their eggs during the breeding season.

Drop a bit of meat or fish near an undisturbed crayfish in an aquarium. What appendages are used to pick up and handle food? Notice the activity of the small mouth appendages as the animal feeds.

External Features

👉 Place a preserved crayfish (or lobster) in a dissecting pan and add water to the pan.

The **exoskeleton** is a cuticle secreted by the epidermis and hardened with a nitrogenous polysaccharide called **chitin,** with the addition of mineral salts such as calcium carbonate. The cuticle must be shed or **molted** (ecdysis) several times while the crayfish is growing, each time being replaced by a new soft exoskeleton that soon hardens. During and immediately after molting crayfish spend a great deal of time immobile and hiding. Why? _____

The **cephalothorax** is covered by a hard **carapace.** A transverse **cervical groove** marks the head-thorax fusion line. Posterior to this groove are two grooves that separate the median middorsal cardiac area of the carapace, which covers the heart, from the broad lateral extensions of the carapace, which cover the gills. These lateral extensions are also called **branchiostegites** ("gill-cover"). Lift up the edge of the branchiostegite to disclose the **gill chamber** and the feathery **gills.**

Extending anteriorly in the head region is the pointed **rostrum** (L., bill, snout), and coming from under the rostrum are the stalked **eyes** and two pairs of

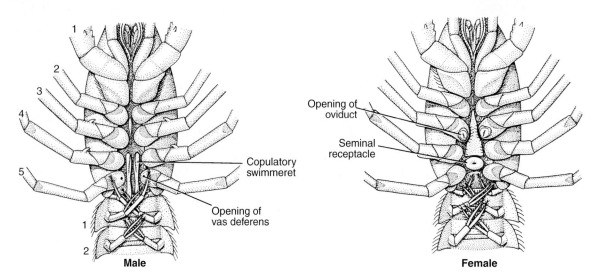

Figure 12-2
Ventral views of crayfish, male and female.

antennae (the second pair of antennae are also called antennules).

On the ventral side the five fused somites of the head bear the two pairs of antennae and three pairs of small mouthparts. The thorax, made up of eight somites, bears three pairs of **maxillipeds** and five pairs of walking legs.

The **abdomen** is made up of six somites, the first five of which bear the **swimmerets** (Figure 12-1B). The sixth is a flat process, the **telson** (Gr., extremity), which bears the **anus** on the ventral side and a pair of fan-shaped **uropods.**

Genital Openings. In the male the genital openings of the sperm ducts are located medially at the base of each of the fifth walking legs (Figure 12-2). In the female the genital openings of the oviducts are located at the base of each third walking leg. Where is the opening of the seminal receptacle? _____
During copulation the male turns the female over and presses the sperm along grooves of the specialized first (copulatory) swimmerets into the seminal receptacle of the female.

Dissection of the Appendages

As you dissect the crayfish appendages, you will see how they illustrate the principle of **serial homology.** Arthropods descended from an annelid-like ancestor bearing **biramous** (two-branched) appendages. These appendages were at first unspecialized, much like the swimmerets of crayfish (Figure 12-3). Because they are all derived from a common form and are similar in development they are considered **homologous.** However, in crayfish and their relatives, the common biramous appendage became modified into mouthparts, walking legs, chelipeds, and swimmerets with different

functions. During evolution some of the biramous branches became lost or modified, and new parts were added. This condition is called **serial homology,** the evolution of a series of structures all homologous to each other but modified for different functions. The crayfish and its relatives possess the best examples of serial homology in the animal kingdom.

☞ From the animal's left side, remove and study each of the appendages described in the next paragraph. However, do not remove any appendages until you have read the instructions for doing so.

Although appendages are numbered consecutively beginning at the anterior end of the animal, it is easier to remove them by beginning at the **posterior** end and proceeding forward. To remove an appendage, grasp it at the base with forceps as near the body as possible and work it loose gradually by gently manipulating the forceps back and forth. Some of the appendages are quite small and feathery, and some are attached to gills. *Be very careful to remove all of the parts of each appendage together* (Figure 12-3). As you remove each one, identify its **medial** and **lateral** sides. Pin the appendages in order on a sheet of paper in the bottom of a dissecting pan. Keep covered with water. Alternatively, the appendages may be glued (e.g., with Elmer's Glue-All) in order on a piece of white cardboard and labeled.

Uropods. The broad uropods (Gr. *oura,* tail, + *pous,* foot) on the most posterior segment are biramous, and together with the medial telson they make up the strong **tail fan** that is used in the rapid backward movements so important in escape. The tail fan also helps protect the eggs and young on the female's swimmerets.

Appendage	Function

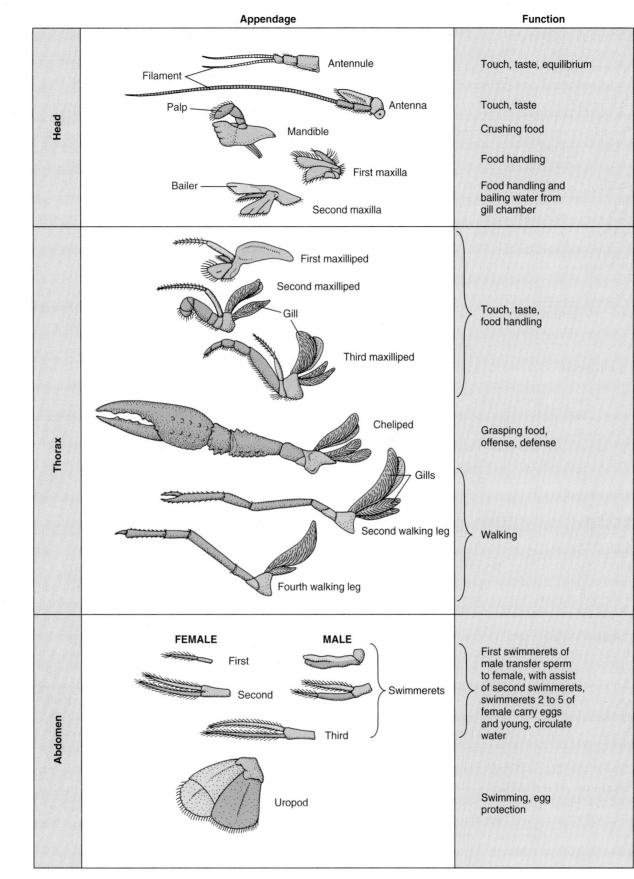

Head

Antennule — Touch, taste, equilibrium

Filament

Antenna — Touch, taste

Palp

Mandible — Crushing food

First maxilla — Food handling

Bailer

Second maxilla — Food handling and bailing water from gill chamber

Thorax

First maxilliped

Second maxilliped

Gill

Third maxilliped — Touch, taste, food handling

Cheliped — Grasping food, offense, defense

Gills

Second walking leg

Fourth walking leg — Walking

Abdomen

FEMALE · MALE

First

Second

Third — Swimmerets — First swimmerets of male transfer sperm to female, with assist of second swimmerets, swimmerets 2 to 5 of female carry eggs and young, circulate water

Uropod — Swimming, egg protection

Figure 12-3

Appendages of the crayfish. Protopod in pink; endopod in purple; exopod in yellow.

Swimmerets (Pleopods). The swimmerets illustrate the primitive biramous plan.

The first two pairs in the male are modified as **copulatory organs.** They are large and grooved and used to direct sperm onto the female (See Figure 12-2). The first pair in the female is usually reduced in size. What is the function of the swimmerets? _____

The part of the appendage attached to the body is called the **protopod.** The medial branch is the **endopod,** and the lateral branch is the **exopod.**

Walking Legs (Pereiopods). The first (most anterior) pair of walking legs are the **chelipeds,** with enlarged claws **(chelae)** used in defense. The other four pairs are used in walking and food handling. There is no exopod so they are called **uniramous** (single branched). The first four pairs are attached to gills. Which of the four pairs of walking legs bear chelae? _____

Note the **genital pores** on the medial side of the protopods of the male's fifth walking legs. Where are the genital pores of the female located? _____

Maxillipeds. Each of the three paired maxillipeds (L. *maxilla,* jaw, + *pes,* foot) has an endopod and a very slender exopod. Be sure to obtain both in removing the appendage. Remove the third (most posterior) maxilliped and its attached gill.

The second and first maxillipeds are similar to the third but smaller. The first maxilliped bears no gill. The maxillipeds are food handlers that break up food and move it to the mouth.

Maxillae. The maxillae anterior to the maxillipeds are foliaceous (thin and leaflike) and direct food toward the mouth. On the second maxilla the protopod is expanded into four little processes, the endopod is small and pointed, and the exopod forms a long blade called the **gill bailer,** which beats to draw currents of water from the gill chambers.

Mandibles. The mandibles on either side of the mouth are heavy triangular structures, bearing teeth on their inner edges. These are the protopods. A little palp folded above each tooth margin represents the endopod. The exopod is absent. The mandibles work from side to side to direct food into the mouth and to hold it while the maxillipeds tear it up. Pry the mandibles apart and remove the left mandible carefully. A strong mandibular muscle attached to the base may tear off with the mandible.

Antennae. The endopod of each antenna is a very long, many-jointed filament. The exopod is a broad, sharp, movable projection near the base. On its broad protopod on the ventral side of the head is the **renal opening** from the excretory gland. Antennae and antennules are sensitive to touch, vibrations, and the chemistry of the water (taste). When you fed the crayfish how were the antennae used? _____

Antennules. Each antennule has a three-jointed protopod as well as two long, many-jointed filaments. The antennules are also concerned with equilibrium.

Branchial (Respiratory) System

Remove part of the carapace on the animal's right side and note the feathery **gills** lying in the branchial chamber. You have already seen that some of the gills are attached to certain appendages. These outer gills are called the foot gills. How many appendages have gills attached? Move the appendages to determine this. Separate the gills carefully, laying aside the foot gills. Another row of gills underneath is attached to membranes that hold the appendages to the body. These gills are called the joint gills. Some genera, but not *Cambarus,* have a third row of gills (side gills) attached to the body wall.

☞ Remove a gill, place it in a watch glass, cover with water, and examine with a hand lens. Now cut the gill in two and look at one of the cut ends.

The **central axis** bears **gill filaments** that give it a feathery appearance. Notice that the central axis and also the little filaments contain canals. These represent blood vessels (afferent and efferent) that enter and leave the filaments. What do the terms afferent and efferent mean? _____ What happens to the blood as it passes through the filaments? Water enters the gill chamber by the free ventral edge of the carapace and is drawn forward over the gills by the action of the gill bailer of the second maxilla, facilitated by movements of the other appendages.

Internal Structure

☞ Remove the dorsal portion of the exoskeleton as follows: insert the point of the scissors under the posterior edge of the lateral carapace about 1.3 cm to one side of the medial line. Cut forward to a point about 1 cm posterior to the eye. Do the same on the other side, thus loosening a dorsal strip about 2.5 cm wide. Carefully remove this center portion of the carapace, a little at a time, being careful not to remove the underlying **epidermis** and **muscles,** which cling to the carapace, especially in the head region. Loosen such tissue with your scalpel and push it carefully back into place.

☞ Remove the dorsal portion of the abdominal exoskeleton in the same way, uncovering each somite carefully so as not to destroy the long **extensor muscle** lying underneath.

The thin tissue covering the viscera is the **epidermis,** which secretes the exoskeleton. Notice the position of the pinkish portion of the epidermis, which lies in the

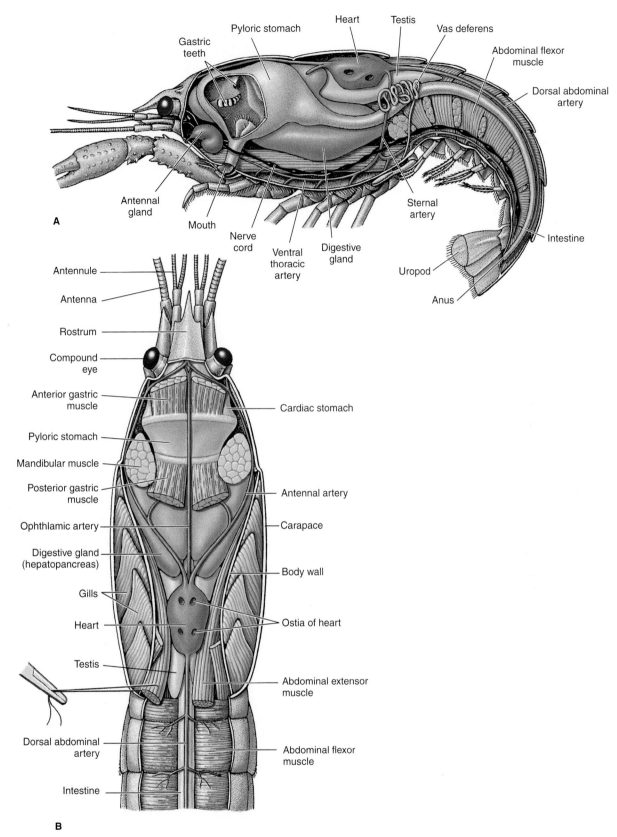

Figure 12-4

Internal structure of a male crayfish. **A,** Lateral view. **B,** Dorsal view. What median artery extends forward from the heart to supply the cardiac stomach, esophagus, and head? _____ What pair of arteries pass diagonally forward and downward over the digestive gland to supply the stomach, antennae, antennal glands, and part of the head? _____ What artery leaves the posterior of the heart? _____

same position as the cardiac region of the carapace. This covers the **pericardial sinus** containing the **heart.** The sinus may be filled with colored latex if the circulatory system has been injected for easy identification.

☞ Remove the epidermis carefully with your forceps to expose the viscera.

The large **stomach** lies in the head region, anterior to the heart (Figure 12-4). On each side of the stomach and heart are large cream-colored lobes of the **hepatopancreas** (digestive gland). They extend the full length of the thorax. This gland is the largest organ in the body.

Muscular System. As you removed the carapace, you noticed the tendency of certain muscles in the head region to cling to it. These are **gastric muscles,** which attach the stomach to the carapace (Figure 12-4B). On each side of the stomach lies a little mass of muscle, the **mandibular muscles,** which move the mandibles. One of these may have been removed when the mandible was removed. On each side of the thorax a narrow band of muscles runs longitudinally to the distal end of the abdomen. These are **extensor muscles,** which are used to straighten the abdomen. In the abdomen, lying ventral to the extensors and nearly filling the abdomen, are the **flexor muscles,** which flex the abdomen (bend it ventrally). Why are these muscles so large? _____

Circulatory System. The small, angular **heart** is located just posterior to the stomach. If the circulatory system has been injected, the heart may be filled and covered with a mass of colored injection fluid or latex filling the sinus. Remove the latex carefully bit by bit, being careful not to destroy the tiny blood vessels leaving the heart. The heart lies in a cavity called the **pericardial sinus,** which is enclosed in a membrane, the **pericardium.** The pericardium may have been removed with the carapace.

This is an "open" type of circulatory system. The hemolymph leaves the heart in arteries but returns to it by way of venous sinuses, or spaces, instead of veins. Hemolymph enters the heart through three pairs of slit-like openings, the **ostia,** which open to receive the hemolymph and then close when the heart contracts to force out the hemolymph through the arteries.

Your instructor may want to demonstrate the heartbeat of a living crayfish (instructions given on p. 143).

Five arteries leave the anterior end of the heart (Figure 12-4): a median **ophthalmic artery** (Gr. *ophthalamos,* eye) extends forward to supply the cardiac stomach, the esophagus, and the head; a pair of **antennal arteries,** one on each side of the ophthalmic, pass diagonally forward and downward over the digestive gland to supply the stomach, the antennae, the antennal glands, and parts of the head; and a pair of **hepatic arteries** (Gr. *hepatos,* liver) from the ventral surface of the heart supply the hepatopancreas.

Leaving the posterior end of the heart are the **dorsal abdominal artery,** which extends the length of the abdomen, lying on the dorsal side of the intestine, and the **sternal artery,** which runs straight down (ventrally) to beneath the nerve cord, where it divides into the **ventral thoracic** and **ventral abdominal arteries,** which supply the appendages and other ventral structures (Figure 12-4). Do not attempt to find these ventral arteries now. They will be referred to later.

Reproductive System. The **gonads** in each sex lie just under the heart. Their size and prominence will depend on the season in which the animals were killed. To find them, lay aside the heart and abdominal extensor muscle bands. The gonads are very slender organs usually slightly different in color from the digestive glands, lying along the medial line between and slightly above the glands. The gonads are sometimes difficult to distinguish from the digestive glands.

In the female the gonads, or **ovaries,** are slender and pinkish, lying side by side, with the anterior ends slightly raised. In some seasons the ovaries may be swollen and greatly distended with eggs, appearing orange in color. A pair of **oviducts** leaves the ovaries and passes laterally over the digestive glands to the genital openings on the third walking legs.

The male gonads, or **testes,** are white and delicate. The **sperm ducts** pass diagonally over the digestive glands and back to the openings in the fifth walking legs (Figure 12-4A).

Digestive System. The **stomach** is a large, thin-walled organ, lying just back of the rostrum. It is made up of two parts—anteriorly a large firm **cardiac chamber** and posteriorly a smaller soft **pyloric chamber.** These will be examined later.

Sometimes a mass of calcareous crystals (**gastroliths**) is attached to each side of the cardiac chamber near the time of molting. These limy masses are thought to have been recovered from the old exoskeleton by the blood and used in the making of the new exoskeleton.

The **intestine,** small and inconspicuous, leaves the pyloric chamber, bends down to pass under the heart, and then rises posteriorly to run along the abdominal length above the large flexor muscles. The intestine ends at the **anus** on the telson.

The large **hepatopancreas,** also called **liver** or **digestive glands** (Figure 12-4), furnishes digestive secretions that are poured through hepatic ducts into the pyloric chamber. The hepatopancreas also is the chief site of absorption and serves for storage of food reserves.

☞ To see the whole length of the digestive system, remove the left lobe of the hepatopancreas and gonad and push aside the left mandibular muscle.

You can now see the **esophagus,** connecting the stomach and mouth, and the intestine, arising from the stomach and running posteriorly to the abdomen where it lies just ventral to the dorsal abdominal artery. You can also see the **sternal artery** descending ventrally.

☞ Now remove the stomach by severing it from the esophagus and intestine, turn it ventral side up, and open it longitudinally. Wash out the contents, if necessary.

The cardiac chamber contains a **gastric mill,** which consists of a set of three chitinous teeth, one dorsomedial and two lateral, that are used for grinding food. They are held by a framework of ossicles and bars in the stomach and operated by the gastric muscles.

In the cardiac stomach the food is ground up and partially digested by enzymes from the hepatopancreas before it is filtered into the smaller pyloric stomach in liquid form. Large particles must be egested through the esophagus. Rows of setae and folds of the stomach lining strain the finest particles and pass them from the pyloric stomach into the hepatopancreas or into the intestine where digestion is completed.

Excretory System. In the head region anterior to the digestive glands and lying against the anterior body wall are a pair of **antennal glands** (also called **green glands,** though they will not appear green in the preserved material) (Figure 12-4). They are round and cushion shaped. In the crayfish each gland contains an **end sac,** connected by an excretory tubule to a bladder. Fluid is filtered into the end sac by hydrostatic pressure in the hemocoel. As the filtrate passes through the excretory tubule, reabsorption of salts and water occurs, leaving the urine to be excreted. A duct from the bladder empties through a renal pore at the base of each antenna.

How would urine production differ between freshwater crayfish and marine lobster?_____ Why? _____ The role of the antennal glands seems to be largely the regulation of the ionic and osmotic composition of the body fluids.

Excretion of nitrogenous wastes (mostly ammonia) occurs by diffusion in the gills and across thin areas of the cuticle.

Nervous System.

☞ Carefully remove all of the viscera, leaving the esophagus and sternal artery in place. The brain is a pair of **supraesophageal ganglia** that lie against the anterior body wall between the antennal glands. Can you distinguish the three pairs of nerves running from the ganglia to the antennae, eyes, and antennules?

From the brain, two connectives pass around the esophagus, one on each side, and unite at the **subesophageal ganglion** on the floor of the cephalothorax.

☞ Chip away the calcified plates that cover and conceal the double ventral nerve cord in the thorax

and follow the cord posteriorly. By removing the big flexor muscles in the abdomen you can trace the cord for the length of the body. Note the ganglia, which appear as enlargements of the cord at intervals. Observe where the nerve cord divides to pass on either side of the sternal artery. Note the small lateral nerves arising from the cord.

In the annelids there is a ganglion in each segment, but in the arthropods there is some fusion of ganglia. The brain is formed by the fusion of three pairs of head ganglia, and the subesophageal ganglion is formed by the fusion of at least five pairs.

Sense Organs. The crayfish has many sense organs: **tactile hairs** over many parts of the body, **statocysts, antennae, antennules,** and **compound eyes.** The eyes have already been observed.

The tactile hairs are variously specialized for touch reception, detection of water currents, and orientation. A number of chemoreceptors are present on the antennae, antennules, and mouthparts.

A **statocyst** is located in the basal segment of each antennule. The pressure of sand grains against sensory hairs in the statocyst gives the crayfish a sense of equilibrium.

Oral Report

Be prepared to demonstrate any phase of your dissection to your instructor. (1) Locate on the dissection any structure mentioned in the exercise, and (2) explain its function. (3) Explain how the appendages of the crayfish or lobster illustrate the principle of serial homology.

Other Crustaceans

Class Branchiopoda

Fairy Shrimp, *Eubranchipus* or *Brachinecta* (Order Anostraca). Fairy shrimp have 11 pairs of basically similar appendages used for locomotion, respiration, and egg carrying. They swim ventral side up. They have no carapace. Note the dark eyes borne on unsegmented stalks. The females carry their eggs in a ventral brood sac, which can usually be seen when the animal is moving (Figure 12-5).

Water Flea, *Daphnia* (Order Cladocera). Water fleas (Figure 12-5) are common in pond water. Mount a living *Daphnia* on a slide, using enough water to prevent crushing but not enough to allow free swimming. *Daphnia* is 1 to 3 mm long and, except for the head, is covered by a thin transparent carapace. The large biramous second antennae are the chief organs of locomotion. There are five pairs of small leaflike swimmerets on the thorax. These are used to filter microscopic algae from the water for food. The blood in the open system

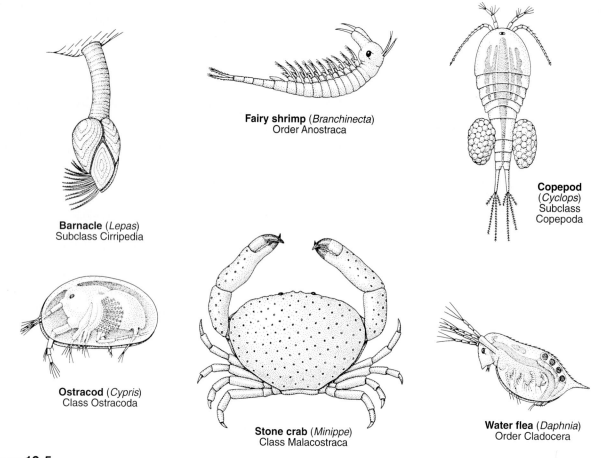

Figure 12-5
Some representative crustaceans.

is red from the presence of hemoglobin. The paired eyes are fused. The brood pouch in the female is large and posterior to the abdomen. They can reproduce parthenogenetically. (See p. 143 for demonstration of heartbeat.)

See the Experimenting in Zoology feature at the end of this chapter for a study of the phototactic behavior of *Daphnia*.

Class Maxillopoda

Subclass Ostracoda. Ostracods with their transparent bivalved carapace resemble tiny clams (1 to 2 mm). They have a median eye and seven pairs of appendages, including the large antennules and antennae (Figure 12-5).

Subclass Copepoda. *Cyclops* and other copepods are found everywhere in fresh and brackish water. *Cyclops* has a median eye near the base of the rostrum and long antennae, modified in the male. The cephalothorax bears appendages; the abdomen has none. The sixth thoracic segment in the female carries large, pendulous egg sacs. The last abdominal segment bears a pair of caudal projections covered with setae (Figure 12-5).

Subclass Cirripedia. The barnacles (all marine) might be mistaken for molluscs because they are enclosed in a calcareous shell. *Balanus*, the acorn barnacle, has a six-piece shell that surrounds the animal like a parapet. The animal protrudes its six pairs of biramous appendages through the opening at the top to create water currents and sweep in plankton and particles of detritus.

The gooseneck barnacle, *Lepas* (Figure 12-5), attaches to floating objects by a long stalk, actually the drawnout front part of the head. The body proper is enclosed in a bivalve carapace strengthened by calcareous plates. Six pairs of delicate filamentous and biramous appendages can be protruded from the carapace. Feathery with long setae, the appendages form an effective net to strain food particles from the water.

Class Malacostraca

Class Malacostraca is a large group. It includes the Isopoda (sow bugs, pill bugs, wood lice, and others with a dorsoventrally flattened body); the Amphipoda (beach fleas, the freshwater *Gammarus*, and others that are laterally compressed); and the Decapoda (shrimps, crabs, crayfish, and lobsters).

☞ Examine a crab (Figure 12-5) and compare its structure and appendages with those of the crayfish.

Crustacean Development, Exemplified by the Brine Shrimp

The brine shrimp, *Artemia salina* (Gr. *Artemis,* a goddess of mythology), which is not really a true shrimp but a member of the more primitive order Anostraca, has a pattern of development typical of most marine crustaceans. If time permits, your instructor may have prepared a series of cultures, started 10 days before the laboratory period and every 2 days thereafter. This should provide you with five larval stages for study. The larvae hatch 24 to 48 hours after the dry eggs are placed in natural or artificial seawater. The stages are classified on the basis of the number of body segments and the number of appendages present. Since growth cannot continue without molting (ecdysis), the larvae will go through a number of molts before reaching adult size. You may find some of the shed exoskeletons in the cultures.

The animals can be narcotized or killed by adding a few drops of ether or chloroform per 10 ml of culture water.

☞ To observe the larvae, mount a drop of the culture containing a few larvae on a slide and cover with a coverslip. Try to identify the stages listed here.

Nauplius. The newly hatched larva, called the nauplius, has a single median ocellus and three pairs of appendages (two pairs of antennae and one pair of mandibles), and the trunk is still unsegmented.

Metanauplius. The first and second maxillae have developed, and there is some thoracic segmentation.

Protozoea. There are now seven pairs of appendages because the first and second pairs of maxillipeds have been added. The compound eyes are now developing.

Zoea. The third pair of maxillipeds have now appeared. The eyes are complete and there are several thoracic segments.

Mysis. Most or all of the 19 body segments and 11 pairs of appendages found in the adult are now present. In these transparent forms you should be able to see the digestive tract and, in live specimens, its peristaltic movements.

The brine shrimp attain adulthood at about 3 weeks and can be maintained on a diet of yeast, fed sparingly. The adults closely resemble their much larger freshwater cousins, the fairy shrimp, *Eubranchipus* (Figure 12-5).

Drawings

✐ Use separate paper for sketching crustaceans other than the crayfish, or the developmental stages of crustacean larvae, as seen in the brine shrimp.

Projects and Demonstrations

1. *Keeping crayfish.* Crayfish are easily maintained in an aquarium tank at room temperature. They will thrive if fed sparingly on raw meat such as small pieces of fish, shrimp, frog, or beef. Crayfish need plenty of oxygen. Keep the water in the aquarium shallow and provide rocks that can serve for cover. Rocks will also enable the crayfish to crawl out of the water. The addition of aquatic plants enhances the appearance of the aquarium and will also serve as food.

2. *Heartbeat of crayfish.* Anesthetize the crayfish in a suitable-sized dish by immersing in 15% ethanol, or club soda, freshly opened. Remove from the anesthetic when the animal fails to respond to prodding; then remove the carapace and observe the beating heart. The crayfish should survive several hours under these conditions.

3. *Heartbeat of* Daphnia. Place a small drop of petrolatum (Vaseline) in the center of a Syracuse watch glass. Fasten *Daphnia* by one valve of its carapace to the petrolatum but be careful that water circulates between the valves. Observe the heartbeat. Start with water at 0° C. As the water slowly rises to room temperature, record the heartbeats at the different temperature. Determine the number of heartbeats in 15 seconds at each 2° or 3° rise in temperature.

EXPERIMENTING IN ZOOLOGY
The Phototactic Behavior of *Daphnia*

Many species of freshwater zooplankton move toward light (positive phototaxis). In nature, movement toward light often means movement toward shallower, warmer water that is rich in algae and microorganisms. However, most zooplankton have few defense mechanisms and are very vulnerable to predators. Movement toward light would seem to make them more visible to potential predators. Thus, natural selection might favor one type of response to light when no predation risk is apparent, and a different response to light when predation risk exists.

Getting Ready

Work with a partner for this exercise. Try to work in a room that can be darkened. Obtain an open-ended glass tube 2 to 3 cm in diameter and 25 to 30 cm long (Figure 12-6). Wash the tube and place a black rubber stopper tightly into one end. Use a wax pencil to mark off three equal sections of tube: left, middle, and right. Fill the tube to within 3 to 4 cm of the top with bottled water or dechlorinated tap water. Have a second rubber stopper available for the open end. Have a stopwatch available for timing the experiment.

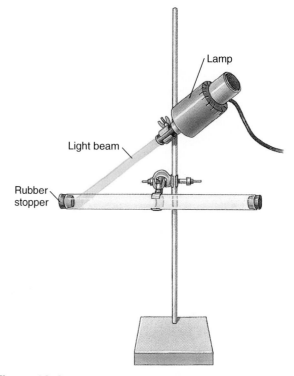

Figure 12-6
Setup for observing the phototactic behavior of *Daphnia*.

How to Proceed

Use a pipette to count out 20 *Daphnia* and add them to the tube. Place the second stopper firmly into the open end. Clamp the tube to the ring stand so that it is horizontal. Turn out the room lights and allow the *Daphnia* to acclimate to the tube for the next 10 minutes. At the end of each minute for the next 10 minutes you will record where the *Daphnia* are distributed in the tube (left, middle, right). If light from the windows is insufficient to see the *Daphnia,* use diffuse lighting from a lamp some distance from the tube. Turn the lamp on just long enough to count the *Daphnia* and then turn it back off until the next observation. After the tenth observation, calculate the mean percentage of *Daphnia* in each section of the tube over the 10 minutes. Wash out the tube and fill it with clean water and 20 new *Daphnia* and repeat the experiment. Calculate the mean distribution of *Daphnia* for the two trials.

After two trials with the *Daphnia* in the dark, repeat the experiment with a dissecting scope illuminator (see Figure 12-6) clamped at an angle so that it illuminates the end of one stopper making it visible to the *Daph-*

nia. Clamp it 4 to 5 cm away from the tube so that it does not heat the water. Again, record the distribution of *Daphnia* in the tube over the next 10 minutes and repeat the experiment. Calculate the mean distribution of *Daphnia* for the two trials.

Compile the data from the entire class and calculate overall means for all control trials and all lighted trials.

Do the *Daphnia* appear to respond to light? Do they exhibit negative or positive phototaxis? In the light trials, do you notice any change in distribution over the 10 minutes?

Questions for Independent Investigation

Recent studies on vertical migrations in zooplankton have found that stimuli from predators can influence phototaxis. Aquatic organisms like *Daphnia* are often very sensitive to chemical stimuli in the water.

1. Design an experiment to test how fish odor might affect phototaxis in *Daphnia*. We suggest adding a small amount of water (40 to 50 ml) from a tank that contains a bass or sunfish to your experimental tube containing the *Daphnia*. What would be a good control for this experiment?

2. Do *Daphnia* respond the same to chemical cues from bass or sunfish as they do to cues from frog tadpoles? Why might there be a difference?

3. *Daphnia* also have many invertebrate predators (e.g., hydra, midge larvae—*Chaoborus*). Do chemical cues from invertebrate predators have an effect on their phototactic behavior?

4. How might the effects of gravity impact phototaxis? Design an experiment to examine both gravity and light on *Daphnia* behavior.

References

DeMeester, L. 1991. An analysis of the phototactic behaviour of *Daphnia magna* clones and their sexual descendants. Hydrobiologia **225:**217–227.

DeMeester, L. 1993. Genotype, fish-mediated chemicals, and phototactic behavior in *Daphnia magna*. Ecology **74:**1467–1474.

Neill, W. 1990. Induced vertical migration in copepods as a defense against invertebrate predators. Nature **345:**524–525.

Parejko, K., and S. Dodson. 1991. The evolutionary ecology of an antipredator reaction norm: *Daphnia pulex* and *Chaoborus americanus*. Evolution **45:**1665–1674.

The Uniramia Arthropods: Myriapods and Insects

The Uniramia include insects and myriapods, a large group of mandibulate arthropods that breathe by means of tracheae and have only one pair of antennae and only uniramous appendages. How many pairs of antennae do crustaceans have? _____ Do crustaceans have uniramous or biramous appendages? _____ There are few aquatic adults, and those are mostly freshwater forms. Some insect young are aquatic and possess gills, but their gills are not homologous to the gills of crustaceans.

EXERCISE 13A
The Myriapods—Centipedes and Millipedes

The term **myriapods** (Gr. *myries,* myriad, numberless, + *podos,* foot) is a common name of convenience for several groups of Uniramia (the Chilopoda, Diplopoda, Pauropoda, and Symphyla). They have two tagmata—head and trunk. There is one pair of antennae, and the trunk bears paired appendages on all but the last segment (Figure 13-1).

Class Chilopoda—The Centipedes

Phylum Arthropoda
    Subphylum Uniramia
        Class Chilopoda
            Order Scolopendromorph
                Genus *Scolopendra* (or *Lithobius*)

Centipedes, or "hundred-leggers," are active predators that live in moist places such as under logs, stones, and bark, where they feed on worms, larvae, and insects.

☞ If living examples are available, watch their locomotion and note the agile use of the body and legs.

Note the general shape of the body and the arrangement of segments and appendages. Is the body circular or flattened in cross section? _____

EXERCISE 13A
The Myriapods—Centipedes and Millipedes
Class Chilopoda—The centipedes
Class Diplopoda—The millipedes

EXERCISE 13B
The Insects—The Grasshopper and the Honey Bee
Romalea, the lubber grasshopper
Apis, the honey bee

EXERCISE 13C
The Insects—The House Cricket
Acheta domesticus, the house cricket

EXERCISE 13D
Collection and Classification of Insects
Key to the Principal Orders of Insects

☞ On a preserved specimen of *Lithobius* (Gr. *lithos,* stone, + *bios,* life) or *Scolopendra* (Gr., a kind of centipede) (Figures 13-1 and 13-2) examine the head.

Some species have simple **ocelli;** others have large faceted **eyes** resembling the compound eyes of insects. Which does your specimen have? _____

☞ Pull aside the large poison fangs to uncover the mouth area and examine under a dissecting microscope.

Find the **antennae;** the **labrum,** anterior to the mouth; the **mandibles** and **first maxillae,** lateral to the mouth; and the **second maxillae,** bearing a long palp and a short labial portion just posterior to the mouth.

The first trunk appendages are the prehensile **maxillipeds,** each bearing a terminal **poison fang.** Do the rest of the trunk appendages bear legs? The last segment bears the **gonopores,** and the **anus** is located on a short telson. Find the **spiracles.** In *Lithobius* they are located near the bases of the legs. In *Scolopendra* the spiracles are located more dorsally and are present on alternate segments.

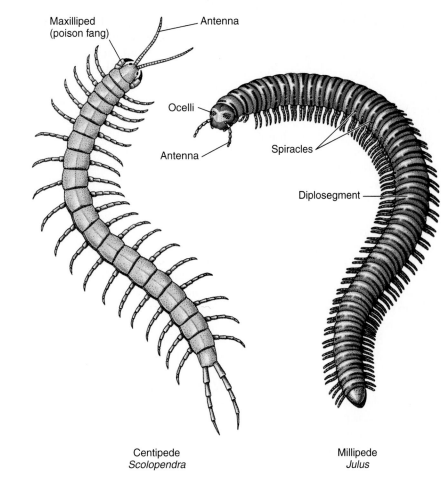

Figure 13-1

Examples of myriapods—centipede and millipede.

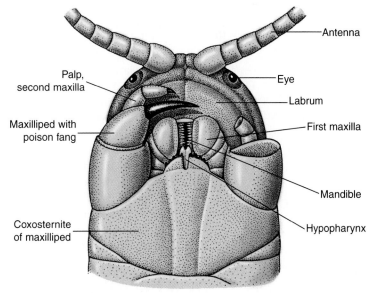

Figure 13-2

Mouthparts of *Scolopendra,* ventral view.

Class Diplopoda—The Millipedes

Phylum Arthropoda
    Subphylum Uniramia
        Class Diplopoda
            Subclass Chilognatha
                Order Juliformia
                    Genus *Julus* or *Spirobolus*

Millipedes, or "thousand foot worms," are found throughout the world, usually hiding in damp woods under bark, leaves, rocks, or logs. They are herbivorous, feeding on decaying wood or leaves. Their most distinguishing feature is the presence of diplosegments, which are double trunk segments probably derived from the fusion of two single segments and each bearing two pairs of legs. The name Diplopoda comes from Greek *diploos,* double, and *pous, podos,* foot. (See Figure 13-1.)

☞ If living millipedes such as *Spirobolus* (L. *spira,* coil, + Gr. *bōlos,* lump) or *Julus* (Gr. *iulus,* plantdown) are available, note the shape of the body, the use of the antennae, the use of the legs, and the animal's ability to roll up.

Notice the rhythm of the power and recovery strokes of the legs with the opposite sides of each segment exactly out of phase. It is similar to that seen in the polychaete *Nereis.* The movements of the legs pass in regular successive waves toward the anterior end. If a specimen has been chilled, you should be able to determine the number of legs working together in each wave and the number of segments between waves.

☞ Study a preserved specimen, and locate on the head the **ocelli, antennae, labrum, mandibles,** and **labium** (fused second maxillae).

Are there appendages on the first trunk segments? _____ How many pairs on the next three segments? _____ The **gonopores** open on the third trunk segment at the bases of the legs. How does this compare with the centipede? _____ The dorsal overlapping of the exoskeletal plates provides full protection for the animal, even when rolled into a ball. Notice that the diplosegments each have two pairs of **spiracles.**

EXERCISE 13B
The Insects—The Grasshopper and the Honey Bee

Insects are by far the largest group of animals. It is estimated that there are nearly 1 million named species of insects—more than all other named animals combined—and that several million species of insects remain to be discovered and described.

Among their chief characteristics are **three pairs of walking legs, one pair of antennae,** body typically divided into **head, thorax,** and **abdomen,** and respiratory system of **tracheal tubes.** Most insects are also provided with one or two pairs of **wings.** They show many variations of adaptive structures according to their habits and life cycles. Their sense organs are often specialized and perhaps account for much of their success in the competition for ecological niches.

Most insects are less than 2.5 cm long, but they range from 1 mm to 20 cm, with the largest insects usually living in tropical areas.

Some insects are highly specialized; others have a generalized plan of body structure. We shall study and compare an example of each. A grasshopper is a fairly generalized type of insect. A honey bee, on the other hand, has become specialized for particular conditions. Not only is its morphology modified for special functions and adaptations, but it belongs to the social insects in which patterns of group organization involve different types of individuals and division of labor.

Romalea, the Lubber Grasshopper

Phylum Arthropoda
    Subphylum Uniramia
        Class Insecta
            Order Orthoptera
                Genus *Romalea*

Behavior of the Grasshopper

☞ If living grasshoppers are available, observe them in a terrarium or place one in a glass jar with moist paper toweling in the bottom and a stick or other object to perch on.

Is the color adaptive given its natural habitat? Can it move its head?

Observe how the grasshopper moves and how it uses its legs. How does it crawl up a stick? How does it use its claws? What position does it assume when quiescent in the jar? How does it jump? How are its legs adapted for jumping? Does it use its wings?

Note movements of the body while it is at rest. Are the movements related to breathing? How does the grasshopper breathe? How does it get air into its body?

What is the common food of grasshoppers? Observe how one eats a piece of lettuce leaf. Watch how it moves its mouthparts.

Take a specimen from the jar. Note that it will regurgitate its greenish digestive juices and food on a glass plate. Is this a defensive adaptation?

Written Report

 Record your observations on p. 153.

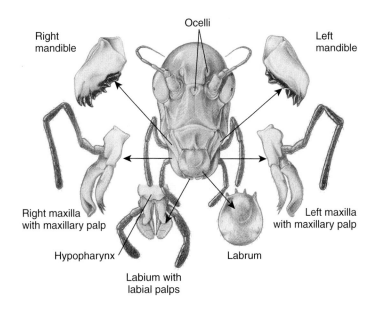

Right mandible

Ocelli

Left mandible

Right maxilla with maxillary palp

Left maxilla with maxillary palp

Hypopharynx

Labrum

Labium with labial palps

Figure 13-3
Head and mouthparts of the grasshopper.

From Peter H. Raven and George B. Johnson, Biology, *4th ed. Copyright © 1996 Times Mirror Higher Education Group, Inc., Dubuque, Iowa. Reprinted by permission. All rights reserved.*

External Structure

☞ Study a preserved specimen of the grasshopper. It is a model of compactness, so use a dissecting microscope or hand lens to observe smaller structures.

Note the division of the body into three tagmata: **head, thorax,** and **abdomen.** Is the grasshopper segmented throughout, or is segmentation more apparent in certain regions of the body? _____

The chitinous **exoskeleton** is secreted by the underlying epidermis. It is composed of hard plates, called **sclerites,** that are bounded by sutures of soft cuticle.

The Head of the Grasshopper. The head of the grasshopper is freely movable. Notice the **compound eyes,** the **antennae,** and three **ocelli,** one dorsal to the base of each antenna and one in the groove between them. Lift the movable bilobed upper lip, or **labrum** (L., lip) (Figure 13-3) and observe the toothed **mandibles.** The mouth contains a membranous **hypopharynx** for tasting food. The bilobed lower lip, or **labium** (L., lip), is the result of the fusion of the second maxillae. The labium bears on each side a three-jointed labial palp. Between the mandible and labium are the paired **maxillae** (L., jawbone), each with a maxillary palp, a flat lobe, and a toothed jaw. Note how the mouthparts are adapted for biting and chewing.

In summary, the insect head bears four pairs of true appendages: the antennae, the mandibles, the maxillae, and the labium (fused second maxillae). There are at least six somites in the head region, although some of them are apparent only in the insect embryo.

☞ After studying the rest of the external features, if time permits, you may be asked by the instructor to use forceps and teasing needles to carefully remove all the mouthparts and arrange them in their relative positions on a sheet of paper.

Thorax of the Grasshopper. The thorax is made up of three somites: **prothorax, mesothorax,** and **metathorax,** each bearing a pair of legs (Figure 13-4). The mesothorax and metathorax also bear a pair of wings. **Spiracles** (external openings of the insect's tracheal system) are located above the legs in the mesothorax and metathorax. Note the leathery **forewings** (on the mesothorax) and the membranous **hindwings** (on the metathorax). Which is more useful for flight? _____ What appears to be the chief function of the forewings? _____ The small **veins** in the wings, really tracheal tubes, are used by entomologists in the identification and classification of insects.

Examine the legs of the grasshopper and identify the basal **coxa** (L., hip), small **trochanter** (Gr., ball of hip joint), large **femur,** slender, spiny **tibia,** and five-jointed **tarsus** with two **claws** and a terminal pad, the **arolium.** Which pair of legs is most specialized, and for what function? _____

Abdomen of the Grasshopper. There are 11 somites in the abdomen of the grasshopper. Notice the large **tympanum,** the organ of hearing, one located on each side of the first abdominal segment. On which of the abdominal segments are the paired spiracles located? _____ In both sexes, somites 2 to 8 are

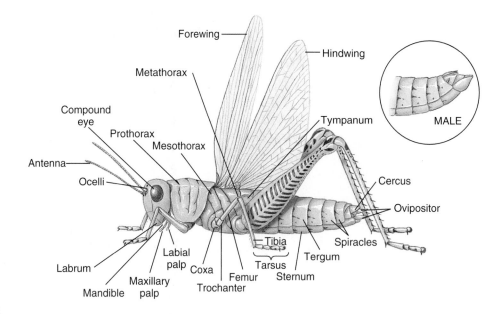

Figure 13-4
External features of a female grasshopper. The terminal segments of a male with external genitalia are shown in the inset.

similar and unmodified and somites 9 and 10 are partially fused.

The eleventh somite forms the genitalia (secondary sex organs). On each side behind the tenth segment is a projection, the **cercus** (pl. **cerci;** Gr. *kerkos,* tail). In the female the posterior end of the abdomen is pointed and consists of two pairs of plates with a smaller pair between, with the whole forming the **ovipositor.** Between the plates is the opening of the **oviduct.** The end of the abdomen in the male is rounded. What is the sex of your specimen? _____

Drawing

Label the external view of the grasshopper on p. 153.

Apis, the Honey Bee

Phylum Arthropoda
    Subphylum Uniramia
        Class Insecta
            Order Hymenoptera
                Genus *Apis*
                    Species *Apis mellifera*

External Structure

Examine a preserved or freshly killed honey bee. Use a dissecting microscope or hand lens to study smaller structures.

The body of the honey bee, like that of the grasshopper, is divided into three tagmata: **head, thorax,** and **abdomen.**

Head of the Honey Bee. Identify **antennae, compound eyes,** and the three **ocelli,** located dorsally

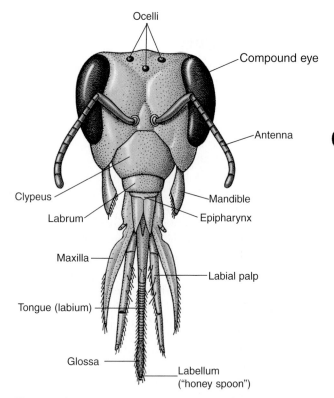

Figure 13-5
Head and mouthparts of a honey bee.

between the compound eyes. The mouthparts can be used for both chewing and sucking (Figure 13-5). Observe the narrow upper lip, or **labrum,** with a row of bristles on its free margin. Below the labrum are the **mandibles,** which are brownish in color. From its mouth projects a sucking apparatus made up of the long slender hairy **tongue** (or **labium**), paired **labial**

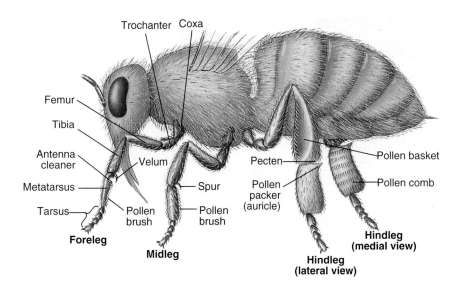

Figure 13-6
Adaptive legs from the left side of a honey bee worker.

palps, one on each side of the tongue, and paired, broad **maxillae,** one on each side of the labial palps.

Thorax of the Honey Bee. The thorax is composed of three somites: **prothorax, mesothorax,** and **metathorax,** but the lines of division between the somites are less distinguishable than on a grasshopper. The honey bee's first abdominal somite is also a part of the thorax, lying anterior to its narrow "waist." Each somite bears a pair of legs, and the mesothorax and metathorax each bear a pair of wings. Note that both pairs of wings are thin and membranous, unlike the grasshopper in which only the hindwings are membranous. How would this difference affect the wings' function? _____

Notice small hooks on the front margin of the hindwing. During flight the hooks catch hold of a groove near the margin of the forewing. The wings of a bee may vibrate 400 times or more per second during flight.

The legs of the honey bee have the same segments as those of the grasshopper but are highly adapted for specific functions.

☞ Use your hand lens, or remove the legs from one side of the bee and examine under a dissecting microscope.

Note hairs on the **foreleg.** Are they branched? _____ These hairs serve to collect pollen. The **pollen brush** (Figure 13-6) consists of long hairs on the proximal end of the tarsus. The pollen brushes on the forelegs and middle legs brush pollen off the body hairs and deposit it on the pollen combs of the hindlegs (Figure 13-6).

At the distal end of the tibia is a movable spine, the **velum** (L., covering). The velum covers a **semicircular notch** that bears a row of stiff bristles, the **antenna comb.** The velum, notch, and comb together make up

the **antenna cleaner.** The antenna is freed from pollen as it is drawn through this antenna cleaner.

The **middle leg** has a long, sharp **spur** projecting from the end of the tibia that is used to remove pollen from the pollen basket of the leg behind. The middle leg also bears a pollen brush.

The **hindleg** is the largest of the legs and the most specialized. One of its striking adaptations is the **pollen basket,** a wide groove with bristles on the outer surface of the tibia. By keeping these bristles moist with mouth secretions the bee can use the basket for carrying pollen. On the inner surface of the metatarsus are **pollen combs,** which are composed of rows of stout spines. The large spines found along the distal end of the tibia and the proximal end of the metatarsus make up the **pollen packer.** The pecten (pollen rake) removes the pollen from the pollen brush of the opposite leg, and then, when the leg is bent, the auricle packs it into the pollen basket. The bee carries her baskets full of protein-rich pollen back to the hive and pushes it into a cell, where it will be cared for by other workers.

Abdomen of the Honey Bee. The abdomen of the bee has 10 segments, the first of which is really part of the thorax, as mentioned before. The last three segments are modified and hidden within the seventh segment. Can you identify five pairs of spiracles on the abdomen? _____

The Honey Bee Sting. The amazingly intricate **sting** is a modified ovipositor that has evolved for defense. It consists of a large **poison gland** that receives secretions (a mixture of proteins, peptides, and other compounds) from two acid glands by way of a common duct (Figure 13-7). The poison sac discharges into the cavity of a sting bulb; from there the poison passes through the canal of the sting shaft and into the

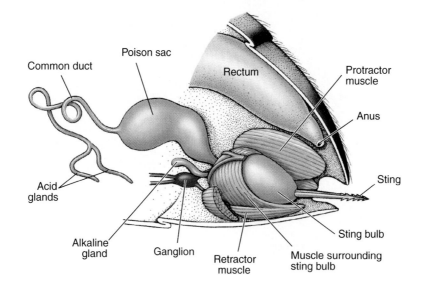

Figure 13-7
Sting of the worker honey bee.

wound. The sting bulb also receives a short alkaline gland of uncertain function.

When a honey bee stings, it thrusts the barbs of the sting into the victim's flesh. The entire sting apparatus is torn out of the bee's abdomen, fatally wounding the bee. The barb actually consists of a sting shaft together with two barb lancets that are moved rapidly back and forth on the stylet by the action of protractor and retractor muscles. Therefore even after the bee has hurriedly departed, the self-sacrificial sting continues to work in deeper and poison continues to be injected into the wound for 30 to 60 seconds.

The sting of the queen bee is used only against rival queens and can be withdrawn and reused many times. It is much more firmly attached within the queen's abdomen and the lancets have fewer and smaller barbs.

The Honey Bee as a Social Animal. In the hive of the honey bee there are three **castes: workers, queen,** and **drones. Workers** are sexually inactive genetic females and make up most of the society. They do most of the work of the hive, except lay eggs. They collect food, clean out the hive, make honey and wax, care for the young, guard the hive, ventilate the hive, and so on.

The single **queen** is a sexually mature female that lays the eggs, which may or may not be fertilized by sperm stored in her spermatheca.

Drones are males; all develop from unfertilized eggs and consequently are haploid (workers and the queen are diploid). There are usually only a few hundred drones in a hive. Their main duty is to fertilize the queen during the nuptial flight, although usually only one is required to provide the queen enough sperm to last her lifetime.

Internal Structure of an Insect

Study of the internal structures of a grasshopper is not always satisfactory because the various organs appear somewhat poorly defined. We suggest that internal structures are more easily studied in the anesthetized cricket, instructions for which are given in Exercise 13C.

Oral Report

Be prepared to (1) identify the major external features of the grasshopper and/or honey bee, (2) explain the adaptations of the honey bee worker appendages, and (3) explain the sting of the worker honey bee.

Lubber Grasshopper

External view of female (male abdomen at bottom left).

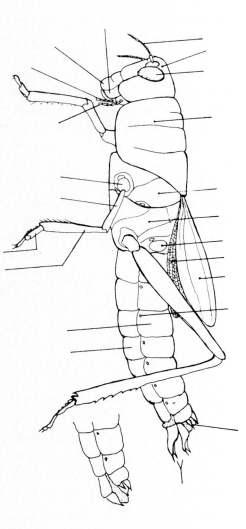

Behavior of Grasshoppers

Comparing Characteristics of Annelida and Arthropoda

	Annelida	Arthropoda
Segmentation		
Body covering		
Main body cavity		
Appendages		
Circulatory system		
Respiration		
Excretory organs		
Sense organs		
Division of animal kingdom (protostome/deuterostome)		

EXERCISE 13C

The Insects—The House Cricket*

Acheta domesticus, the House Cricket

Phylum Arthropoda
    Subphylum Uniramia
        Class Insecta
            Order Orthoptera
                Genus *Acheta*
                    Species *Acheta domesticus*

The house cricket, or gray cricket, commercially available as fish bait, is a European import and is not native to this country. Although crickets have functional wings, the flight muscles do not completely develop and crickets never fly. Adult house crickets live an average of two months, during which time the female will deposit up to 2000 eggs. Only the adult male sings, and one of the functions of his singing is to attract a female. At a rearing temperature of 30° C the eggs hatch in 13 days. There are eight larval instars (growth stages) before the final ecdysis (molt) to the adult 48 days later.

External Structure

☞ Examine several alcohol-preserved crickets. Determine the sex of the crickets (Figures 13-8E and 13-9C), and determine which are larval instars and which are adults.

Only adult crickets have fully developed wings. The last two or three larval instars of insects in the superorder Exopterygota (which includes crickets of the order Orthoptera) have external wing pads (buds). This is one important characteristic that distinguishes them from the Endopterygota, in which the last two or three instars have internal wing pads (e.g., butterflies, moths, flies, and bees). Crickets, like all other exopterygote insects, have **gradual metamorphosis.** The animal grows by successive molts (ecdysis), with each stage called a larval instar; after the last molt, the insect is called an adult.

☞ Pin an adult male or female preserved cricket lateral side up in the dissecting dish.

Note that the basic 18 metameres (segments) of the insect are functionally organized into three body regions, or tagmata: **head** (five segments), **thorax** (three segments), and **abdomen** (10 segments). Feeding and sensory organs are on the head, locomotory organs (wings and legs) are on the thorax, and digestive and reproductive organs are in the abdomen. All head seg-

---

* Exercise contributed by J. P. Woodring of Louisiana State University.

ments are fused into a unit head capsule, and the mouthparts represent the modified appendages. The three thoracic segments are the **prothorax, mesothorax,** and **metathorax.** The abdominal segments are simply numbered 1 through 9 (the cerci represent segment 10). The entire dorsum is called the **tergum** (L., back) (or notum) and any one specific segment of the dorsal plate is called a **tergite.** The lateral body surface is the **pleuron** (Gr., side) (pleurite for one segment), and the ventral body surface is the **sternum** (L., breastbone) (sternite for one segment). Find and identify all of these parts.

☞ Cut off a prothoracic leg and identify all of the segments and the tympanic membrane (Figure 13-8). Cut off the mesothoracic wing of a male cricket (Figure 13-8A), and determine how male crickets are able to sing. Place a cricket ventral side up, bend the head back, fold out the labrum, and pin the head in this position (Figure 13-8C). Identify all the mouthparts by moving each with fine forceps. Carefully remove one maxilla and the labium to see the **hypopharynx.**

The hypopharynx acts as a tongue and bears the openings of the **salivary glands.**

Functional Observations

☞ Students should work in pairs. Anesthetize two crickets with carbon dioxide. Immobilize one cricket ventral side up by crossing insect pins over the insect (do not stick the pins through the cricket). When the cricket recovers from the anesthetic, feed it some colored, moistened food and observe the action of the mouthparts under the dissecting microscope.

The heart is an almost transparent tube visible through the intersegmental membranes along the middorsal line. Cut the wings off the other cricket and immobilize it *dorsal side up* with insect pins crossed over the body. When the cricket recovers from the anesthetic, determine the ventilation rate (abdominal contractions) and the heart rate in ventilations and beats per minute.

Internal Anatomy

☞ If necessary, reanesthetize the cricket from which the wings have been removed, and pin the animal *dorsal side up* through the head and epiproct (Gr., *epi,* upon, + *proktos,* anus; see Figure 13-9C and D) onto the wax in the dish. Flood the entire animal with saline and cut the cricket open with fine-tipped scissors. Pin open the cricket with insect pins.

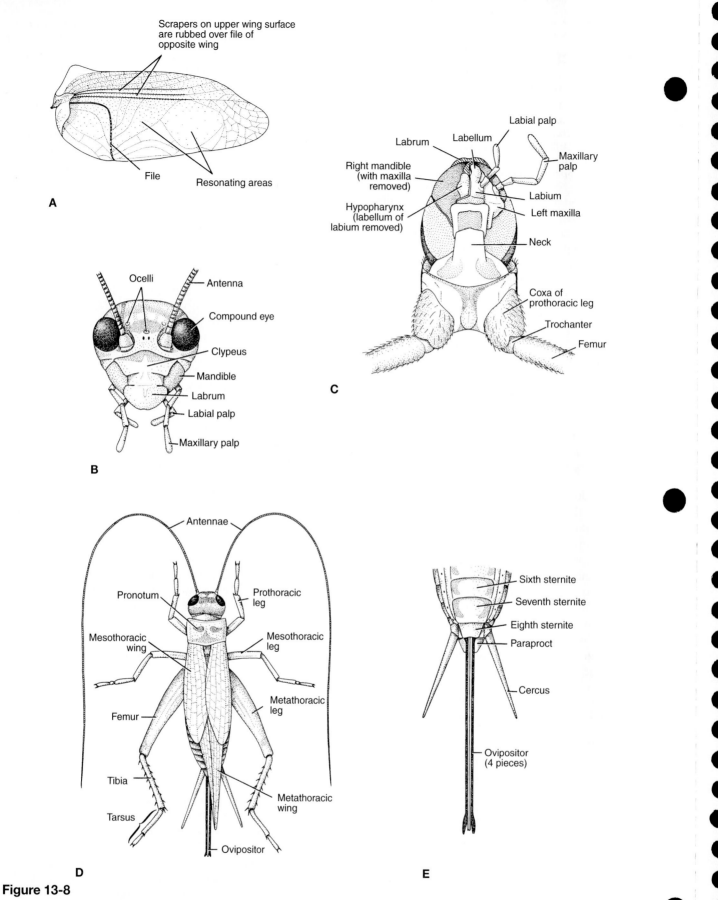

Figure 13-8

External structure of a cricket. **A,** Undersurface of male mesothoracic wing. **B,** Frontal view, and **C,** Ventral view of head. **D,** Dorsal aspect of adult female. **E,** Ventral aspect of adult female abdomen.

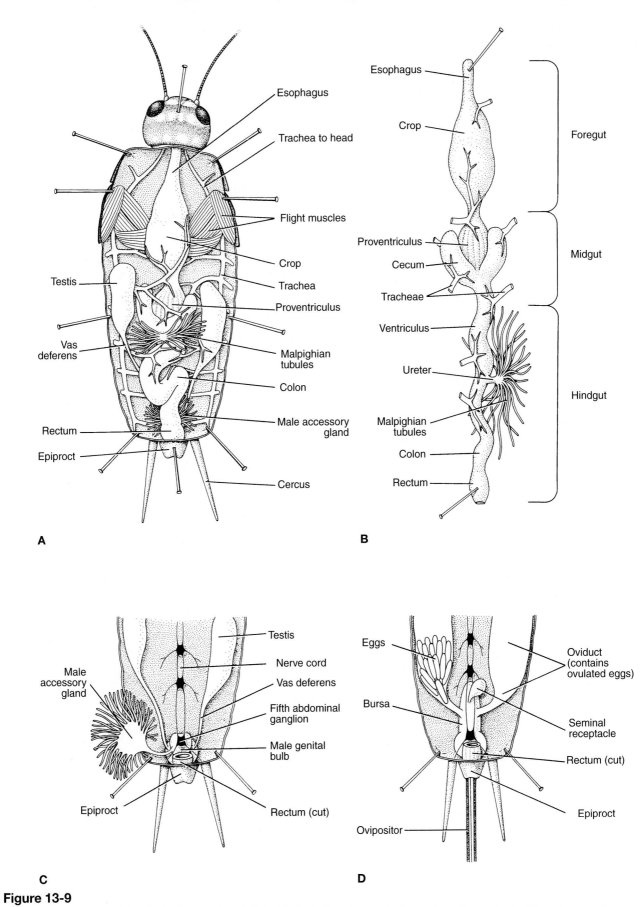

Figure 13-9

A, Internal anatomy of a cricket (fat bodies not shown). **B,** Isolated gut. **C,** Male reproductive anatomy. **D,** Female reproductive anatomy.

Observe **peristalsis** of the gut and movement of the **malpighian tubules.** The cricket will remain alive and the organ systems will function normally for several hours under these conditions. The chalky white **fat body,** which functions as a liver, is spread throughout the body and will vary greatly in amount and location according to the age and diet of the cricket. Note the **tracheal tubes,** which appear silvery because of contained air. These branch throughout the body.

☞ Sever the esophagus and rectum. Grasp the esophagus with forceps and gradually lift the entire digestive tract out by carefully cutting each tracheal connection to the gut (use the dissecting microscope). Stretch and straighten out the gut and pin it as illustrated to one side of the dish (Figure 13-9B).

Identify all structures labeled in Figure 13-9. Note that the **ureter** arises in the middle of the hindgut.

Both male and female reproductive systems are best seen after the digestive tract has been removed. Identify the gonads **(testes** or **ovaries),** gonoducts **(vas deferens** or **oviducts), accessory gland, bursa** or **genital bulb,** and the genitalia. The genitalia are the **aedeagus** (e-de′a-gus; Gr. *aidoia,* genitals) of the male and the **ovipositor** (L. *ovum,* egg, + *ponere,* to place) of the female.

☞ Pour off the saline from your specimen but leave it pinned in place. Rinse out the debris and fat body with a mild jet of water from a plastic squeeze bottle. Pour off all the water and add several crops of methylene blue dye to the tissues to stain the nerve cord and ganglia. Wait one minute, then cover the specimen with water. Pick away tissue covering parts of the nervous system. Rinse and restain if necessary.

A double ventral **nerve cord** passes back through the body. In the thorax are three pairs of large, fused **segmental ganglia.** The nerve cords are widely separated. In the abdomen are five smaller ganglia; the fifth ganglion (overlying the genital bulb in the male and the bursa in the female) is larger than the rest and supplies all the posterior end of the body (Figure 13-9C and D).

Two cerebral ganglia lie in the head, but are not visible without dissection of the head.

Oral Report

Be prepared to locate on your dissection any structure mentioned in the exercise and state its function.

EXERCISE 13D
Collection and Classification of Insects

Where to Collect Insects

1. Using a sweep net, sweep over grass, alfalfa, or weed patches.
2. Spread a cloth, newspaper, or inverted umbrella under a bush or shrub; beat or shake the plant vigorously.
3. Overturn stones, logs, bark, leaf mold, and rubbish; look under dung in pastures.
4. Watch for butterflies to alight, then a drop a net over them.
5. Look around outdoor lights at night.
6. At night, suspend a sheet from a limb or clothesline with the lower part of the sheet spread on the ground. Direct automobile headlights or a spotlight on the sheet. The insects will be attracted to the white light, hit against the sheet, and drop onto the cloth below.
7. Moths may be baited by daubing a mixture of crushed banana or peach and molasses or sugar on the bark of trees; visit the trees at night with a flashlight to collect the moths.
8. Locate soil insects by placing humus and leaf matter in a Berlese funnel.
9. Water insects may be seined with a water net. Aquatic insect larvae may be found in either quiet or running water, attached to plants and leaves or under stones, or sieved out of bottom mud.
10. In early spring the sap exuding from stumps or tree trunks attracts various insects.

Key to the Principal Orders of Insects*

This key will enable you to place your insects in the correct order. The use of a two-choice, or dichotomous, key such as this one is explained in Exercise 2. This key is designed for use with adult (final instar) insects and is therefore not suitable for identification of insect larvae. If you wish to carry identification to family, genus, or species, consult one of the references listed on page 164.

Terms Used in the Key

Cercus (pl., cerci) One of a pair of jointed anal appendages.

Tarsus (pl., tarsi) The leg segment distal to the tibia, consisting of one or more segments or subdivisions.

Lepidoptera

1   With functional wings 2
     Without functional wings, or with
     forewings thickened and concealing
     membranous hindwings 15

Diptera

2 (1)   Wings covered with minute scales;
     mouthparts usually a coiled tube
     (butterflies, moths) . . . . **Lepidoptera**
     Wings usually clear, not covered with
     scales; mouthparts not a coiled tube 3

Thysanoptera

3 (2)   With one pair of wings (true flies). . . . **Diptera**
     With two pairs of wings 4

4 (3)   Wings long, narrow, fringed with long
     hairs, body length 5 mm or less
     (thrips) . . . . **Thysanoptera**
     Wings not narrow and fringed, body
     usually longer than 5 mm 5

Ephemeroptera

5 (4)   Abdomen with two or three threadlike
     "tails"; hindwings small (mayflies). . . . .
     . . . . **Ephemeroptera**
     Abdomen with only short filaments or
     none; hindwings larger. 6

6 (5)   Forewings clearly longer and with
     greater area than hindwings 7
     Forewings not longer, or only slightly
     longer than hindwings, and with same or
     less area than hindwings 9

Trichoptera

7 (6)   Forewings noticeably hairy; antennae as
     long or longer than body
     (caddis flies). . . . . **Trichoptera**
     Wings transparent or translucent, not
     hairy; antennae shorter than body 8

continued

*Key adapted from Borror, D. J., and White, R. E.: 1970, A field guide to the insects of America north of Mexico. Boston, Houghton Mifflin Co.

8 (7)   Tarsi 2-segmented or 3-segmented; body
        not wasplike or beelike 14
        Tarsi 5-segmented; usually wasplike or
        beelike (sawflies, ichneumons, winged
        ants, wasps, bees) . . . . **Hymenoptera**

Tarsi

Hymenoptera

9 (6)   Head prolonged ventrally into a beaklike
        structure (scorpionflies) . . . . **Mecoptera**
        Head not prolonged ventrally 10

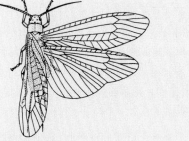

Mecoptera

10 (9) Antennae very short and bristlelike; eyes
        large; abdomen long and slender
        (dragonflies, damselflies) . . . . **Odonata**
        Antennae not short and bristlelike; eyes
        moderate to small 11

Odonata

11 (10) Hindwings broader than forewings; cerci
        present (stoneflies) . . . . **Plecoptera**
        Hindwings little if any broader than
        forewings; cerci absent 12

Plecoptera

12 (11) Mothlike; wings noticeably hairy and
        opaque; antennae as long or longer than
        body (caddis flies) . . . . **Trichoptera**
        Not mothlike; wings not noticeably hairy,
        usually clear; antennae shorter than
        body. 13

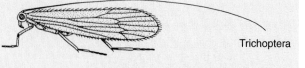

Trichoptera

13 (12) Wings with few cross veins; tarsi
        4-segmented; length to 8 mm
        (termites). . . . . **Isoptera**

Isoptera

Wings with numerous cross veins;
tarsi 5-segmented; length to 75 mm
(fishflies, dobsonflies, lacewings,
ant lions) . . . . **Neuroptera**

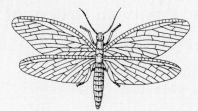

Neuroptera

14 (8)   Mouthparts sucking, beak arising from
rear of head (cicadas, hoppers,
aphids). . . . . **Homoptera**
Mouthparts chewing, beak absent; body
length less than 7 mm (book lice,
bar lice) . . . . **Pscoptera**

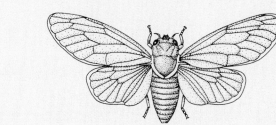

Homoptera

15 (1)   Wings entirely absent 16
Wings modified, forewings hard and
leathery and covering hindwings 27

16 (15) Narrow-waisted, antlike (ants, wingless
wasps) . . . . **Hymenoptera**
Not narrow-waisted or antlike 17

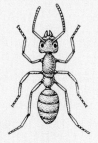

Hymenoptera

17 (16) Body rarely flattened laterally; usually do
not jump. 18
Body flattened laterally; small jumping
insects (fleas). . . . . **Siphonaptera**

Siphonaptera

18 (17) Parasites of birds and mammals; body
nearly always flattened dorsoventrally 19
Never parasitic; body usually not flattened . . 20

19 (18) Head as wide as or wider than thorax
(chewing lice). . . . . **Mallophaga**

Mallophaga

Head narrower than thorax
(sucking lice) . . . . **Anoplura**

Anoplura

continued

20 (18)   Abdomen with stylelike appendages
or threadlike tails (silverfish,
bristletails) . . . . **Thysanura**
Abdomen with neither styles nor tails 21

Thysanura

21 (20)   Abdomen with a forked tail-like jumping
mechanism (springtails) . . . . **Collembola**
Abdomen lacking a jumping
mechanism 22

Collembola

22 (21)   Abdomen usually with two short tubes;
small, plump, soft-bodied (aphids,
others) . . . . **Homoptera**
Abdomen without tubes; usually not
plump and soft-bodied 23

23 (22)   Lacking pigment, whitish; soft-bodied 24
Distinctly pigmented; usually
hard-bodied 25

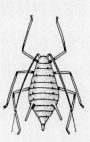

Homoptera

24 (23)   Antennae long, hairlike; tarsi 2-segmented
or 3-segmented (psocids) . . . . **Psocoptera**
Antennae short, breadlike; tarsi
4-segmented (termites) . . . . **Isoptera**

25 (23)   Body shape variable; length over
5 mm 26
Body narrow; length less than 5 mm
(thrips) . . . . **Thysanoptera**

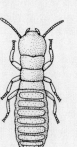

Isoptera

26 (25)   Antennae 4-segmented or 5-segmented;
mouthparts sucking
(wingless bugs) . . . . **Hemiptera**

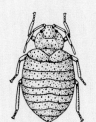

Hemiptera

Antennae many-segmented; mouthparts
chewing (some cockroaches,
walkingsticks) . . . . **Orthoptera**

Orthoptera

27 (15) Abdomen with forcepslike cerci
(earwigs) . . . . **Dermaptera**
Abdomen lacks forcepslike cerci 28

28 (27) Mouthparts sucking; beak usually
elongate 29
Mouthparts chewing. 30

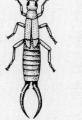

Dermaptera

29 (28) Forewings nearly always thickened at
base, membranous at tip; beak rises from
front or bottom of head (true bugs)
. . . . **Hemiptera**

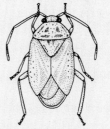

Hemiptera

Forewings of uniform texture throughout;
beak arises from hind part of head
(hoppers) . . . . **Homoptera**

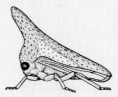

Homoptera

30 (28) Forewings with veins, at rest held rooflike
over abdomen or overlapping
(grasshoppers, crickets, cockroaches,
mantids) . . . . **Orthoptera**

Orthoptera

Forewings without veins, meeting in a
straight line down back
(beetles) . . . . **Coleoptera**

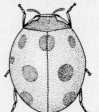

Coleoptera

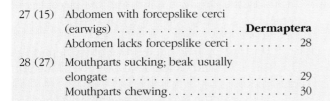

References

Note that in addition to the general guidebooks listed here, regional guides are often useful for surrounding areas.

Arnett, R. H. 1985. American insects: a handbook of the insects of America and Mexico. New York, Van Nostrand Reinhold Company.

Borror, D. J., and R. E. White. 1970. A field guide to the insects of America north of Mexico. Boston, Houghton Mifflin Co.

Chu, H. F. 1949. How to know the immature insects. Dubuque, Wm. C. Brown Co., Publishers.

Claassen, P. W. 1931. Plecoptera nymphs of America (north of Mexico). Springfield, IL, Pub. of the Thomas Say Foundation, by C. C. Thomas.

Covell, C. V., Jr. 1984. A field guide to the moths of eastern North America. Boston, Houghton Mifflin Co.

Dillon, E. S., and L. S. Dillon. 1972. A manual of common beetles of eastern North America. New York, Dover Publications.

Edmunds, G. F., Jr., S. L. Jensen, and L. Berner. 1976. The mayflies of North and Central America. Minneapolis, University of Minnesota Press.

Ehrlich, P. R., and A. H. Ehrlich. 1961. How to know the butterflies. Dubuque, Wm. C. Brown Co., Publishers.

Harris, J. R. 1952. An angler's entomology. New York, F. A. Praeger.

Holland, W. J. 1968. The moth book: a popular guide to a knowledge of the moths of North America. New York, Dover Publications, Inc.

Jaques, H. E. 1947. How to know the insects, ed. 2. Dubuque, Wm. C. Brown Co., Publishers.

Jaques, H. E. 1951. How to know the beetles. Dubuque, Wm. C. Brown Co., Publishers.

Jewett, S. G. 1959. The stoneflies (Plecoptera) of the Pacific Northwest. Corvallis, Oregon State College.

Klots, A. B. 1951. A field guide to the butterflies of North America east of the great plains. Boston, Houghton Mifflin Co.

LaFontaine, G. 1981. Caddisflies. New York, Lyons & Burford.

Lehmkahl, D. M. 1979. How to know the aquatic insects. The Pictured Key Nature Series. Dubuque, Wm. C. Brown Co.

McCafferty, W. P. 1981. Aquatic entomology: the fisherman's and ecologist's illustrated guide to insects and their relatives. Boston, Science Books International.

McPherson, J. E. 1982. The Pentatomoidea (Hemiptera) of Northeastern North America. Carbondale, IL, Southern Illinois University Press.

Merritt, R. W., and K. W. Cummins. 1984. An introduction to the aquatic insects of North America, ed. 2. Dubuque, Kendall/Hunt Publishing Co.

Miller, P. L. 1984. Dragonflies. New York, Cambridge University Press.

Milne, L., and M. Milne. 1980. The Audubon Society field guide to North American insects and spiders. New York, Alfred A. Knopf, Inc.

Needham, J. G., and M. J. Westfall, Jr. 1975. A manual of the dragonflies of North America (Anisoptera): including the Greater Antilles and the provinces of the Mexican border. Berkeley, University of California Press.

Opler, P. A. 1992. A field guide to eastern butterflies. The Peterson Field Guide Series. Boston, Houghton Mifflin Co.

Otte, D. 1981–84. The North American grasshoppers. Vol. 1: Acrididae (Gomphocerinae and Acridinae); Vol. 2: Acrididae (Oedipodinae). Cambridge, MA, Harvard University Press.

Pyle, R. M. 1981. The Audubon Society field guide to North American butterflies. New York, Alfred A. Knopf, Inc.

Sborboni, V., and S. Forestiero. 1998. Butterflies of the world. Buffalo, NY, Firefly Books Inc.

Schuh, R. T., and J. A. Slater. 1995. True bugs of the world. Ithaca, Cornell University Press.

Scott, J. A. 1986. The butterflies of North America: a natural history and field guide. Stanford, CA, Standard University Press.

Smart, P. 1975. The international butterfly book. New York, Crowell.

Stehr, F. W. (ed.). 1991. Immature insects. Vol. 1 & 2. Dubuque, Kendall/Hunt Publishing Co.

Swan, L. A., and C. S. Papp. 1972. The common insects of North America. New York, Harper & Row, Publishers.

White, R. E. 1983. A field guide to the beetles of North America. Boston, Houghton Mifflin Co.

Wiggins, G. B. 1977. Larvae of the North American caddisfly genera (Trichoptera). Toronto, University of Toronto Press.

The Echinoderms
Phylum Echinodermata
A Deuterostome Group

Echinoderms are an all-marine phylum that comprises sea lilies, sea stars, brittle stars, sea urchins, sand dollars, and sea cucumbers. They are a strange group, strikingly different from any other animal phylum, a group that abandoned the adaptive advantages of bilateral symmetry to become radial.

Echinoderms are deuterostomes, thus sharing with two other phyla—Hemichordata and Chordata—several embryological features that set them apart from all the rest of the animal kingdom: anus developing from or near the blastopore and mouth developing elsewhere, enterocoelous coelom, radial and regulative cleavage, and mesoderm derived from enterocoelous pouches. Thus all four phyla are presumably derived from a common ancestor.

It is the echinoderm's **dermal endoskeleton** of calcareous plates and spines, often fused into an investing armor, that provides the group with their name (Gr. *echinos,* hedgehog, sea urchin, + *derma,* skin). There are other characteristics, unique to echinoderms, that both define the group and limit its evolutionary potential. They have a **water-vascular system** that powers a multitude of tiny tube feet used for locomotion and food gathering. Many are invested with pincerlike **pedicellariae** that snap at creatures that would settle on them. To breathe, many echinoderms rely on numerous **dermal branchiae** (skin gills) that project delicately through spaces in their skeletal armor. As if to emphasize their uniqueness, echinoderms lack a definite head, their nervous system and and sense organs are primitive, locomotion is slow, and they lack segmentation. Of all their distinguishing features, however, none delineates the group more conspicuously than its **pentaradiate symmetry:** body parts always arranged radially in five or multiples of five, no matter how the body plan has become adapted to different feeding strategies. This radial symmetry is, however, secondarily acquired because their larvae are unmistakably bilaterally symmetrical (Figure 14-1).

Classification

This classification is abbreviated.

Classification
Phylum Echinodermata

EXERCISE 14A

Class Asteroidea—The Sea Stars
Asterias

EXERCISE 14B

Class Ophiuroidea—The Brittle Stars
Brittle stars

EXERCISE 14C

Class Echinoidea—The Sea Urchin
Arbacia

EXERCISE 14D

Class Holothuroidea—The Sea Cucumber
Sea cucumber

Phylum Echinodermata

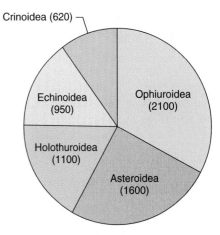

Crinoidea (620)
Echinoidea (950)
Ophiuroidea (2100)
Holothuroidea (1100)
Asteroidea (1600)

Phylum Echinodermata

Class Crinoidea (cry-noi′ de-a) (Gr. *krinon,* lily, + *eidos,* form, + *ea,* characterized by). Sea lilies and feather stars. Aboral attachment stalk of dermal ossicles. Anus on oral surface; five branching arms with pinnules; ciliated ambulacral groove on

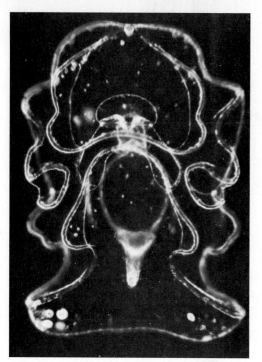

Figure 14-1
Bilaterally symmetrical *Auricularia* larva of the sea cucumber, ×100.

oral surface with tentacle-like tube feet for food collecting; spines, madreporite, and pedicellariae absent. Examples: *Antedon, Florometra.*

Class Asteroidea (as′ter-oi′de-a) (Gr. *aster,* star, + *eidos,* form, + *ea,* characterized by). Sea stars. Star shaped, with arms not sharply marked off from central disc; ambulacral grooves open, with tube feet on oral side; tube feet often with suckers; anus and madreporite aboral; pedicellariae present. Example: *Asterias.*

Class Ophiuroidea (o′fe-u-roi′de-a) (Gr. *ophis,* snake, + *oura,* tail, + *eidos,* form). Brittle stars and basket stars. Star shaped, with arms sharply marked off from central disc; ambulacral grooves closed, covered by ossicles; tube feet without suckers and not used for locomotion; pedicellariae absent. Examples: *Ophiura, Gorgonocephalus.*

Class Echinoidea (ek′i-noi′de-a) (Gr. *echinos,* sea urchin, hedgehog, + *eidos,* form). Sea urchins, heart urchins, and sand dollars. More or less globular or disc shaped, with no arms; compact skeleton, or test, with closely fitting plates; movable spines; ambulacral grooves closed and covered by ossicles; tube feet with suckers; pedicellariae present. Examples: *Arbacia, Strongylocentrotus, Lytechinus, Mellita.*

Class Holothuroidea (hol′o-thu-roi′de-a) (Gr. *holothourion,* sea cucumber, + *eidos,* form). Sea cucumbers. Cucumber shaped; with no arms; spines absent; microscopic ossicles embedded in thick muscular wall; anus present; ambulacral

grooves closed; tube feet with suckers; circumoral tentacles (modified tube feet); pedicellariae absent; madreporite plate internal. Examples: *Thyone, Parastichopus, Cucumaria.*

EXERCISE 14A

Class Asteroidea—The Sea Stars

Asterias

Phylum Echinodermata
  Subphylum Eleutherozoa
    Class Asteroidea
      Order Forcipulatida
        Genus *Asterias*

Sea stars are the "prima donnas" of echinoderms, familiar to many people as beautifully symmetrical symbols of marine life. Those commonly seen are intertidal species, especially along rocky coastlines. Many live on high-energy beaches that receive full force of the surf. Other species inhabit a variety of benthic habitats, often at great depths in the ocean. Sea stars, like other echinoderms, are strictly bottom dwellers. Some are particle feeders, but most are predators of slow-moving prey, such as molluscs (their favorite food), crabs, corals, and worms, since sea stars are themselves slow-moving animals. Adult sea stars seem to have few enemies, suggesting that they produce something that discourages potential predators.

Behavior

☞ Examine a living sea star in a dish of seawater. Using these suggestions, observe its behavior.

Note the general body plan of the star with its **pentaradial** (Gr. *pente,* five, + L. *radius,* ray) symmetry, its five **arms,** or **rays,** its **oral-aboral flattening,** and the **mouth** on the underside. Lift up the dish and look at this oral surface. Notice the rows of **tube feet.** How are they used? How are the ends of the tube feet shaped? What is the sequence of action of a single tube foot as the animal is moving? Tube feet are filled with fluid and are muscular, providing the necessary components for a hydraulic skeleton. When a foot contracts, water flows into a bulblike **ampulla** inside the arm. Tube feet and ampullae are parts of the **water-vascular system.**

Tilt the dish to one side (pour out a little water if necessary) and watch the animal's reaction. Does it move up or down the inclined plane? Now tilt the dish in the opposite direction. Does the animal change direction? Is it positively geotactic (moves toward the earth) or negatively geotactic?

Place a piece of fresh seafood (oyster, fish, or shrimp) near one of the arms. Is there any reaction? Are the arms flexible? If the sea star makes no move toward the food, touch the tip of an arm with it, or slip it under the end of an arm. Hold the dish up and look under-neath. How is the food grasped? How is it moved toward the mouth? What position does the animal assume when feeding?

Examine the aboral surface with a hand lens or dis-secting microscope. The **epidermis** is ciliated. Place a drop of carmine suspension (in seawater) on the exposed surface and see which direction the ciliary cur-rents take. Notice the calcareous **spines** protruding through the skin (Figure 14-2). These are extensions of skeletal ossicles. Do the spines move? Small fingerlike bulges in the epidermis are **skin gills** (also called **dermal branchiae,** and **papulae** (sing. **papula;** L., pim-ple), concerned with gaseous exchange. Around the spines you will see small **pedicellariae** (L., *pediculus,* little foot, + *aria,* like) (Figure 14-2). These are calcare-ous two-jawed pincers that are modified spines and are concerned with capturing tiny prey and protecting the dermal branchiae from collecting sediment or small par-asites. Touch with a small camel's hair brush and observe the pincer action.

Look at the tip of each arm to see a small red **eye-spot** and elongate tube feet modified as **sensory tenta-cles** (Figure 14-2).

Written Report

Record your observations on separate paper.

External Structure

☞ Place a preserved sea star in a dissecting pan and cover with water. If your observations of exter-nal structure are on a living sea star, place the animal in a clean pan or culture dish and cover with seawater.

The star-shaped body is composed of a **central disc** and five **rays** (arms). Are all the rays alike? What would account for some of the rays being shorter than others? _____ Compare your specimen with those of your neighbors. Preserved specimens may seem rigid but live stars can bend their arms by means of muscles.

Aboral Surface. The central disc bears a small, porous **madreporite plate** (Fr. *madrépore,* reef-building coral, + *ite,* suffix for body part) composed of calcium carbonate (Figure 14-3). It allows seawater to seep into an intricate **water-vascular system,** which provides the means of locomotion. The rays on each side of the madreporite are called the **bivium;** the other three are the **trivium.** The **anus** opens in the center of the cen-tral disc, but it is probably too small to see.

☞ Submerge one of the arms in water and examine the dorsal body wall under a dissecting micro-scope. Compare with a piece of dried-up test.

Figure 14-2
Distal portion of the ray of a living sea star (*Asterias*).

Labels: Sensory tentacles; Tube foot; Eyespot; Spine surrounded by pedicellariae; Skin gills

The body is covered with a thin, ciliated **epidermis,** through which white calcareous **spines** extend from the endoskeleton beneath. Are the spines movable? _____ Notice that around the base of each spine is a raised ring of skin bearing tiny, calcareous pincerlike **pedicellariae.** Some are also found between the spines. Sometimes pedicellariae can be seen more easily on a dried-up piece of test. Some pedicellariae have straight jaws and others have curved jaws. They are moved by tiny muscles. What is the function of the pedicellariae? _____ Between the spines are soft, transparent fingerlike projections, the **skin gills,** or **dermal branchiae.** These are hollow evaginations of the coelomic cavity. What is the function of the dermal branchiae? _____

Drawings

✎ Sketch some pedicellariae in your notebook.

A small, pigmented **eyespot** is located at the tip of each ray.

Oral Surface. The **ambulacral** (L. *ambulare,* to walk) groove of each ray contains rows of **tube feet (podia).** The **ambulacral spines** bordering the groove are movable and can interlock when the groove is con-tracted to protect the tube feet. Note the size, shape, number, and arrangement of the tube feet. How many rows of tube feet are there? _____

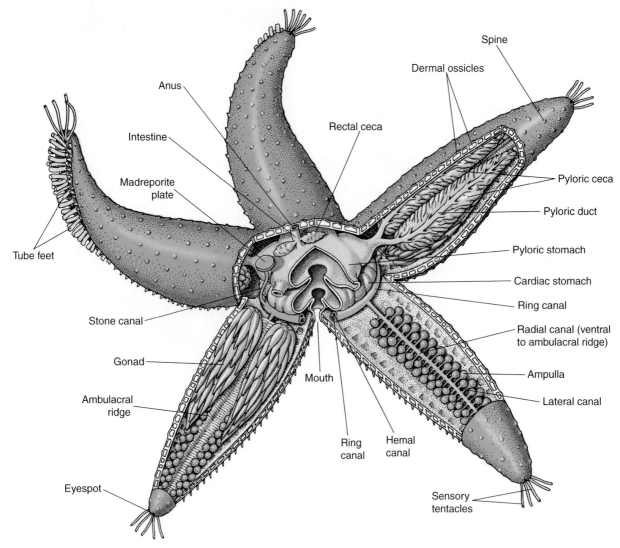

Figure 14-3
Anatomy of the sea star.

 Scrape away some tube feet and note arrangement of the pores through which they extend. If you are making observations on a living sea star, do not scrape off tube feet but instead brush them with a camel's hair brush, noting the response of both tube feet and ambulacral ossicles.

The tube feet are a part of the water-vascular system. They are hollow, and their tips form suction discs for attachment (Figures 14-3 and 14-4). These are effective not only in locomotion but also in opening bivalves for food.

The central **mouth** is surrounded by five pairs of movable spines. Push the spines outward, bend the arms back slightly, and note the thin **peristomial membrane,** which surrounds the mouth. Sometimes the mouth is filled with part of the everted **stomach.**

Endoskeleton

Echinoderms are the first of the invertebrates to have a mesodermal endoskeleton. It is formed of calcareous plates, or **ossicles,** bound together by connective tissue.

 If you are studying a preserved sea star, cut off part of one of the rays of the bivium, remove the aboral wall from the cut-off piece, and study its inner surface. Compare with a piece of dried body wall, which is excellent for examining the arrangement of ossicles.

Note the skeletal network of irregular ossicles. Now look at the cut edge of the body wall and identify the outer layer of **epidermis,** the thicker layer of **dermis,** or connective tissue in which the ossicles are embedded, and the thin inner layer of ciliated **peritoneum.** Do sea stars have a true coelom? _____ The dermis and

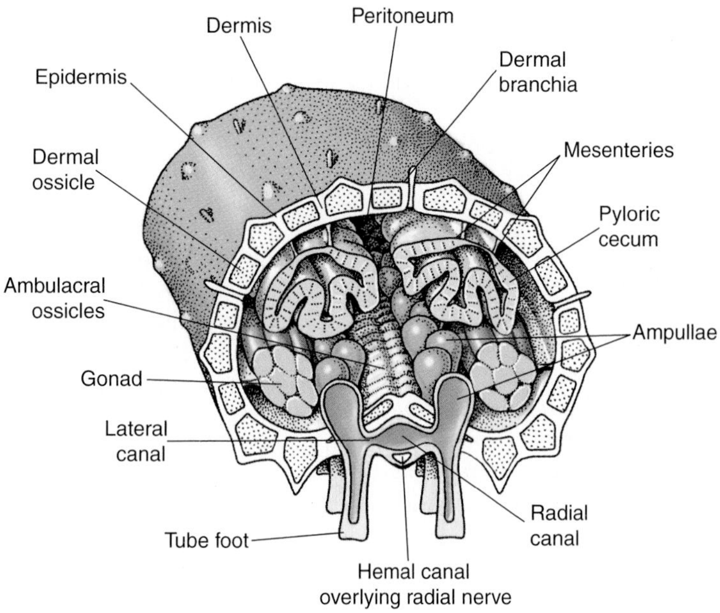

Figure 14-4
Cross section of a sea star arm.

peritoneum are mesodermal in origin. Are the spines part of the endoskeleton? _____ Hold the piece up to the light. The thin places you see between the ossicles are where the dermal branchiae extend through the connective tissue.

Look at the oral surface of the cut-off ray. Note how the **tube feet** extend up between the **ambulacral ossicles,** emerging on the inside surface of the wall as bulb-like **ampullae** (am-pool'ee; sing. **ampulla;** L., flask). Compress some of the ampullae and note the effect on the tube feet. Press on the tube feet and see the effect on the ampullae. Both are muscular and can regulate the water pressure by contraction. Scrape away some of the ampullae and tube feet and examine the shape of the ossicles in the ambulacral groove. How do they differ from the ossicles elsewhere? _____ Note the alternating arrangement of the **ambulacral pores,** through which the tube feet extend. Whereas the ambulacral ossicles form a groove on the oral surface of each arm, they form an **ambulacral ridge** on the inner surface.

Drawings

In your notebook sketch a series of ambulacral ossicles, showing the arrangement of the openings for the tube feet.

Internal Structure (Dissection)

Place the specimen aboral side up in a dissecting pan and cover with water. Select the three rays of the trivium and snip off their distal ends. Insert a scissors' point under the body wall at the cut end of one of the rays. Carefully cut along the dorsolateral margins of each ray to the central disc. Lift up

the loosened wall and carefully free any clinging organs. Uncover the central disc, but cut around the madreporite plate, leaving it in place. Be careful not to injure the delicate tissue underneath. The tissue around the anal opening in the center of the disc will cling. Cut the very short intestine close to the aboral wall before lifting off the body wall.

The **coelomic cavity** inside the rays and disc contains **coelomic fluid,** which bathes the visceral organs.

Digestive System. A pentagonal **pyloric stomach** (Gr. *pylōros,* gatekeeper) (see Figure 14-3) lies in the central disc, and from it a **pyloric duct** extends into each arm, where it divides to connect with a pair of large, much-lobulated **pyloric ceca** (digestive glands) (see Figure 14-3). A very short **intestine** leads up from the center of the stomach to the anus in the center of the disc. Attached to the intestine are two **rectal ceca,** small, branched sacs of uncertain function. Below (ventral to) the pyloric stomach is the larger, five-lobed **cardiac stomach,** which fills most of the central disc (see Figure 14-3). Each lobe of the stomach is attached to the ambulacral ridge of one of the arms by a pair of **gastric ligaments,** which prevent too much eversion of the stomach.

When a sea star feeds on a bivalve, it folds itself around the animal, attaches its tube feet to the valves, and exerts enough pull to cause the shell to gape a little. Then, by contracting its body walls, (causing pressure of the coelomic fluid), it everts its stomach and inserts it into the slightly opened clam shell. There it digests the soft parts of the clam with juices from the pyloric ceca. Partly digested material is drawn up into the stomach and pyloric ceca, where digestion is completed. There is little waste fecal matter. When the sea star is finished with feeding, the stomach withdraws into the coelom by contraction of stomach muscles and relaxation of the body wall, which allows coelomic fluid to flow back into the arms.

Many stars feed on small bivalves by engulfing the entire animal, digesting out its contents, then casting the shell out through the mouth.

Reproductive System. Sexes are separate (dioecious) in sea stars. Remove the pyloric ceca from one arm to find paired **gonads** attached to the sides of the ray where the ray joins the disc (see Figure 14-3). During the breeding season the gonads are larger than at other times. Each gonad opens aborally to the exterior at the point of attachment by a very small **reproductive duct** and **genital pore.** The sex can rarely be determined by simple inspection of the gonads, although the female gonads may be a little coarser in texture and more orange than the male gonads.

☞ Make a wet mount of a mashed bit of gonad and examine with a microscope.

In the ovary, eggs with large nuclei will be found; in the testes, many small sperm are to be found. In early summer large streams of eggs and sperm are shed into the water, where fertilization occurs externally.

Nervous System. The nervous system of a sea star is somewhat primitive and actually consists of three interrelated systems. Foremost of these is the **oral** system, consisting of a **nerve ring** around the mouth in the peristomial membrane, and a **radial nerve** to each arm running along the ambulacral groove to the eyespot.

☞ To find the nerve ring, remove the tube feet and movable spines around the mouth and expose the peristomial membrane.

The nerve ring is a whitish thickening on the outer margin of this membrane. To see one of the **radial nerves,** bend an arm aborally and look along the oral surface of the ambulacral groove for a whitish cord (Figure 14-4). Trace the nerve from the ring to its termination in the arm.

The other two systems are an aboral system lying near the upper surface of the sea star and a deep system positioned between the oral and aboral systems. You will be unable to see these. Freely connected to these three systems is the **epidermal nerve plexus.** This consists of a network of nerve cell bodies and their processes lying just beneath the epidermis that coordinate the responses of the dermal branchiae to tactile stimulation.

Sense organs are not well developed. Chemoreceptors and cells sensitive to touch are found all over the surface.

Each pigmented **eyespot** consists of a number of light-sensitive ocelli.

Water-Vascular System. The water-vascular system is found only in echinoderms, which use it for loco-motion and, in the case of sea stars, for opening clam shells. If this system in your specimen has been injected with a colored injection mass, its features can be studied to greater advantage.

☞ Carefully remove the stomach from the central disc.

The **madreporite plate** (orange in life) on the aboral surface (see Figure 14-3) contains ciliated grooves and pores. From it a somewhat curved **stone canal** (yellow in life) leads to a **ring canal** (see Figure 14-3), which is found around the outer edge of the peristomial membrane next to the skeletal region of the central disc. The ring canal may be difficult to find if not injected.

Five **radial canals,** one in each arm, radiate out from the ring canal, running along the apex of the ambulacral groove just below the ambulacral ossicles and above the radial nerve. The position of the radial canal is best seen in a cross section of one of the arms (Figure 14-4). Short **lateral canals,** each with a valve, connect the radial canal with each of the tube feet. Now look on the inside of an arm and study the alternating arrangement of the ampullae. Note how each ampulla connects with a tube foot through a **pore** between the ambulacral plates.

The function of the madreporite plate has been the subject of controversy. The traditional view that the madreporite serves as a point of entry for seawater into the water-vascular system has been confirmed with the use of radioactive isotopes. Seawater that enters the madreporite passes down the stone canal to the ring canals, from there to the radial canals, and finally through the lateral canals to the ampullae and tube feet.

Tube feet have longitudinal muscles; ampullae have circular muscles. When the tube feet are contracted, most of the water is held in the ampullae (Figure 14-4). When the ampullae contract, water is forced into the elastic tube feet, which elongate because of the hydrostatic pressure within them. When the cuplike ends of the extended tube feet contact a hard surface, they attach with a suction force and then contract, pushing the water back into the ampullae and pulling the animal forward. Valves in the lateral canals prevent the backflow of water into the radial canals. Although a single foot is not very strong, hundreds of them working together can move the animal along slowly and can create a tremendous pull on the shell of a mussel. Suckers are of little use on a sandy surface, where the tube feet serve as tiny legs. Some species have no suckers but use the stiff podia like little legs to "walk." Sea stars can travel about 15 cm per minute.

Cross Section of the Arm of a Sea Star (Microslide)

Identify the **epidermis** covering the entire animal, **dermis** containing the **ossicles,** muscular tissue, **peritoneum,** and **coelomic cavity.** The spines have not yet erupted, for these are young sea stars. Observe the **dermal branchiae** projecting from the coelom through the body wall. **Pyloric ceca** hang from the aboral wall by **mesenteries.** Notice the ampullae in the coelom and their connection to the **tube feet** and **lateral canals.** The canals may not always be seen. Why? Locate the **ambulacral ossicles** with the pores through which tube feet extend. Find the **radial canal,** a small tube under the **ambulacral groove,** and the **radial nerve** beneath the radial canal (Figure 14-4).

EXERCISE 14B
Class Ophiuroidea—
The Brittle Stars

Brittle Stars

Phylum Echinodermata
　　Subphylum Eleutherozoa
　　　　Class Ophiuroidea
　　　　　　Order Ophiurida
　　　　　　　　Available genera

Where Found

Brittle stars, most agile of echinoderms, are widely distributed in all oceans and at all depths. In many habitats they are also the most abundant echinoderms, yet are seldom seen by casual observers. They are secretive animals that hide under and between intertidal and subtidal rocks by day to escape predation by fish. Even at night they may extend only their arms from hiding places to feed. If a diver should expose their retreat, they will quickly scuttle to safety, using rowing movements of their arms. Should a fish (or a human) catch a brittle star by one arm, the animal usually will simply cast off the arm (autotomize), leaving the predator with a wiggling arm while its erstwhile owner scurries to safety beneath a nearby rock. It is common to find brittle stars with missing or partly regenerated arms.

External Features

General Body Form. Note that the arms of brittle stars are sharply marked off from the central disc—a characteristic of ophiuroids (Figure 14-5). The arms are more flexible than those of the sea stars. Does their appearance give a clue as to why? _____
In both appearance and function the arms resemble vertebral columns. In fact, internally they consist of a series of calcareous vertebral **ossicles,** each joined to the next by two pairs of muscles. Externally the arm is encased in a series of aboral, lateral, and oral **plates.**

☞ Pull an arm off a preserved specimen and examine it in cross section. Locate the muscles and note the vertebral articulations.

Are there spines on the arms? _____ How do the spines compare with those of asteroids? _____ Do you find any pedicellariae or skin gills? _____

Oral Surface. On the oral side note five triangular **jaws** around the mouth. Find the five **oral shields** (also called **buccal shields**), which are oval plates located on the interradial area between the rays (Figure 14-6). One of these is slightly modified as a **madreporite plate,** and its tiny pores connect with a madreporite canal inside. Compare the location of the madreporite plate in asteroids and ophiuroids.

Distal to the oral shields and close to each arm is a pair of grooves, representing the openings of the **bursae.** (In *Ophioderma* a second pair is located distal to the first.) The bursae, comprising 10 saclike cavities within the disc, are peculiar to ophiuroids. Water is pumped in and out of them for respiratory purposes, and the gonads discharge their products into them. In some species they also serve as brood pouches, but in *Ophioderma* development is external.

The tube feet (podia) are often called tentacles in ophiuroids. They are small, do not have suckers, and project laterally between the skeletal plates. They are largely sensory in function but may also assist in locomotion. Examine the rough spines used in gripping the substrate. How do they compare with those of asteroids?

Behavior

Avoid rough handling of a brittle star because this may cause it to "freeze," becoming immobile, or even to cast off one or more of its arms.

Locomotion. Brittle stars "walk" by twisting, highly flexible arm movements. How are the arms used in locomotion? Watch their movements carefully. What provides the actual force for forward movement? Do the arms push or pull the brittle star? _____ Do the arms work in pairs? _____ How are the podia (tube feet) used—or are they used at all in locomotion? _____ Does one arm always lead or follow? _____ Note that locomotory patterns differ somewhat in different species of brittle stars.

Turn the brittle star over and watch its righting response. If placed on a sandy substrate, will the brittle star burrow? _____

Feeding. Most ophiuroids are either active predators or selective deposit feeders. Deposit feeders capture organic material with the podia that secrete copious amounts of mucus. Mucus and food are rolled into a ball by the podium, then transferred to a small scale that lies adjacent to the podium. The next podium picks up the food and transfers it to the next scale, and so on until the food ball reaches the mouth. Active predators such as species of the genus *Ophioderma* capture benthic (bottom dwelling) organisms by curling an arm around the prey and sweeping it into the mouth.

Figure 14-5
A, The brittle star *Ophiopholis aculeata* from the Gulf of Maine. **B,** Basket star, *Astrophyton muricatum,* on octocoral. Ophiuroids have sharply marked-off arms and are fragile.

☞ Drop very small bits of fish or shrimp into the aquarium near (but not touching) some brittle stars to see the method of feeding.

Do the animals appear to sense the presence of food before touching it? Is the species you are studying an active predator or a selective deposit feeder? _____ The digestive system is much reduced as compared to sea stars. The mouth leads by a short esophagus to a stomach, the site of digestion and absorption. There is no intestine, anus, or hepatic ceca.

Reactions to Other Stimuli. Note how the animal reacts to mechanical stimulation. Touch the tip of an arm. Does it retreat or advance toward the source of the stimulus? Touch a more proximal part of the arm. Is the reaction the same or different? Stimulate the base of an arm or the central disc. What happens?

Can you determine whether ophiuroids—at least the species you are studying—respond positively or negatively to light?

The water-vascular, hemal, and nervous systems are on a plan similar to that of asteroids.

Written Report

✍ Record your behavioral observations on separate paper.

Table of comparison

Complete a table of comparative characteristics of the external anatomy of the sea star and brittle star on p. 179. Include such items as symmetry, shape, integument, ambulacra, tube feet, skin gills, pedicellariae,

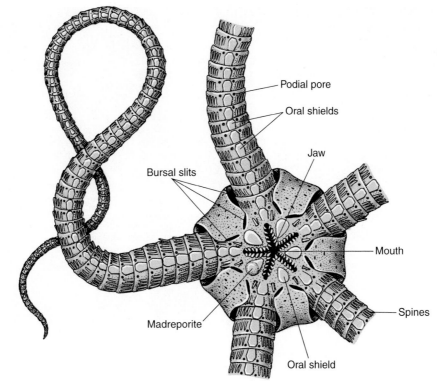

Figure 14-6
Oral view of the central disc of a brittle star.

skeleton, and digestive system. After you have observed the sea urchin and sea cucumber, add these to the table.

EXERCISE 14C

Class Echinoidea— The Sea Urchin

Sea urchins, like sea stars, are familiar denizens of seashores. Lacking arms, and with the body enclosed within a globose shell, or test, of interlocking plates which bear movable spines, sea urchins have evolved a body design quite unlike that of sea stars and brittle stars. Nevertheless, they bear the typical pentamerous plan of all echinoderms, with a water-vascular system and other characteristics that set the strange echinoderms apart from any other animal phylum. Sea urchins and their kin, heart urchins and sand dollars, all bear a spiny armament suggesting *echinos,* the Greek word for hedgehog, from which the echinoids get their scientific name. Unlike the carnivorous sea stars and brittle stars, sea urchins are herbivores that scrape encrusting algae from rock surfaces, nibble on plants, or trap and eat drifting food. Sea urchins are "regular" echinoids: radially symmetrical, globose in shape, and armed with long spines. Other echinoids, such as sand dollars and heart urchins are "irregular" echinoids: bilaterally symmetrical

with bodies variably shaped, such as the flattened sand dollars or the heart-shaped heart urchins.

Arbacia

Phylum Echinodermata
    Subphylum Eleutherozoa
        Class Echinoidea
            Genus *Arbacia*
                Species *Arbacia punctulata*

Where Found

Like sea stars and brittle stars, echinoids are strictly marine, benthic animals widely distributed in all seas, from the intertidal to deep sea. Some favor rocky, high-energy coastlines with pounding surf, but most echinoids are subtidal, grazing in turtle-grass buds or on coral reefs or (especially the irregular echinoids) burying themselves in sandy bottoms where they feed on microscopic organic matter. *Arbacia punctulata* (Gr. *Arbaces,* first king of Media*), the purple sea urchin on the East Coast, is found from Cape Cod to Florida and Cuba. Other common species include *Strongylocentrotus drobachiensis* (Gr. *strongylos,* round, + *kentron,* spine), the green sea

---

*The name of the genus *Arbacia,* bestowed by British zoologist John Edward Gray in 1835, is an example of a "nonsense" name that lacks any descriptive value. Gray apparently chose the name after reading Lord Byron's poem *Sardanapalus* concerning Arbaces, who, according to legend, founded the Median empire (now part of northern Iran) about 830 B.C.

Figure 14-7
The green sea urchin *Strongylocentrotus drobachiensis*. Note the slender, suckered tube feet. Urchins often attach bits of shell, marine algae, and other debris to themselves for camouflage. Small, stalked, white-tipped pedicellariae can be seen surrounding the bases of spines near the center of the photograph.

urchin of both East and West coasts of North America; *Strongylocentrotus purpuratus,* purple urchin of the West Coast; and species of *Lytechinus* (Gr. *lytos,* broken, + *echinos,* urchin) (Figure 14-7) on both coasts.

Behavior and External Structure

The external features of a sea urchin are best observed in a living animal. However, if living forms are not available, submerge a preserved specimen in a bowl of water and use this account, directed toward live urchins, to identify the external structures.

☞ Place a living sea urchin on a glass plate, submerge in a bowl of seawater, and observe under a dissecting microscope.

Spines. Examine the long, movable **spines,** each attached at a ball-and-socket joint by two sets of ring muscles. Remove a spine (instructor's option) and note that its socket fits over a rounded **tubercle** on the test (Figure 14-8). An inner ring of **cog muscles** holds the spine erect. Hold the tip of a spine and try to move it. Do you feel the locking mechanism of the cog muscles? Of what advantage is such a locking mechanism? _____ Now with a probe touch the epidermis near a spine. Does the spine move? In which direction? The outer ring of muscles is responsible for directional movement. Are all the spines the same length? _____ Are they all pointed on the distal ends?

Tube Feet. Notice that the tube feet all originate from rows of perforations in the five **ambulacral**

regions. The podia can be extended beyond the ends of the spines. Do any of them possess suckers? Are any of them suckerless? _____ Do they move? What happens when you touch one? When you jar the bowl?

Mouth and Peristome. Examine the oral side of the urchin and find the **mouth** with its five converging **teeth** and collarlike **lip** (Figure 14-8). The teeth are part of a complex chewing mechanism called **Aristotle's lantern,** which is operated internally by several sets of muscles.* The lip contains circular, or "purse-string," muscles.

The membranous **peristome** surrounding the mouth is perforated by five pairs of large oral tube feet called **buccal podia.** Do these podia have suckers? _____ They are probably sensitive to chemical stimuli. The peristome also bears some small spines.

Pedicellariae. Notice on the peristome a number of three-jawed **pedicellariae** on the ends of long, slender stalks. Smaller but more active pedicellariae are located among the spines. Stimulate some of them by touching gently with a camel's hair brush. You may want to pinch off some of the pedicellariae to examine more closely on a slide, particularly if you are using a preserved specimen. Their function is to discourage intruders and to help keep the skin clean.

Locomotion. Note how the urchin uses its spines and tube feet in locomotion. Carefully, so as not to injure its tube feet, turn the urchin over (oral side up). Does it

---

* The curious name for the protrusible chewing mechanism of sea urchins derives from a passage in the writings of Aristotle (384–322 B.C.) where he compared the apparatus to the frame of a lantern.

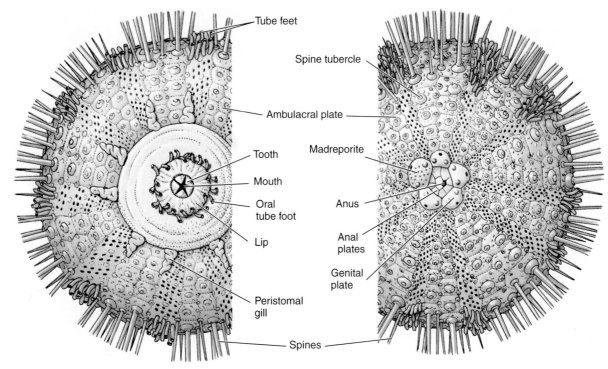

Figure 14-8
External structure of the sea urchin. The oral (*left*) and aboral (*right*) surfaces are shown with the spines partly removed.

use its spines or tube feet to right itself? _____
Notice which ambulacra turned first and mark that row by removing some of its spines or by marking some of the spines with thread; then turn the animal over again and see if the same ambulacra turn first. Repeat once more.

Tilt the glass plate to determine if the urchin moves up or down. Tilt in the opposite direction and see what happens. Is it positively or negatively geotactic? _____

Some sea urchins are adapted for burrowing into rock or other hard material by using both their spines and chewing mechanisms. *Strongylocentrotus purpuratus,* common on the North American Pacific coast, excavates cup-shaped depressions in stone.

Most echinoids have tiny modified spines called **sphaeridia** (Gr. *sphaira,* sphere,+ *idion,* dim. suffix), believed to be organs of equilibrium. In *Arbacia* these are minute glassy bodies located one in each ambulacrum close to the peristome. Try to find and remove the sphaeridia to see whether their removal affects the urchin's righting reaction or its geotactic responses.

Direct a beam of bright light toward an urchin. How does it react? _____

Epidermis. The test, podia, pedicellariae, and spines are covered with **ciliated epidermis,** although the epidermis may have become worn off from the exposed spines. Drop a little carmine suspension on various parts of the sea urchin and find the direction of the ciliary currents. Of what advantage are such currents to the urchin? _____

Gills. At the outer edge of the peristome, between the ambulacra, find five pairs of branching peristomial **gills,** which open into the coelomic cavity.

Written Report

Record your observations on sea urchin behavior on separate paper.

Test.

☞ Examine a dried sea urchin test from which the spines have been removed and also a dried sand dollar and/or heart urchin.

The test, or endoskeleton, is composed of calcareous plates (ossicles) that are symmetrically arranged and interlocked or fused so as to be immovable. Note the tubercles to which the spines were attached. Note the arrangement of the plates into 10 meridional double columns—five double rows of **ambulacral plates** alternating with double rows of interambulacral plates. What is the function of the perforations in the ambulacral plates? _____ In asteroids the tube feet were extended between the plates rather than through them.

Examine the test of a sand dollar or a heart urchin or both. Can you find perforations in them similar to those of the urchin? _____ Do they have ambulacra, and is their arrangement pentamerous? _____
The ambulacra in echinoids are homologous to the ambulacra of asteroids and ophiuroids.

On the aboral surface, note the area that is free from spines, the **periproct.** The **anus** is centrally located, surrounded by four (sometimes five) valvelike **anal plates** (Figure 14-8). Around the anal plates are five **genital plates,** so called because each bears a **genital pore.** Note that one of the genital plates is larger than the others and has many minute pores. This is the **madreporite plate,** which has the same function in sea urchins as it does in sea stars, for the echinoids have a **water-vascular system** in common with all echinoderms.

The test grows both by the growth of plates and by the production of new plates in the ambulacral area near the periproct.

Table of comparison ◄

Complete a table of the comparative characteristics of the external anatomy of the sea star, brittle star, and sea urchin on p. 179. Include such items as symmetry, shape, integument, ambulacra, tube feet, skin gills, pedicellariae, skeleton, digestive system, and so on. After you have observed the sea cucumber, add that to the table.

EXERCISE 14D
Class Holothuroidea— The Sea Cucumber

Sea Cucumber

Phylum Echinodermata
    Subphylum Eleutherozoa
        Class Holothuroidea
            Order Dendrochirotida
                Genus *Cucumaria* (or *Thyone*)
                or
            Order Aspidochirotida
                Genus *Parastichopus*

Where Found

Sea cucumbers, perhaps the oddest members of a phylum distinguished by strange animals, look remarkably like their vegetable namesake. They are characterized by an elongate body, leathery body wall with warty surface, absence of arms, and with mouth and anus located at opposite poles of the animal. They are benthic (bottom-dwelling), slow-moving animals found in all marine habitats. Two common genera on the East Coast of the United States are *Thyone* and *Cucumaria* (Figure 14-9); *Parastichopus* is a familiar genus on the West Coast.

Behavior and External Structure

☞ If possible, study living specimens in an aquarium or in a bowl of seawater containing a generous

A

B

Figure 14-9
Two sea cucumbers from the Pacific coast of North America. **A,** *Parastichopus californicus* is a deposit feeder that grazes the bottom with its tentacles. **B,** *Cucumaria miniata* is a suspension feeder that traps planktonic organisms on its extended, mucus-coated tentacles. When loaded, the tentacles are pushed one by one into the mouth and the food is ingested.

layer of sand, and then examine a preserved specimen. Sea cucumbers are slow to react. They should be left undisturbed for some time before the laboratory period if you are to see them relaxed and feeding.

Note that the holothurian, unlike the other echinoderms, is orally-aborally elongated and has a cylindrical body with the **mouth** encircled by **tentacles** at one end and the **anus** at the other.

A more detailed description and behavior of the animal will depend somewhat on the species you happen to be observing. Notice the tentacles. Are they branched and extensible (as in *Cucumaria* and *Thyone*), or short and shield shaped (as in *Parastichopus*), or of some other type? _____ The tentacles, which are modified tube feet, are hollow and a part of the **water-vascular system;** they are connected internally with the radial canals. The type of tentacle structure is related to feeding habits. *Cucumaria* and *Thyone* are suspension feeders that stretch their mucus-covered tentacles into the water or over the substrate until they are covered with tiny food organisms, then they thrust the tentacles into the mouth, one by one, to lick off the food. Can you observe these actions? *Parastichopus* and some others are deposit feeders that simply shovel mud and sand into the mouth, digest out organic particles, and void the remainder.

Have you noticed any rhythmic opening and closing of the anus? This is a respiratory movement coordinated with the pumping action of the cloaca, which pumps water into and out of the **respiratory trees** (internal respiratory organs).

Does the animal try to burrow into the sand? Does it cover itself completely, or leave the ends exposed? *Thyone* may take 2 to 4 hours to bury its middle by alternate circular and longitudinal muscle contractions. It is likely to be more active in the late afternoon and night than in the morning. If you are watching *Parastichopus* move, what does its muscular action remind you of? Are there waves of contraction?

Does the sea cucumber react to mechanical stimulus? _____ Try touching a tentacle. Does one or more than one tentacle react? _____ Touch several tentacles. What happens when you stroke the body or gently pick the animal up? _____ Did it expel water when you picked it up? Having observed its movement and reactions, can you see the advantages of the hydraulic skeleton? What other phyla used the hydraulic skeleton to advantage? _____

Note the **podia.** Are they scattered all over the body or arranged in ambulacral rows? _____ This pattern differs among different species. Are the podia all alike; that is, are ventral podia any different from dorsal podia? _____ Are any of them suckered? If the pentamerous arrangement of ambulacra is evident in your specimen, how many rows make up the ventral **sole?** _____ How many are on the dorsal surface? _____ If you place the animal on a solid surface, do the podia attach themselves? _____ Can the animal right itself if turned over? _____ Are the podia involved in the righting action? _____ Are muscles involved?

Does the animal show any geotactic reaction if placed on a vertical or sloping surface? _____ Some burrowing forms are positively geotactic and move downward; other species are negatively geotactic and climb upward. *Thyone* gives no geotactic response.

Do you find any pedicellariae or skin gills? _____ Do you feel the presence of a test under the epidermis? That is because the skeleton of the holothurian is usually limited to microscopic ossicles embedded in the tough, leathery body wall.

Written Report

Record your observations on sea cucumber behavior on separate paper.

Internal Structure

Locate the five longitudinal ambulacral areas of the sea cucumber. The sole (the ventral side applied to the substrate) has three ambulacra with well-developed podia. The dorsal side has two ambulacra with smaller podia (in some species podia are absent on the dorsal surface). With scissors, open the body by making a longitudinal incision on the ventral (sole) side of the animal, between the central and right ambulacra. Pin down the walls and cover with water.

Digestive System. Note the large **coelomic cavity.** Just behind the mouth is the **pharynx,** supported by a ring of calcareous plates (Figure 14-10). It is followed by a short muscular **stomach** and a long convoluted **intestine,** held in place by mesenteries and expanding somewhat at the end to form a **cloaca,** which empties at the **anus.**

Respiratory System. Two branched **respiratory trees** are attached to the cloaca; they serve as both respiratory and excretory organs. They are aerated by a rhythmic pumping of the cloaca. Several inspirations 1 minute or more apart are followed by a vigorous expiration that expels all the water.

Note the muscles between the cloaca and the body wall. Long **retractile muscles** (How many? _____) run from the pharynx to join the longitudinal **muscle bands** in the body wall.

Water-Vascular System. A **ring canal** surrounds the pharynx. One or more rounded or elongated sacs called **polian vesicles** hang from the ring canal into the coelom and open into the ring canal by a narrow neck. Polian vesicles are believed to function as expansion chambers in maintaining pressure within the water-vascular system. One or more **stone canals** also open into the ring canal from the body cavity. In adult sea cucumbers the water-vascular system has usually lost contact with the seawater outside. Coelomic fluid, rather than seawater, enters and leaves the system.

Five **radial canals** extend from the ring canal forward along the walls of the pharynx to give off branches to the tentacles, which are actually modified tube feet. From there the radial canals run back along the inner surface of the ambulacra, where each gives off **lateral canals** to the **podia** and **ampullae.** Valves in the lateral canals prevent backflow. Note the ampullae along the ambulacra in the inner body wall.

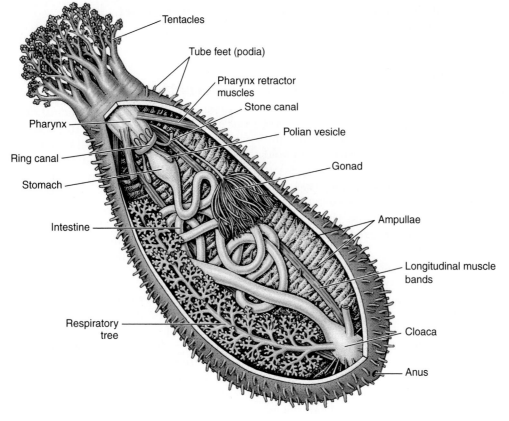

Figure 14-10
Internal anatomy of a sea cucumber.

Reproductive System. The **gonad** consists of numerous tubules united into one or two tufts on the side of the dorsal mesentery. These become quite large at sexual maturity. A **gonoduct** passes anteriorly in the mesentery to the **genital pore.** The sexes are separate, and fertilization is external.

Endoskeleton. The **endoskeleton** consists largely of tiny calcareous ossicles scattered in the dermis. These can be seen on a prepared slide.

Table of comparison

Complete the comparative table (p. 179) of the likenesses and differences between the external anatomy of the sea star, brittle star, sea urchin, and sea cucumber, mentioning such things as shape, ambulacra, tube feet, skin gills, skeleton, integument, and so on.

Comparative Table of Echinoderm Characteristics

Name_____

Date_____

Section_____

LAB REPORT

Characteristic	Sea star	Brittle star	Sea urchin	Sea cucumber

Phylum Chordata:
A Deuterostome Group
The Protochordates
Subphylum Urochordata
Subphylum Cephalochordata

What Defines a Chordate?

The chordates show a remarkable diversity of form and function, ranging from the **protochordates** to humans. Although all organ systems are well developed in this group, the nervous system has been chiefly responsible for giving the phylum its eminence among animals. Most chordates are vertebrates, but the phylum also includes a few invertebrate groups. All animals that belong to the phylum Chordata must have at some time in their life cycle these characteristics.

1. **Notochord.** The notochord (Gr. *nōton,* back, + L. *chorda,* cord) (Figure 15-1) is a slender rod of cartilage-like connective tissue lying near the dorsal side and extending most of the length of the animal. It is regarded as an early endoskeleton and has the functions of such. In most vertebrates it is found only in the embryo.

2. **Pharyngeal Gill Slits.** The pharyngeal gill slits (Figure 15-3) are a series of paired slits in the pharynx, serving as passageways for water to the gills. In some vertebrates they appear only in the embryonic stages.

3. **Dorsal Tubular Nerve Cord.** A dorsal tubular nerve cord, with its modification, the brain, forms the central nervous system. It lies dorsal to the alimentary tract and has a fluid-filled cavity, in contrast to the invertebrate nerve cord, which is ventral and solid.

4. **Postanal Tail.** A postanal tail projects beyond the anus at some stage and serves as a means of propulsion in water. It may or may not persist in the adult. Along with body muscles and stiffened notochord, it provides motility for a free-swimming existence.

These features vary in chordates. Some ancestral chordates have all of these structures throughout life. In

---

What Defines a Chordate?
Classification
Phylum Chordata

EXERCISE 15A
Subphylum Urochordata—*Ciona,* an Ascidian
Ciona

EXERCISE 15B
Subphylum Cephalochordata—Amphioxus
Amphioxus

---

many chordates having more derived characteristics, the gill slits never break through from the pharynx but merely form pouches that have no function, the notochord is replaced by the vertebral column, and only the dorsal nerve cord actually persists in the adult as a diagnostic chordate character. The **protochordates** demonstrate each of the chief chordate characteristics at some point in their life cycle.

Classification

Phylum Chordata

Group Protochordata (pro′to-kor-da′ta) **(Acrania).** Chordates with no cranium or vertebral column.

Subphylum Urochordata (u′ro-kor-da′ta) (Gr. *oura,* tail, + L. *chorda,* cord) **(Tunicata).** Tunicates. Only larval forms have all chordate characteristics; adults sessile, without notochord and dorsal nerve cord; body enclosed in tunic. Example: *Molgula,* a sea squirt.

Subphylum Cephalochordata (sef′a-lo-kor-da′ta) (Gr. *kephalē,* head, + L. *chorda,* cord). Lancelet. Notochord and nerve cord persist throughout life; lance shaped. Example: *Branchiostoma* (amphioxus).

Phylum Chordata

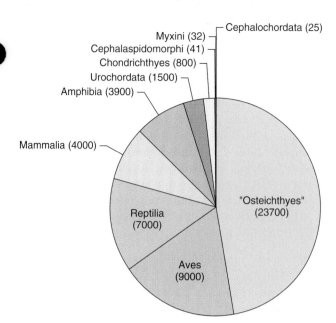

- Cephalochordata (25)
- Myxini (32)
- Cephalaspidomorphi (41)
- Chondrichthyes (800)
- Urochordata (1500)
- Amphibia (3900)
- Mammalia (4000)
- "Osteichthyes" (23700)
- Reptilia (7000)
- Aves (9000)

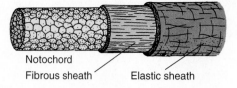

- Notochord
- Fibrous sheath
- Elastic sheath

Figure 15-1

Structure of the notochord and its surrounding sheaths. Cells of the notochord proper are thick walled, pressed together closely, and filled with semifluid. Stiffness is caused mainly by turgidity of fluid-filled cells and surrounding connective tissue sheaths. The primitive type of endoskeleton is characteristic of all chordates at some stage of the life cycle. The notochord provides longitudinal stiffening of the main body axis, a base for trunk muscles, and an axis around which the vertebral column develops.

Group Craniata. Animals with a cranium and vertebral column; that is, the vertebrates.

Subphylum Vertebrata (ver′te-bra′ta) (L. *vertebratus,* backboned). Enlarged brain enclosed in cranium; nerve cord surrounded by bony or cartilaginous vertebrae; notochord in all embryonic stages and persists in adults of some fishes; typical structures include two pairs of appendages and body plan of head, trunk, and postanal tail.

Superclass Agnatha (ag′na-tha) (Gr. *a,* without, + *gnathos,* jaw). **(Cyclostomata).** No jaws or ventral fins; notochord persistent.

Class Cephalaspidomorphi (sef-a-lass′pe-do-mor-ph′e) (Petromyzontes). Lampreys.

Class Myxini (mik-sy′ny). Hagfishes.

Superclass Gnathostomata (na′tho-sto′ma-ta) (Gr. *kephalē,* head, + *aspidos,* shield, + *morphē,* form). Jaws present; usually paired limbs; notochord persistent or replaced by vertebral centra.

Class Chondrichthyes (kon-drik′thee-eez) (Gr. *chondros,* cartilage, + *ichthys,* a fish). Sharks, skates, rays, and chimaeras.

Classes Actinopterygii (ak′ti-nop-te-rij′ee-i) and **Sarcopterygii** (Sar-cop-te-rij′ee-i), formerly grouped within the class Osteichthyes (os′te-ik′thee-eez) (Gr. *osteon,* bone, + *ichthys,* a fish). Bony fishes.

Class Amphibia (am-fib′e-a) (Gr. *amphi,* both or double, + *bios,* life). Amphibians. The frogs, toads, and salamanders.

Class Reptilia (rep-til′e-a) (L. *repere,* to creep). Reptiles. The snakes, lizards, turtles, crocodiles, and others.

Class Aves (ay′veez) (L. pl. of *avis,* bird). Birds.

Class Mammalia (ma-may′lee-a) (L. *mamma,* breast). Mammals.

EXERCISE 15A
Subphylum Urochordata— *Ciona,* an Ascidian

Ciona

Phylum Chordata
  Subphylum Urochordata
    Class Ascidiacea (sea squirts)
      Order Enterogona
      Family Cionidae
        Genus *Ciona*
          Species *Ciona intestinalis*

The Urochordata (Gr. *oura,* tail, + L. *chorda,* cord) are commonly called tunicates because of their leathery covering, or tunic. They are divided into three classes: Ascidiacea, the sea squirts; Thaliacea, the salpians; and Larvacea, the appendicularians. The largest group is the ascidians (Gr. *askidion,* leather bag or bottle), which are also the most generalized. They are called sea squirts because of their habit, when handled, of squirting water from the excurrent siphon. Any of the small translucent ascidians may be used for this exercise. Adult tunicates are all sessile, whereas the larvae undergo a brief free-swimming existence.

Where Found

Tunicates are found in all seas and at all depths. Most of them are sessile as adults, although some are pelagic (found in the open ocean). *Ciona intestinalis* (Gr. *Chionē,* demigoddess of mythology) is a cosmopolitan species common in shallow water on wharf pilings, on anchored and submerged objects, and on eelgrass. It grows to 15 cm in its largest dimension. Although sea squirts (ascidians of the class Ascidiacea) are common

Figure 15-2
Ciona intestinalis, a solitary tunicate, showing its siphons in use. It remains anchored throughout life to one spot on the seafloor. Its free-swimming larva bears all the chordate hallmarks: notochord, gill slits, dorsal nerve cord, and postanal tail.

they also are commonly overlooked in the marine environment. Nearly all shallow-water sea squirts are found attached to almost any available rigid surface: wharf pilings, rocks, shells, and ship bottoms. One of the best places to see sea squirts is on pilings where they may cover the surface.

External Features and Behavior

Sea squirts are fairly hardy in the marine aquarium.

☞ Examine, in a fingerbowl of seawater, a living solitary tunicate such as *Ciona* or *Molgula,* or a portion of a colony of *Perophora* or other ascidians as available.

Observe the use of the two openings, or **siphons.** When fully submerged and undisturbed the siphons are open, and respiratory water, kept moving by ciliary action, enters the more terminal siphon (called the **incurrent,** or **oral siphon**) at the mouth, circulates through a large pharynx, and leaves through the **excurrent,** or **atrial, siphon** on one side (the dorsal side) (Figures 15-2 and 15-3).

☞ Release a little carmine suspension near the animal to verify this.

The outer covering of a tunicate is called the **tunic,** or **test** (Figure 15-3). It is secreted by the **mantle,** which lies just inside it, and it contains cellulose—an uncommon substance in animals. The mantle contains the muscle fibers by which the body can contract. If the tunic is translucent enough and the light is properly adjusted, you may be able to see some of the internal structure.

Neuromuscular System. The tunicate nervous system is reduced and not well understood. There is a cerebral ganglion (closely associated with a sub-neural gland of uncertain function) located between the siphons. The tunic probably has no sensory nerves, but pressure on the tunic may be transmitted to nerves in the mantle. Both direct and crossed reflexes have been observed in some ascidians and can be tested in a living *Ciona* or other tunicate by touching selected areas with the tip of a glass, rod, or dissecting needle.

Direct reflexes result from mechanical stimulation of the **outer** surface of the siphons or tunic.

☞ *Gently* stimulate various areas of the tunic and note the response. Touch gently the outer surface of one of the siphons, note the response, and then stimulate the other siphon.

What areas of the body are most sensitive? **Crossed reflexes** result from mechanical stimulation of the **inner** surface of the siphons.

☞ Gently touch the inner surface of the oral siphon. What happens? Try a stronger stimulus of the same siphon. Do you get the same response? How do these responses differ from those in which the outer surface was stimulated? Repeat with the atrial siphon.

Of what protective value would these reflexes be to the animal? _____

Internal Structure

☞ Internal structure can be observed on either a living or preserved specimen. Use fine scissors to slit the tunic longitudinally, beginning at the incurrent siphon and continuing the cut to the base of the pharynx. Be *very* careful to cut *only* the tunic and not the mantle beneath it. Slip the animal out of its tunic, then return the animal to the bowl of seawater and study with a dissecting microscope.

Respiratory System. The **branchial sac,** or **pharynx,** is the largest internal structure, and the space between it and the mantle is the **atrium** (Figure 15-3). The pharyngeal wall is perforated with many pharyngeal slits through which water passes into the atrium to be discharged through the excurrent siphon. The vascular wall of the pharynx serves as a gill for gaseous exchange.

Circulatory System. In a living specimen, with test removed and a light properly adjusted, you should be able to see the beating of the **heart,** located near the posterior end on the right side. The tubular heart empties

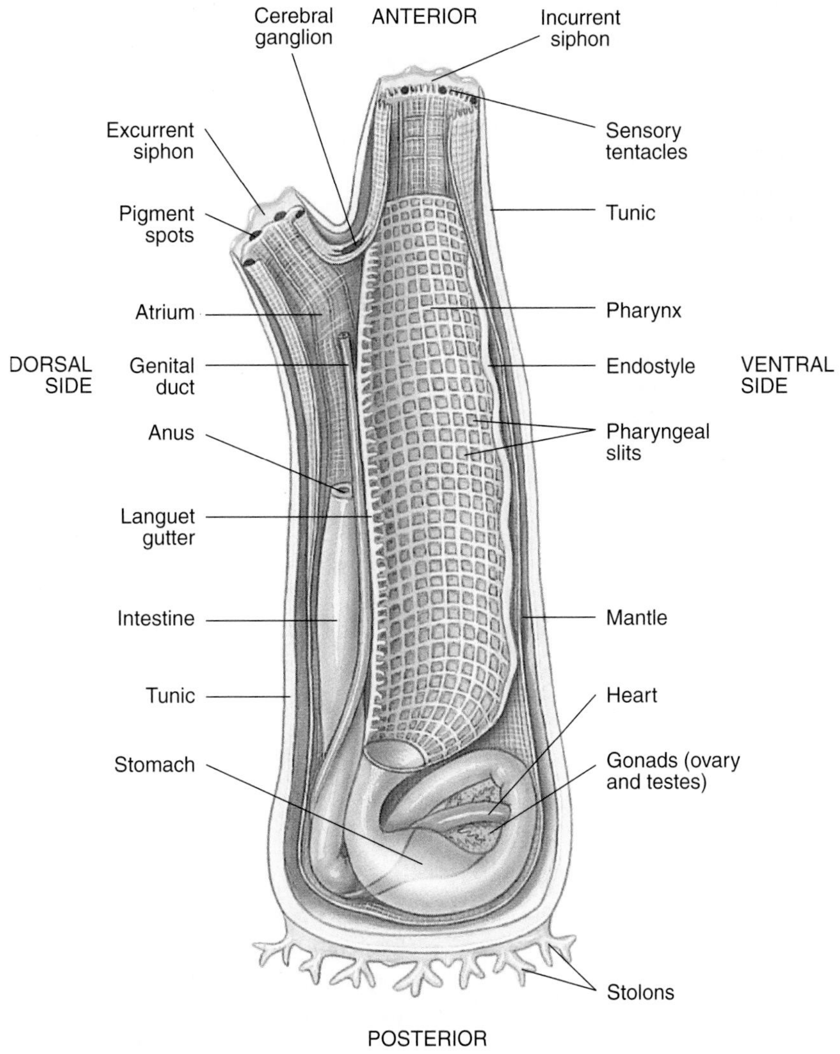

Figure 15-3

Structure of a solitary sea squirt, *Ciona*.

into two vessels, one at each end. Its peristaltic waves are of unusual interest, because they send the blood in one direction for awhile and then reverse direction and pump the blood in the opposite direction. Apparently there are two pacemakers that initiate contractions, one at each end of the heart, and they alternate in dominance over each other. The tunicate vascular system is an open type of system. Blood cells are numerous and colorful. There are no respiratory pigments.

Digestive System. At the junction of the **mouth** and the **pharynx** is a circlet of **tentacles** that form a grid which screens the incurrent water. By dropping a grain of sand into the incurrent siphon of a living tunicate, you may be able to observe the ejection reflex.

Inside the pharynx along the midventral wall is the **endostyle,** which is a ciliated groove that secretes a great deal of mucus. Cilia on the walls of the pharynx distribute the mucus. Food particles become tangled in the mucus and are propelled by cilia to a dorsal gutter (Figure 15-3) in which the mucus with its trapped food

becomes rolled into a compact cord. (In *Ciona* and a few other tunicates the gutter is lined with a row of curved, ciliated, fingerlike processes called languets [F. *languette,* small tongue].) The cord is propelled posteriorly to the esophagus and stomach.

☞ If living ascidians are available, examine one that has been submerged for some time in a suspension of carmine particles in seawater. Open the pharynx by cutting through the incurrent siphon and downward, a little to one side of the midventral line. Then cut around the base and lay the animal open in a pan of water. You should be able to see a concentration of carmine particles in the middorsal area. Cut out a small piece of the pharyngeal wall (free from the tunic and mantle) and mount on a slide to observe gill slits and beating cilia.

It may be difficult to differentiate the **esophagus, stomach,** and **intestine.** The **anus** empties into the atrium near the excurrent siphon.

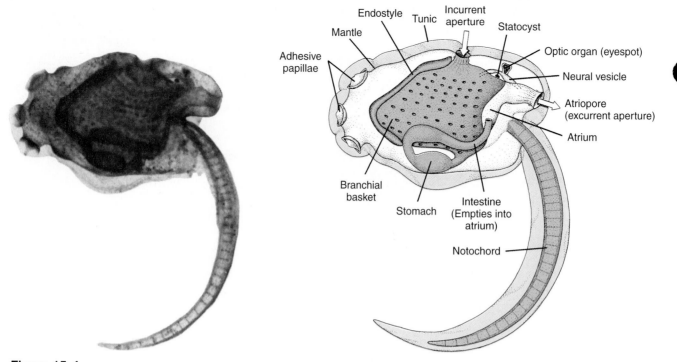

Figure 15-4
Tadpole larva of a tunicate, photographed from a stained slide specimen.

Excretion. A ductless structure near the intestine is assumed to be a type of nephridium and to be excretory in function.

Reproduction. *Ciona,* like most tunicates, is hermaphroditic and bears a single ovary and testis, each with a gonoduct that opens into the atrium.

Some tunicates are colonial. In some the zooids are separate and attached by a stolon, as in *Perophora.* In others the zooids are regularly arranged and partly united at the base by a common tunic; all the incurrent siphons are at one side and the excurrent siphons at the other. Another type has a system of zooids that share a common atrial (excurrent) chamber. Colonies are formed asexually by budding.

Solitary tunicates generally shed their eggs from the excurrent siphon, and development occurs in the sea. Colonial species usually brood their eggs in the atrium, and the microscopic larvae leave by the excurrent siphon. For a very brief period the larvae are nonfeeding and live a planktonic free-swimming existence; then they settle down, attach to the substrate, and metamorphose.

Ascidian Larvae (Study of Stained Slides)

Ascidian larvae are free-swimming. (Figure 15-4). They do not look like sessile adult sea squirts and are actually more characteristic of the chordates than are the adults. They possess not only gill slits, but also a notochord, dorsal tubular nerve cord, and a tail, structures that have

been lost in the adult. You may be able to identify some of these structures on a stained slide.

Adhesive papillae at the anterior end of the larva are used to attach to some object during metamorphosis, which occurs within a short time after hatching. The **notochord** can be identified in the long tail. This character, which becomes lost in the adult, gives the subphylum Urochordata ("tail-cord") its name. A **nerve cord** dorsal to the notochord enlarges anteriorly into a neural vesicle. Can you identify a pigmented, photoreceptive **eyespot?** A smaller pigmented area anterior to the eye is a **statocyst.** What is the function of the statocyst? _____ At metamorphosis these portions of the nervous system degenerate, and a ganglion serves as the nerve center.

Look for the anterior **oral (incurrent) aperture** and the more posterior **atriopore (excurrent aperture).** Perhaps you can identify the **branchial basket** (pharynx) with **gill slits, stomach, intestine, atrium, and endostyle.**

Exercise 15B
Subphylum Cephalochordata— Amphioxus

Amphioxus
Phylum Chordata
    Subphylum Cephalochordata
        Genus *Branchiostoma* (= *Amphioxus*)
            Species *Branchiostoma lanceolatus*

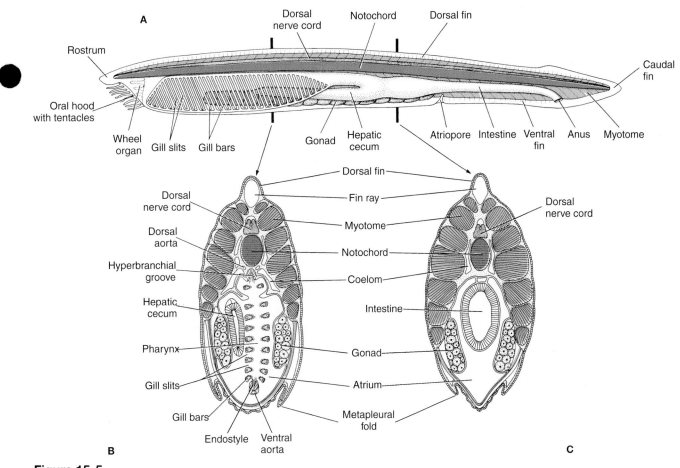

Figure 15-5
Structure of amphioxus, showing external structure and transverse sections through the pharynx and intestine.

The little lancelet, *Branchiostoma* (Gr. *branchia,* gills, + *stoma,* mouth), commonly called amphioxus, illustrates basic chordate structure and is considered similar to the ancestor of the vertebrates. Besides the basic characteristics—**notochord, pharyngeal gill slits, dorsal hollow nerve cord,** and **postanal tail**—it also possesses the beginning of a **ventral heart** and a **metameric arrangement** of muscles and nerves. There are only two genera of Cephalochordates—*Asymmetron* (Gr. *asymmetros,* ill-proportioned) and *Branchiostoma.*

Where Found

Branchiostoma is common along the southern California and southern Atlantic coasts of the United States, as well as the coasts of China and the Mediterranean Sea. On sandy bottoms it dives in headfirst, then twists upward so that the tail remains buried in sand and the anterior end is thrust upward into the water.

External Structure

 Place a preserved mature specimen in a watch glass and cover with water. Do not dissect or mutilate the specimen.

How long is it? _____ Why is it called a lancelet? _____ Does it have a distinctive head? _____ Observe the **dorsal fin,** which broadens in the tail region **(caudal fin)** and continues around the end of the tail to become the **ventral fin.**

The anterior tip is the **rostrum** (Figure 15-5A). With the hand lens, find the opening of the **oral hood,** which is fringed by a number of slender oral tentacles, also called buccal cirri, that strain out large particles of sand and are sensory in function.

On the flattened ventral surface are two **metapleural folds** of skin extending like sled runners to the ventral fin. Find the **atriopore,** which is anterior to the ventral fin. The atriopore is the opening of the atrium (L., entrance hall), a large cavity surrounding the pharynx. The **anus** opens slightly to the left of the posterior end of the ventral fin.

In mature specimens, little blocklike **gonads** (testes or ovaries) lie in the atrium anterior to the atriopore and just above the metapleural folds on each side. They can be seen through the thin body wall.

Study of the Whole Mount

 Examine with low power a stained and cleared whole mount of an immature specimen.

The chevronlike **myotomes** (Gr. *mys,* muscle, + *tomos,* slice) along the sides of the animal are segmentally arranged muscles. Does the myotome of an amphioxus zigzag more or less than the myomere of a bony fish? _____ How might the differences in musculature help us predict whether the amphioxus or the fish would be the better swimmer? _____ Identify the various parts of the **fin** and note its skeletal support, the transparent **fin rays.** You may have to reduce the light to see the fin rays.

Beneath the rostrum is a large chamber called the **buccal cavity,** which is bounded laterally by fleshy, curtainlike folds and is open ventrally. The rostrum and lateral folds together make up the **oral hood** (Figure 15-5A). The roof of the oral hood bears the notochord, which may have a supporting function in spreading the hood open. Each of the oral tentacles is stiffened by a skeletal rod of fibrous connective tissue. Behind the buccal cavity is an almost perpendicular membrane, the **velum** (L., veil), pierced ventrally by a small opening, the true **mouth,** which is always open and leads into the **pharynx.** On the walls of the buccal cavity, projecting forward from the base of the velum, are several fingerlike ciliated patches that compose the **wheel organ.** The rotating effect of its cilia helps maintain a current of water flowing into the mouth. Around the mouth, projecting posteriorly from the velum, are about a dozen delicate **velar tentacles,** also ciliated. Both oral tentacles and velar tentacles have chemoreceptor cells for monitoring the incurrent water.

The large **pharynx** narrows into a straight **intestine** extending to the **anus** (Figure 15-5). The sidewalls of the pharynx are composed of a series of parallel, oblique **gill bars,** between which are **gill slits.** Just posterior to the pharynx is a diverticulum of the intestine called the **hepatic cecum,** or liver, which extends forward along one side of the pharynx. Surrounding the pharynx is the **atrium,** a large cavity that extends to the **atriopore.** Water entering the mouth filters through the gill slits into the atrium and then out the atriopore.

Cephalochordates, like sponges, clams, and tunicates are filter-feeders. They use a mucus-ciliary method, feeding on minute organisms. As the animal rests with its head out of the sand, the ciliated tentacles, wheel organ, and gills draw in a steady current of food-laden water, from which the cirri and velar tentacles strain out large or unwanted particles. On the floor of the pharynx is an **endostyle** (Figure 15-5) consisting of alternating rows of ciliated cells and mucus-secreting cells, and in the roof of the pharynx is a ciliated **hyperbranchial groove** (= epipharyngeal groove). Particles of food entangled in the stream of mucus secreted by the endostyle are carried upward by cilia on the inner surface of the gill bars, then backward toward the **intestine** by cilia in the hyperbranchial groove. Digestion occurs in the intestine.

Oxygen–carbon dioxide exchange occurs in the epithelium covering the gill bars. The **notochord** just dorsal to the digestive system is transversely striated and is best seen in the head and tail regions. It provides skeletal support and a point of attachment for the muscles. Note that it extends almost to the tip of the rostrum. Above and parallel to the notochord is the **dorsal nerve cord** (Figure 15-5). The row of black spots in the nerve cord are pigmented **photoreceptor cells.** Chemoreceptors are scattered over the body but are particularly abundant on the oral and velar tentacles. Touch receptors are located over the entire body.

Circulatory System

Amphioxus does not have a heart; peristaltic contractions of the **ventral aorta** keep the colorless blood in motion, sending it forward and then upward through **afferent branchial arteries** (Figure 15-6) to capillaries in the gill bars for gas exchange. Blood is carried from the gills by **efferent branchial arteries** up to a pair of **dorsal aortas.** These join posterior to the gills to form a **median dorsal aorta,** which gives off **segmented arteries** to the capillaries of the myotomes and to the capillaries in the wall of the intestine. From the intestinal wall, blood, now rich in digested food nutrients, is picked up by the **subintestinal vein** and is carried forward to the **hepatic portal vein,** which enters the **hepatic cecum.** In the capillaries of the liver, nutrients are removed and stored in liver tissue or processed and returned to the blood as needed. Blood returns to the ventral aorta by way of the **hepatic vein.** Blood with waste products from the muscular walls returns by way of left and right **precardinal** and **postcardinal veins,** which empty into left and right **ducts of Cuvier** and then into the ventral aorta.

Oral Report

Compare circulation in amphioxus with that in the earthworm and crayfish. Be able to trace a drop of blood to various parts of the body and back to the ventral aorta, and explain what the blood gains and loses in each of the capillary beds through which it passes on its journey. Locate as many of the main vessels as you can on any transverse sections you may have.

Cross Section—Stained Slide

Look at a cross section through the pharynx with the unaided eye and understand how the section is cut with reference to the whole animal. Note the **dorsal fin,** its supporting **fin ray,** and the ventral **metapleural folds** (Figure 15-5). With the microscope, examine the **epidermis,** a single layer of columnar epithelial cells, and the **dermis,** a gelatinous connective tissue layer. The large

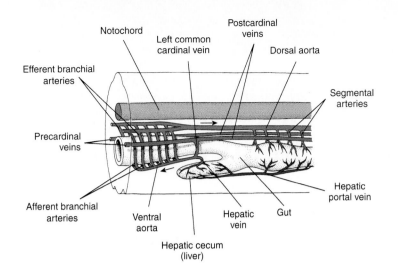

Figure 15-6

Major circulatory vessels of amphioxus. The ventral aorta pumps the blood forward and then upward to the gill capillaries via what arteries? _____ What arteries come from the dorsal aorta and feed the capillaries of the myotomes and intestinal wall? _____ Blood from the intestine, now rich in digested food nutrients, is carried forward to the hepatic cecum via what vein? _____ Blood returns to the ventral aorta by way of what vein? _____

myotomes, or muscles, are paired, but members of a pair are not opposite each other. The myotomes are separated by connective tissue, **myosepta.**

The **nerve cord,** enclosed within the **neural canal,** has in its center a small **central canal,** which is prolonged dorsally into a slit. In some sections dorsal **sensory nerves** or ventral **motor nerves** may be seen. These are given off alternately from the cord to the myotomes. The large, oval **notochord** with vacuolated cells is surrounded by the **notochordal sheath.**

The cavity of the **pharynx** is bounded by a ring of triangular **gill bars** separated by **gill slits** that open into the surrounding **atrium** (See Figure 15-5B). From your study of the whole mount, why do you think the cross section shows the gill bars as a succession of cut surfaces? The somewhat rigid gill bars contain blood vessels and are covered by ciliated **respiratory epithelium,** where gaseous exchange is made. On the dorsal side of the pharynx, find the cili-

ated **hyperbranchial groove** and on the ventral side, the **endostyle.** The latter secretes mucus in which food particles are caught. What is the function of the cilia in the grooves?

The **gonads** (ovaries or testes) lie on each side of the atrial cavity. The reduced **coelom** consists of spaces, usually paired, on each side of the notochord and the hyperbranchial groove, as well as on each side of the gonads and below the endostyle. In favorable specimens little **nephridial** tubules may be found in the dorsal coelomic cavities (See Figure 15-5). What is their function?

Ventral to the notochord are the paired **dorsal aortas.** The **ventral aorta** lies ventral to the endostyle.

Drawings

> On separate paper, make drawings of such sections of amphioxus as your instructor requests.

The Fishes—Lampreys, Sharks, and Bony Fishes

EXERCISE 16A

Class Cephalaspidomorphi (= Petromyzontes)— The Lampreys (Ammocoete Larva and Adult)

Lamprey

Phylum Chordata
   Subphylum Vertebrata
      Superclass Agnatha
         Class Cephalaspidomorphi (= Petromyzontes)
            Genus *Petromyzon*
               Species *Petromyzon marinus*

Hagfishes and lampreys are the only living descendants of the earliest known vertebrates, a group of Paleozoic jawless fishes collectively called ostracoderms. Hagfishes and lampreys are conventionally grouped together in the superclass Agnatha ("without jaws") and share certain primitive characteristics including the absence of jaws, internal ossification, scales, and paired fins. In other respects, however, hagfishes and lampreys are radically different from each other.

Lampreys have a worldwide distribution and most are **anadromous** (an-ad′ruh-mus; Gr. *anadromos*, running upward), meaning that they ascend rivers and streams to spawn. The species *Petromyzon marinus* (Gr. *petros*, stone, + *myzon*, sucking, referring to its habit of holding its position in a current by grasping a stone with its mouth) is the sea lamprey that lives in the Atlantic drainages of Canada, the United States, Iceland, and Europe and is landlocked in the Great Lakes. It grows to be 1 m long and can live both in fresh water and in the sea. It is a marine species that migrates up freshwater streams to spawn. The young larvae, known as **ammocoetes** (sing, **ammocoete**; Gr. *ammos*, sand, + *koitē*, bed, Fr. *keisthai*, to lie, referring to the preferred larval habitat), live in sand for 3 to 5 years and then metamorphose rapidly and become parasites of fishes. Attaching themselves with their suckerlike mouth, they rasp away the fish's flesh with their horny teeth and suck out blood and body fluids. Adults grow rapidly for a year,

EXERCISE 16A

Class Cephalaspidomorphi (= Petromyzontes)— The Lampreys (Ammocoete Larva and Adult)
Lamprey
Ammocoete larva
Adult lamprey

EXERCISE 16B

Class Chondrichthyes—The Cartilaginous Fishes
Squalus, the dogfish shark
Demonstrations

EXERCISE 16C

Class Osteichthyes—The Bony Fishes
Perca, the yellow perch

EXPERIMENTING IN ZOOLOGY

Aggression in Paradise Fish, *Macropodus opercularis*

EXPERIMENTING IN ZOOLOGY

Analysis of the Multiple Hemoglobin System in *Carassius auratus,* the Common Goldfish

spawn in the winter or spring, and soon die. After invading the Great Lakes in the nineteenth century, sea lampreys devastated the important commercial fisheries there. Wounding rates of fishes are lower now, but sea lampreys remain a threat to the commercial fishing trade.

Freshwater lampreys, known as brook or river lampreys, belong to the genera *Lampetra* (L. lambo, to lick or lap up) and *Ichthyomyzon* (Gr. *ichthyos*, fish, + *myzon*, sucking), of which there are about 33 species. They have larval habits similar to the marine form, but adults in about half the species are not parasitic. Nonparasitic forms do not eat as adults and live only a month or so after emerging from the sand to spawn.

Ammocoete Larva

Although the ammocoete larva resembles amphioxus in appearance, life habit, and many anatomical details, it bears several characteristics that anticipate the vertebrate body plan and that are lacking in amphioxus.

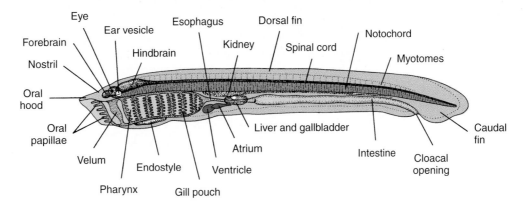

Figure 16-1

Ammocoete larvae, sagittal section.

How is blood propelled in amphioxus? _____
_____

For example the ammocoete has a two-chambered heart, two median eyes each with lens and receptor cells, a three-part brain, thyroid and pituitary glands, and a pronephric kidney. Instead of the numerous gill slits found in the amphioxus, the ammocoete larva has only seven pairs.

☞ Examine a preserved ammocoete larva as well as a stained whole mount of a small specimen.

Study of the Preserved Larva

☞ Cover the preserved larva with water in a water glass. Use a hand lens or dissecting microscope.

List two differences between the preserved specimen and a mature amphioxus. _____
_____ Note the **myotomes** (segmental muscles) appearing faintly on the surface (Figure 16-1). Do the myotomes have the same arrangement as those of the amphioxus? _____ Note the **oral hood** with **oral papillae** attached to the roof and sides of the hood. They are used, as in amphioxus, for filter feeding. The **lateral groove** on each side contains seven small **gill slits.** The anus, or **cloacal opening,** is just anterior to the **caudal fin.** Are the **caudal** and **dorsal fins** continuous? Note the **chromatophores** scattered over the body.

Study of the Stained Whole Mount

☞ With low power, examine a stained whole mount of an ammocoete larva.

On the whole mount, find the darkly stained, dorsal, hollow **nerve cord,** enlarged anteriorly to form the **brain.** Immediately below it is the lighter **notochord** (Figure 16-1).

The oral hood encloses a **buccal cavity,** to the back and sides of which are attached oral papillae. Pos-

terior to the buccal cavity is the **velum,** a large pair of flaps that create water currents. The large **pharynx** has **internal gill slits,** which open into **gill pouches.** The gill pouches open to the outside by small **external gill slits.** How many pairs of gill slits are there in the ammocoete? _____ Using low power, focus upward onto the outer surface of the animal to see the row of small external gill slits. Between the internal gill slits are cartilaginous rods, the **gill bars,** which strengthen the pharynx walls. Note the **gill lamellae** on the pharynx walls. They are rich in capillaries, in which the blood gives up its carbon dioxide and takes up its oxygen from the water.

The ammocoete is a filter-feeder. Was amphioxus also a filter-feeder? _____ Water is kept moving through the pharynx by muscular action of both the velum and the whole branchial basket. This contrasts with the amphioxus, in which water is moved by ciliary action.

The **endostyle** (subpharyngeal gland) in the floor of the pharynx is a closed tube the length of four gill slits. It secretes mucus by a duct into the pharynx. Food particles brought in by water currents are trapped in the mucus and carried by ciliary action to the **esophagus.** During metamorphosis of the larva a portion of the endostyle becomes a part of the thyroid gland of the adult.

The narrow esophagus widens to become the **intestine,** the posterior end of which, called the **cloaca,** also receives the kidney ducts. The **anus** opens to the outside a short distance in front of the postanal tail.

The **liver** lies under the posterior end of the esophagus, and embedded in it is the **gallbladder,** which appears as a clear, round vesicle. The two-chambered **heart** lies under the forepart of the esophagus.

Over the heart and around the esophagus is the **pronephric kidney,** consisting of a number of small tubules that empty into the cloaca by pronephric ducts (not easily distinguished on the whole mounts). Later a mesonephric kidney will develop above the intestine, using the same ducts, and the pronephros will degenerate.

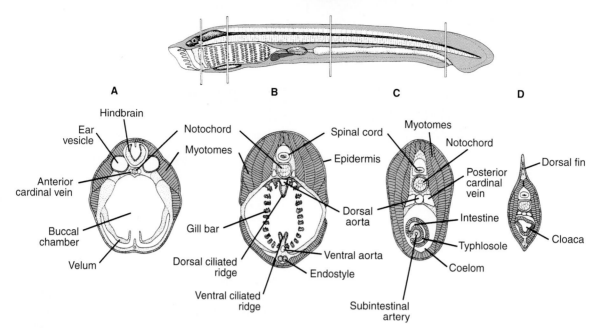

Figure 16-2

Transverse sections of ammocoete larva. **A,** Posterior part of buccal chamber, **B,** pharynx, **C,** intestine, **D,** cloaca.

The tubular dorsal **nerve cord** enlarges anteriorly into a three-lobed **brain,** visible in most slides. How does the nerve cord differ in location from that of the flatworms? _____ How does the nerve cord differ from those of annelids and arthropods? _____ The **forebrain** contains the **olfactory lobe.** In front of the forebrain is the **nasohypophyseal canal,** opening dorsally to the outside by a median **nostril.** The darkly pigmented **eyes** connect with each side of the **midbrain.** At this stage the eyes are covered with skin and muscle and have little sensitivity to light; however, the ammocoete larva does have photoreceptors in the tail. Find the **hindbrain** and, with careful focusing, try to see one of the clear oval **ear vesicles** that flank each side of it. Which lobe of the brain is chiefly concerned with the sense of smell? _____ Of sight? _____ Of hearing and equilibrium? _____ How is the nervous system of this agnathan advanced over that of the cephalochordate amphioxus? _____
_____

Transverse Sections of Ammocoetes

☞ Your slide may contain four typical sections—one each through the brain, the pharynx, the intestine, and the postanal tail. Or it may contain 15 to 20 sections through the body, arranged in sequence. As you study each section, refer to the whole mount again to interpret relationships. Use the low power of the microscope.

Sections Anterior to the Pharynx. Sections through the **forebrain** may include the **oral papilla** and **oral hood.** Sections through the **midbrain** may include the **eyes,** the **buccal chamber,** and portions of the **velar flaps.** Sections through the **hindbrain** may include the **ear (otic) vesicles** and buccal chamber (Figure 16-2A) or the forepart of the pharynx. Compare the size of the brain with that of the spinal cord in more posterior sections. Do you find a **notochord** lying just below the brain in any of the sections? Why is the protochord considered a primitive type of endoskeleton?
_____
_____

Sections Posterior to the Pharynx. Choose a section through the trunk posterior to the pharynx and identify the following (Figure 16-2C): **epidermis; myotomes** (lateral masses of muscles), **nerve cord** surrounded by the **neural canal** and containing a cavity, the **neurocoel; notochord** with large vacuolated cells; **dorsal aorta** (probably contains blood cells); and **posterior cardinal veins,** one on each side of the aorta (blood cells are usually present). You may find the cardinals joining to form the duct of Cuvier. The **coelomic cavity,** lined with peritoneum, contains the visceral organs. What distinguishes a true coloem from other cavities in the body? _____
_____

The visceral contents of the coelomic cavity will vary according to the location of the section.

1. Just behind the pharynx you will find the **esophagus,** of columnar epithelium; the paired **pronephric kidneys,** appearing as sections of small tubules; and chambers of the **heart.**

2. Sections cut posterior to the heart will show the dark **liver** and possibly the hollow **gallbladder**

ventral, or lateral, to the esophagus, with sections of pronephric kidneys or their ducts located under the posterior cardinals.

3. Sections through the intestine will reveal it as a large tube of columnar epithelium with a conspicuous infolding, the **typhlosole,** carrying the **subintestinal artery** (Figure 16-2C). Above the intestine you may find the **mesonephric kidneys,** with their tubules, and a small **gonad** between the kidneys. The **cloaca** is located farther back (Figure 16-2D).

4. In sections posterior to the anus (postanal tail), identify the caudal fins. What other structures can you identify?

Sections through the Pharynx. The body wall, nerve cord, and notochord will be similar to the preceding sections. The central part of the section is taken up by the large pharynx, on whose walls are the **gills,** with their platelike **gill lamellae,** extending into the pharynx or into the **gill pouch,** depending on how the section has been cut. Each lamella has lateral ridges. Some sections, such as that in Figure 16-2B, may show only the gill bars without the feathery lamellae. The gills are liberally supplied with blood vessels. Outside the gill chambers you will find sections through cartilage rods that give support to the branchial basket.

In the middorsal and midventral regions of the pharynx are **ciliated ridges** bearing grooves. The cilia are concerned with the movement toward the esophagus of mucous strands in which food particles are caught. Below the pharynx in certain sections is the bilobed **subpharyngeal gland,** whose function is probably the secretion of mucus. Later, certain portions of this gland are incorporated into the adult thyroid gland. Between the ventral ciliated ridge and the subpharyngeal gland is the single or paired **ventral aorta.** Note the **gill pouches** lateral to the pharynx, the **lateral groove,** and in some sections the **external gill slits** to the outside. Locate the **anterior cardinal veins** on each side of the notochord and the **dorsal aorta** beneath it.

Drawings

✎ On separate paper draw such transverse sections as are required by your instructor. Label fully. How many features of the amphioxus adult do you find repeated in the ammocoete, the larval form of a vertebrate? _____
Do you think this might be interpreted as an example of homology? _____ Keep in mind the structure of these early chordate forms and be able to compare them with those of the fish, amphibian, and mammal forms that you will study later.

Adult Lamprey
External Structure

☞ Examine a preserved specimen of an adult lamprey.

It will be immediately evident that the adult lamprey differs in many anatomical details from the ammocoete, in addition to the obvious difference in their size. During the dramatic metamorphosis from larva to adult, the body becomes rounder and shorter, the pharynx becomes divided longitudinally, the larval hood is replaced by an oral disc with teeth, the eyes enlarge, and the nostril shifts to the top of the head. These changes are essential to the shift from a larval life habit of filter feeding to an adult existence as a parasite of fish.

Note the eel-like shape of the lamprey and its tough, scaleless skin. Among the epithelial cells are numerous gland cells that produce a protective slime. Identify the two **dorsal fins** and the **caudal fin** (Figure 16-3A). There are no paired appendages. What does the lack of paired fins tell you about how a lamprey might swim as compared to a bony fish? _____

Examine the hood-shaped **buccal funnel** supported by a cartilaginous ring that serves as a sucking disc for attachment to the host. The opening is fringed by numerous fingerlike sensory papillae, and the interior of the funnel bears horny "teeth"; these are actually epidermal thickenings and not homologous to true vertebrate teeth, which are derived from mesoderm. Locate the **mouth** at the back of the buccal funnel and dorsal to the **tongue.** The tongue also bears sharp, horny teeth used for rasping.

A single **nostril** located middorsally on top the head opens into an olfactory sac (the latter visible in the sagittal section, below). Just behind the nostril is a small oval area marking the position of the so-called third eye, the **pineal organ.** It is not an eye in the true sense but it does contain photoreceptors that detect changes in illumination and serve to adjust internal activities of the lamprey. The pineal organ is present in most fishes, but it is better developed in lampreys than in any other living vertebrate except certain reptiles. Note the functional, lidless **eyes** and the seven **external gill slits** just behind the eye on each side of the head.

The **lateral line system,** characteristic of nearly all fishes, consists of specialized receptors located in small patches on the head and trunk of the lamprey. They can be located as groups of pores extending below and caudally from the eye on either side of the head. These receptor cells are sensitive to currents and water movement. How would they be useful to the lamprey? _____ In the lamprey the receptors are open to the exterior and not protected within canals as they are in bony fishes.

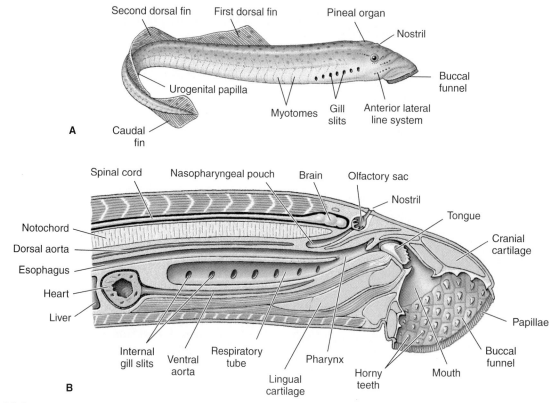

Figure 16-3
Adult lamprey. **A,** External structure. **B,** Sagittal section.

Find the **urogenital* sinus** with its projecting urogenital papilla and the ventral juncture of the trunk and the tail. There is an **anal opening** in front of the urogenital sinus.

How do the structures of the myotomes compare to those of amphioxus? _____

_____

EXERCISE 16B
Class Chondrichthyes— The Cartilaginous Fishes

Squalus, the Dogfish Shark

Subphylum Vertebrata
  Superclass Gnathostomata
    Class Chondrichthyes
      Subclass Elasmobranchii
        Order Squaliformes
          Genus *Squalus*

The cartilaginous fishes are a compact, ancient assemblage of about 815 species that are characterized by their cartilaginous skeletons, powerful jaws, and well-

developed sense organs. They include the sharks, skates, rays, and chimaeras. The subclass Elasmobranchii (e-laz'-mo-bran'kee-i; Gr. *elasmos,* metal plate, + *branchia,* gills) embraces the sharks, skates, and rays—cartilaginous fishes with exposed gill slits opening separately to the outside. Most are carnivores and many are top predators. The dogfish shark is an excellent example of the generalized body plan of early jawed vertebrates.

Where Found

Dogfish sharks are small marine sharks that grow to about 1 m in length (females are slightly larger than males). Two species commonly studied belong to the genus *Squalus* (L. *squalus,* a kind of sea fish), the spiny dogfishes: *Squalus acanthias* of the North Atlantic and *Squalus suckleyi* of the Pacific coast. The two species are morphologically similar. Spiny dogfishes are distinguished by a spine on the anterior edge of both dorsal fins. Spiny dogfishes gather in huge schools of up to 1000 individuals of both sexes when immature, but only of one sex when they are mature. In detecting and capturing their main foods, bottom-dwelling fish and crabs, dogfishes are assisted by their ability to sense weak electrical fields that surround all living animals, using specialized sense organs on the head, the ampullae of Lorenzini. Spiny dogfishes are ovoviviparous (L. *ovum,* egg, + *vivus,* living, + *parere,* to bring forth); that is, they give birth to living young without dependence on

---

* Urogenital = urinogenital. The stem **uro** as used in urogenital derives from the Greek *ouron,* meaning urine. Unfortunately, the stem is used in other terms to mean "tail" (Gr. *oura,* tail) as, for example, in *Urodela, urostyle,* and *uropod.*

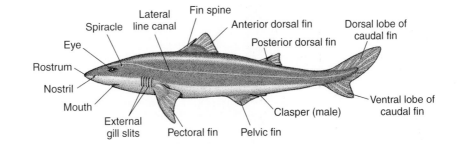

Figure 16-4

External anatomy of the dogfish shark, *Squalus* sp.

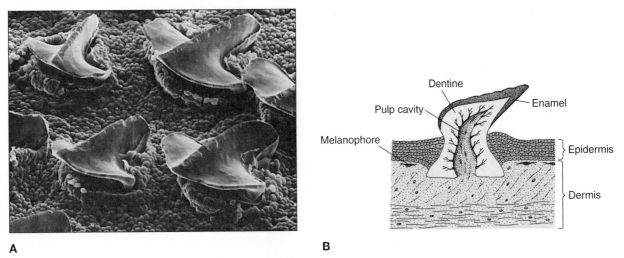

Figure 16-5

A, Surface view of shark skin showing dermal denticles, or placoid scales, SEM, ×195. **B,** Lateral section of a single denticle.

placental nourishment. The embryos develop in an egg capsule in the oviduct until they hatch in the mother just before birth.

External Structure

☞ Examine an intact, preserved dogfish to identify the following features.

The body is divided into the **head** (to the first gill slit), **trunk** (to the cloacal opening), and **tail.** The **fins** include a pair of **pectoral fins** (anterior), which control changes in direction during swimming; a pair of **pelvic fins,** which serve as stabilizers and which in the male are modified to form claspers used in copulation; two medium **dorsal fins,** which also serve as stabilizers; and a **caudal fin,** which is **heterocercal** (asymmetric dorsoventrally) (Figure 16-4).

Identify the **mouth** with its rows of **teeth** (modified placoid scales), which are adapted for cutting and shearing; two ventral **nostrils,** which lead to olfactory sacs and which are equipped with folds of skin that allow continual in-and-out movement of water; and the lateral **eyes,** which lack movable eyelids but have folds of skin that can cover the eyeballs. The part of the head anterior to the eyes is called the **rostrum** (snout).

A pair of dorsal **spiracles** posterior to the eyes are modified gill slits that open into the pharynx. They can be closed by folds of skin during part of the respiratory cycle to prevent the escape of water. The spiracles serve for water intake when the shark is feeding. Five pairs of **gill slits** are the external openings of the gill chambers.

The **pharynx** is the region in back of the mouth into which the gill slits and spiracles open. A **lateral line,** appearing as a white line on each side of the trunk, represents a row of minute, mucus-filled sensory pores used to detect differences in the velocity of surrounding water currents, and thus to detect the presence of other animals, even in the dark. Note the **cloacal opening** between the pelvic fins.

The leathery skin consists of an outer layer of epidermis covering a much thicker layer of dermis densely packed with fibrous connective tissue. Draw your finger lightly over the shark's skin to feel the spines of the **placoid scales** (Gr. *placos,* tablet, plate) (Figure 16-5). Each scale is anchored in the dermis and is built much like a tooth (the shark's teeth are, in fact, modified placoid scales). The scale contains a pulp cavity and a thick layer of dentine, both derived from the dermis, and is covered with hard enamel, derived from the epidermis. The spiny scales help to

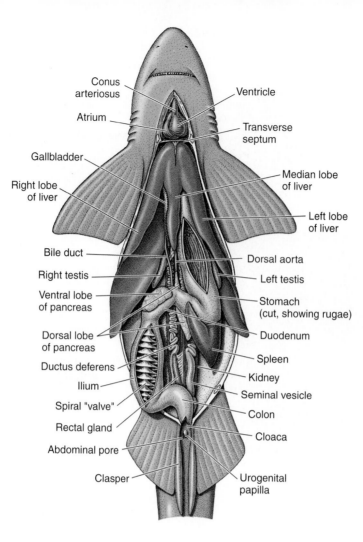

Figure 16-6

Internal anatomy of the dogfish shark, ventral view. What large organ likely assists in keeping the shark buoyant? _____ What structure slows the passage of food through the gut? _____ What organ is formed around the posterior end of the J-shaped stomach? _____ What gland is specific to the cartilaginous fishes and helps regulate the fishes' salt imbalance? _____ What organ functions as the primary source of red blood cell production? _____

reduce friction-producing turbulence as the shark swims, thus lessening drag.

With its dark dorsal and light ventral surfaces **(countershading),** the animal's coloration is protective, whether viewed from above or below.

Internal Structure

If a dissection is not to be made, examine longitudinal and transverse sections of the shark. Note the cartilaginous skull and vertebral column. With the help of Figure 16-6, identify as many of the internal structures as possible.

Dissection of the Shark

☞ Open the coelomic cavity by extending a midventral incision posteriorly from the pectoral girdle through the pelvic girdle and then around one side of the cloacal opening to a point just posterior to it. On each side, make a short transverse cut just posterior to the pectoral fins and another

one just anterior to the pelvic fins. Rinse out the body cavity.

The shiny membrane lining the body cavity and its organs is the **peritoneum** (Gr. *peritonaios,* stretched around). The peritoneum lining the inner surface of the body wall is called **parietal peritoneum,** and that covering the visceral organs and forming the double-membraned **mesenteries** that suspend the digestive organs is called **visceral peritoneum.**

Digestive System

Identify the large **liver** (Figure 16-6). It has two large lobes and a small median lobe. Lying along the right margin of the median lobe of the liver is a thin, tubular sac, the **gallbladder.** Bile from the liver is concentrated in the gallbladder and then discharged during meals by way of the common bile duct into the intestine. Dorsal to the liver is the large **esophagus,** which leads from

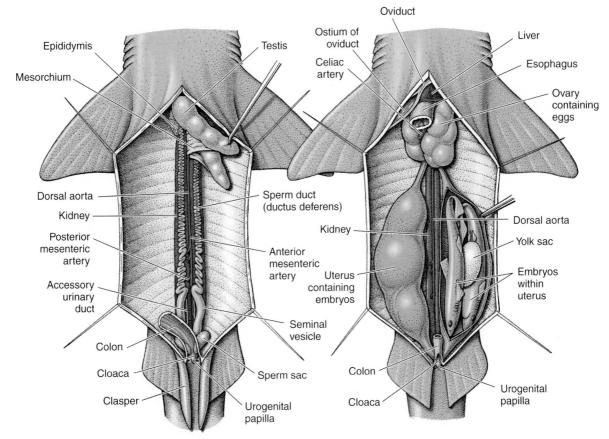

Figure 16-7
Urogenital system of the dogfish shark. **Left,** Male. **Right,** Female.

the pharynx to the J-shaped **stomach.** The digestive tract then turns caudally and gives rise to the **duodenum** (L. *duodeni,* twelve each, socalled because in humans this first part of the intestine is approximately 12 fingerwidths in length). Between the stomach and duodenum is a muscular constriction, the **pyloric valve,** that regulates the entrance of food into the intestine. The **pancreas** lies close to the ventral side of the duodenum, with a slender dorsal portion extending posteriorly to the large **spleen** (not a part of the digestive system). The **valvular intestine** (ileum) is short and wide and contains a **spiral "valve,"** or **ridge.** The short, narrow **rectum** has extending from its dorsal wall a **rectal gland,** which regulates ion balance. The cloaca receives the rectum and urogenital ducts.

👉 Make a longitudinal incision in the ventral wall of the esophagus and stomach. Remove and save the contents of the stomach, if any. Extend the longitudinal incision along the wall of the ileum (taking care not to destroy the blood vessels) to expose the **spiral valve.** Rinse out the exposed digestive tract.

Examine the contents taken from the stomach and compare with those of others being dissected. What do you infer about the shark's eating habits? Examine the

inner surface of the digestive tract. The walls of the esophagus bear large papillae, whereas the walls of the stomach are thrown into longitudinal folds called **rugae** (roog'ee; L. *ruga,* wrinkle). Note the structure of the pyloric valve. Observe the cone-shaped folds of the **spiral valve** and see if you can determine how materials pass through. The spiral valve, not really a "valve" but a fold of tissue that spirals down the ileum much like a spiral staircase, slows the passage of food through the gut. Why might slowing the passage of food be beneficial for the digestive process?

_____

👉 If the intestine has been everted into the cloacal region, carefully pull it back into the body cavity.

Urogenital System

Although the excretory and reproductive systems have quite different functions, they are closely associated structurally, and so are studied together (Figure 16-7).

Male. The soft, elongated **testes** lie along the dorsal body wall, one on each side of the esophagus (Figure 16-7). They are held in place by a mesentery called the **mesorchium** (Gr. *mesos,* middle, + *orchis,* testicle). A number of very fine tubules (vasa efferentia) in the mesentery run from each testis to a much-convoluted

sperm duct (also called Wolffian duct or ductus deferens). The **kidneys** (also called opisthonephroi) are long and narrow and lie behind the peritoneum on each side of the dorsal aorta. They extend from the pectoral girdle to the **cloaca.** The sperm ducts, which serve as both urinary ducts and sperm ducts, take a twisting course along the length of the kidneys to the cloaca and collect wastes from the kidneys by many fine tubules. The sperm ducts widen posteriorly into **seminal vesicles,** which dilate terminally into **sperm sacs** before entering the cloaca. The cloaca is a common vestibule into which both the rectum and the urogenital ducts empty. In the center of the cloaca, dorsal and posterior to the rectum, is a projection called the **urogenital papilla,** which is larger in males than in females. The seminal vesicles empty into the urogenital papilla, which empties into the cloaca.

☞ Slit open the cloaca to see the urogenital papilla.

A groove along the inner edge of each **clasper** is used in conducting spermatozoa to the female at copulation.

Female. A pair of **ovaries** lies against the dorsal body wall, one on each side of the esophagus (Figure 16-7). Enlarged ova may form several rounded projections on the surface of the ovaries. A pair of **oviducts** (Müllerian ducts) run along the dorsal abdominal mesentery. The anterior ends join to form a common opening into the abdominal cavity, called the **ostium tubae.** The oviducts are anteroventral to the liver but may be difficult to find except in large females. Ripened ova leave the ruptured wall of the ovary and enter the abdominal cavity. They are drawn through the ostium into the oviducts, where fertilization may occur. An expanded area of each oviduct dorsal to the ovary is a **shell gland (oviducal gland),** which in *Squalus* secretes a thin membrane around several eggs at a time. The posterior end of each oviduct enlarges into a **uterus,** the caudal end of which opens into the **cloaca.** In immature dogfish both the shell gland and the uterus may not be apparent.

One to six or seven eggs, depending on the species, may develop in each uterus. Vascularized villi on the wall of the uterus come in contact with the yolk sac of the embryo in a placenta-like manner. As mentioned earlier, spiny dogfishes are ovoviviparous because the embryo does not depend on placental nourishment. However, other sharks include some that are dependent on the mother for nourishment through the placental connection (**viviparous;** L. *vivus,* living, + *parere,* to bring forth), and some primitive sharks that lay shelled eggs containing a large amount of yolk (**oviparous;** L. *ovum,* egg, + *parere,* to bring forth). Gestation periods vary from 16 to 24 months, and the young at birth range from 12 to 30 cm in length.

The slender **kidneys** (opisthonephroi) extend the length of the dorsal abdominal wall, dorsal to the peritoneum. A very slender **Wolffian duct** embedded on the ventral surface of each kidney empties into the cloaca through a **urogenital papilla.**

 Slit open the cloaca and identify the urogenital papilla, the entrance of the rectum, and, on the dorsal side, the openings from the uteri.

Circulatory System

Since sharks do not have bone marrow the spleen is the location of red blood cell production. The spleen also functions to filter blood much like the lymph system of other animals. The basic plan of circulation in the shark is similar to that in the ammocoete larva.

☞ Spread the ventral body wall to reveal the cavity containing the heart (**pericardial cavity**). Lift up the **heart** to see the thin-walled, triangular **sinus venosus.**

Blood passes from the sinus venosus to the **atrium,** which surrounds the dorsal side of the muscular **ventricle;** it then flows into the ventricle, which pumps it forward into the **conus arteriosus** (Figure 16-8). **Valves** prevent backflow between compartments.

Venous System.

☞ Slit open the sinus venosus transversely, extending the cut somewhat to the left; wash out its contents.

Look for openings into the sinus venosus of one of each of the following paired veins: (1) **common cardinal** (L. *cardinalis,* chief, principal) **veins (ducts of Cuvier)** (Figure 16-8), which extend laterally and into which empty the large **anterior cardinal sinuses, posterior cardinal sinuses,** and **subclavian veins** (L. *sub,* under, + *clavus,* key); (2) **inferior jugular veins** (L. *jugulum,* collarbone) from the floor of the mouth and gill cavities (this vein is not shown in Figure 16-8); and (3) **hepatic veins,** which empty near the middle of the posterior wall of the sinus venosus and bring blood from the liver.

The **hepatic portal vein** (Figure 16-8) gathers blood chiefly from the digestive system through a system of gastric, pancreatic, and intestinal veins. The hepatic portal vein enters the right lobe of the liver and divides into several small **portal veins** (trace some of these subdivisions); it then divides into a system of capillaries, from which some of the carbohydrates brought from the intestine may be stored in liver cells as glycogen (animal starch) until needed. Blood from capillaries flows into the **hepatic veins** and from there into the sinus venosus.

The **renal portal veins** arise from the **caudal vein** in the tail and carry blood to the capillaries of the kidneys. Many small renal veins carry blood from the kidneys to the posterior cardinal sinuses and from there to the sinus venosus.

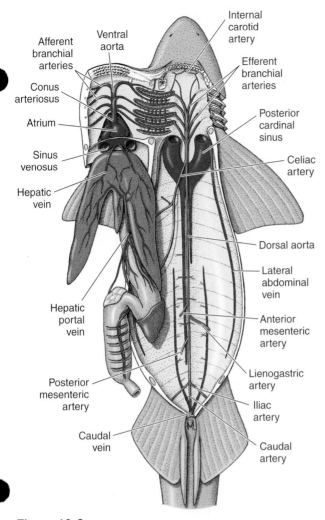

Afferent branchial arteries

Ventral aorta

Internal carotid artery

Conus arteriosus

Efferent branchial arteries

Atrium

Posterior cardinal sinus

Sinus venosus

Celiac artery

Hepatic vein

Dorsal aorta

Lateral abdominal vein

Hepatic portal vein

Anterior mesenteric artery

Posterior mesenteric artery

Lienogastric artery

Iliac artery

Caudal vein

Caudal artery

Figure 16-8
Circulatory system of the dogfish shark.

Arterial System. The arterial system includes (1) afferent and efferent branchial arterial system and (2) the dorsal aorta and its branches.

From the conus arteriosus, trace the **ventral aorta** forward, removing most of the muscular tissue to the lower jaw. The ventral aorta gives off three paired branches, which give rise to **five pairs of afferent branchial arteries** (Figure 16-8). In injected specimens you can follow these arteries into the interbranchial septa, where each gives off tiny arteries to the gill lamellae.

☞ The **efferent branchial arteries** are more difficult to dissect. With scissors, cut through the left corner of the shark's mouth and backward through the centers of the left gill slits, continuing as far as the transverse cut you made earlier at the base of the pectoral fins. Now cut transversely across the floor of the pharynx straight through the sinus venosus, and turn the lower jaw to one side to expose the gill slits and the roof of the pharynx. Locate the spiracle internally. It repre-

sents the degenerated first gill slit. Dissect the mucous membrane lining from the roof of the mouth and pharynx to expose the four pairs of **efferent branchial arteries,** which carry oxygen-rich blood from the gill filaments and unite to form the **dorsal aorta.** By cutting the cartilages under which they pass, trace these arteries back to the gills.

The dorsal aorta extends posteriorly along the length of the body ventral to the vertebral column. It gives rise to **subclavian arteries,** which connect to the pectoral region, a **celiac artery** (Gr. *koilia,* belly) (Figure 16-8), which gives off branches to the intestinal tract and gonads, numerous **parietal arteries** to the body walls, **mesenteric arteries** to the intestine and the rectum, **renal arteries** to the kidneys, and **iliac arteries** (L. *ilia,* flanks) to the pelvic fins. The aorta continues to the tip of the tail as the **caudal artery.**

Respiratory System

In sharks, water taken in through both mouth and spiracles is forced laterally through five pairs of gills and leaves through five pairs of external gill slits (some elasmobranchs have a different number of gills).

☞ On the shark's right (intact) side, separate the gill units by cutting dorsally and ventrally from the corners of each gill slit. Now you can examine the structure of the intact gills on this side and observe the gills in cross section on the other side.

The area between the **external gill slits** and **internal gill slits** comprises the **gill chambers (gill pouches** and **branchial chambers).** The incomplete rings of heavy cartilage supporting the gills and protecting the afferent and efferent branchial arteries are called **gill arches.** Short spikelike projections extending medially from the gill arches are the **gill rakers.** What might be the function of the gill rakers? _____ Cartilaginous **gill rays** fan out laterally from the gill arches to support the gill tissues.

☞ Remove an intact half of a gill arch along with its gill tissue. Examine with a hand lens. Float a small piece of gill in water and examine with a dissecting microscope.

Primary lamellae (gill filaments) are small, platelike sheets of epithelial folds arranged in rows along the lateral face of each gill. Under the microscope the primary lamellae are seen to be made up of rows of tiny plates, called **secondary lamellae,** which are the actual sites of gas exchange. Blood capillaries in the secondary lamellae are arranged to carry blood inward, or in the opposite direction of the seawater, which is flowing outward. This countercurrent flow encourages gas exchange between the blood and water. Gill lamellae are arranged in half-gills, or **demibranchs,** on each side of the branchial arch. The two demibranchs together form the gill unit, or **holobranch.** The spiracles are believed to be remnants of the gill openings found in more primitive chordates. They are usually larger in the slow-moving, bottom-dwelling sharks than in the fast-swimming sharks, in which, because of their motion, there is a more massive flow of water through the mouth.

Oral Report

Be able to identify both the external and internal features of the shark and give the functions of each organ or structure.

Be able to trace the flow of blood from the heart to any part of the body (such as the pectoral region, the kidneys, the tail, and so on) and back to the heart.

Demonstrations

1. Dogfish uterus with developing pups. Or dogfish embryos with attached yolk sac.
2. Various sharks, skates, and rays.
3. Preparation of the skull and/or skeleton of a shark.
4. Corrosion preparation of the arterial system.
5. Shark teeth.
6. Microslides of shark skin.

EXERCISE 16C
Class Osteichthyes— The Bony Fishes

Perca, the Yellow Perch

Subphylum Vertebrata
    Class Actinopterygii
        Superorder Teleostei
           Order Perciformes
                Genus *Perca*
                    Species *Perca flavescens*

Where Found

The yellow perch is a common freshwater fish widely distributed through the lakes of the American Midwest and parts of Canada. A closely related species is found in Europe and Asia.

Characteristics

The Osteichthyes (bony fishes) represents the largest group of vertebrates both in number of species (more than 23,600) and in number of individuals. By adaptive radiation they have developed an amazing variety of forms and structures. They flourish in fresh water or seawater, and in both deep and shallow water. Their chief characteristics usually are **dermal scales, operculum over the gill chamber** of each side, **bony skeleton, terminal mouth, swim bladder, homocercal tail,** and both **median** and **paired fins.**

External Structure

☞ Obtain a preserved fish and, after you have studied its external anatomy, compare it with living fishes in the aquarium. What can their structure tell you of their living habits?

The body of the perch is fusiform, or torpedo shaped. Is it compressed in any of its planes? Identify the **head,** which extends to the posterior edge of the **operculum;** the **trunk,** which extends to the anus; and the **tail** (Figure 16-9). Identify the **pectoral, pelvic, anal, dorsal,** and **caudal fins.** How many of each are there? Which of the fins are paired? _____ Note the **fin rays** which support the thin membrane of each fin. Some of the rays are soft and some are spiny. How would you expect these spiny rays to look when the fish is threatened by a predator? _____ The caudal fin is **homocercal** (Gr. *homos,* same, + *kerkos,* tail), meaning that the upper and lower halves are equal.

The terminal **mouth** is adapted for overtaking prey while swimming. Fishes with superior mouths (those facing upward) are usually surface feeders, whereas those with inferior mouths (facing downward) are usually bottom feeders. Do the **eyes** have lids? _____ Could the perch have binocular vision? Why? _____ On each side in front of the eye a pair of **nostrils** open into an olfactory sac. Water enters the sac through the anterior aperture, which is provided with a flaplike valve, and leaves through the posterior aperture. The **ears** are located behind the eyes, but they are not visible externally. The **lateral line** along the side of the body is a row of small pores or tubules connecting with a long tubular canal bearing sensory organs. These are sensitive to pressure and temperature changes and so are responsive to water currents. Many microscopic sense organs are found in the skin.

Lift a gill cover, or **operculum** (L., cover), and study its structure. Along the ventral margin of the operculum, find a membrane supported by bony rays. This membrane fits snugly against the body to close the branchial cavity during certain respiratory movements. With your probe, examine the **gills** beneath the operculum.

Find the **anus** near the base of the anal fin and the small, slitlike **urogenital opening** just posterior to the anus.

Note the arrangement of the **scales.**

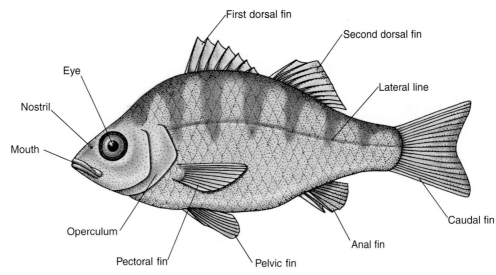

First dorsal fin

Second dorsal fin

Eye

Lateral line

Nostril

Mouth

Caudal fin

Operculum

Anal fin

Pectoral fin

Pelvic fin

Figure 16-9

Yellow perch, external features.

☞ Remove a scale from the lateral line region, mount in water on a slide, and examine with low power.

The anterior, or embedded, side of the perch scale has radiating grooves. The posterior, or free, edge has very fine teeth. These are **ctenoid scales** (ten'oid; Gr., *kteis, ktenos,* comb). Note the fine concentric lines of growth. The scales are covered with a very thin epidermis that secretes mucus over the scales. This reduces friction in swimming and makes capture by a predator more difficult. Ctenoid scales are usually found on fishes with spiny rays in the fins, whereas the soft-rayed fishes usually have cycloid scales, which lack marginal teeth.

Skeletal System

The bony skeleton of the perch consists of an **axial skeleton** (which includes the bones of the skull, vertebral column, ribs, and medial fins) and an **appendicular skeleton** (which includes the pectoral girdle and fins and the pelvic girdle and fins). Examine the mounted perch skeletons on display and compare them with skeletons of other fishes and of amphibians, birds, and mammals. Do you see any basic similarity?

Muscular System

Although the muscles of the perch are less complex than those of land vertebrates, they make up a much larger mass in relation to body size. Tetrapod locomotion results largely from direct action of muscles on bones of the limbs, but fish locomotion results from the indirect action of the segmental muscles—**myomeres**—on the vertebral column, a method by which a large muscle mass produces a relatively small amount of action. This type of movement is efficient in

a water medium, but it would be less effective on land. The myomeres (derived from the embryonic myotomes) consist of blocks of longitudinal muscle fibers placed on each side of a central axis, the vertebral column. Their contraction, therefore, bends the body, and the action passes in waves down the body, alternating on each side.

☞ After cutting off the sharp dorsal and ventral spines, skin one side of the body and note the shape of the myomeres.

They resemble W's that are turned on their sides and stacked together. A horizontal septum of connective tissue divides the muscles into dorsal **epaxial muscles** (Gr. *epi,* upon, + *axis,* axle, meaning above the axis, or vertebral column) and ventral **hypaxial muscles** (Gr. *hypo,* under, + *axis,* axle, meaning below the vertebral column) (Figure 16-10). Posteriorly both epaxial and hypaxial muscles are active in locomotion, but anteriorly the hypaxial muscles serve more for support of body viscera than for locomotion. Try to separate some of the myomeres. Observe the direction of the muscle fibers. Do they run zigzag as the myomeres seem to? Or are they all directed horizontally—or vertically?

☞ Now watch the swimming motions of fishes in the aquarium and try to visualize the use of the body muscles in locomotion. What part do the fins play in locomotion?

Dissection of individual muscles in the fish is difficult and will not be attempted here. Muscles operating the jaws, opercula, and fins are often named according to their function and, as in other vertebrates, include adductors, abductors, dilators, levators, and so on (see Exercise 20).

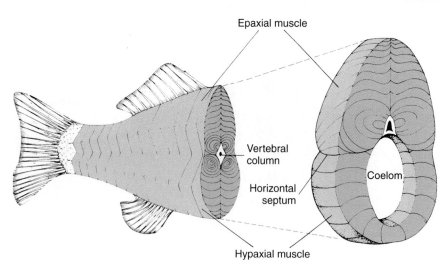

Figure 16-10
Diagram of the skeletal musculature of a teleost fish.

Mouth Cavity, Pharynx, and Respiratory System

Before starting the dissection, it is well, if you have not already done so, to cut off the sharp dorsal and ventral fins to protect the hands.

☞ Cut away the operculum from the left side, exposing the gill-bearing bars, or arches. How many arches are there?

Cut one gill arch, place in water, and examine with the hand lens or dissecting microscope.

If injected, the branchial arteries will be colored. Note the **gill filaments** borne on the posterior, or aboral, side of the arch (Figure 16-11A). Are the filaments arranged in a single or double row? _____ It is in these filaments, containing capillaries from the branchial arteries, that exchange of gases takes place. **Gill rakers** on the oral surface of each gill bar strain out food organisms and offer some protection to the gill filaments from food passing through the pharynx.

☞ Cut through the angle of the left jaw, continuing the cut through the middle of the left gill arches to expose the mouth cavity and pharynx.

Open the mouth wide and note the **gill slits** in the pharynx. In the mouth, locate the fine **teeth.** What would be the main function of these teeth? _____ Just behind the teeth, across the front of both the upper and lower jaw, find the **oral valves.** These are transverse membranes that prevent the outflow of water during respiration. An inflexible **tongue** is supported by the hyoid bone. Explore the **spacious** pharynx, noting the size and arrangement of the gill bars and gill slits. The gills separate the **oral cavity** from the **opercular cavity.**

The mechanics of water movement across the gills involve the combined pumping action of both the oral and opercular cavities—a "double pump" system. The volume of the **oral pump** (mouth cavity) can be changed by raising and lowering the jaw and floor of the mouth. The volume of the **opercular pump** (opercular cavity) can be enlarged and decreased by muscles that swing the operculum in and out. Valves guard the opercular clefts, preventing the backflow of water. The action of the two pumps creates a pressure differential that maintains a smooth flow of water across the gills throughout nearly the entire breathing cycle.

☞ Now watch the fishes in the aquarium, observing their respiratory movements until you understand the sequence.

Water movement across the gills is actually much smoother and less pulsatile than it appears from watching a fish respire. The reason is that the pressure in the opercular cavity is maintained *lower* than the pressure in the mouth cavity for about 90% of the respiratory cycle, and this provides the pressure that drives the water across the gills.

Abdominal Cavity

☞ Starting near the anus and being careful not to injure the internal organs, cut anteriorly on the midventral line to a region anterior to the pelvic fins. Now on the animal's left body wall, make a transverse cut, extending dorsally from the anal region; make another cut dorsally between the pectoral and pelvic fins; then remove the left body wall by cutting between these two incisions. On the right side, make similar transverse cuts so that the right wall can be laid back, but do not remove it.

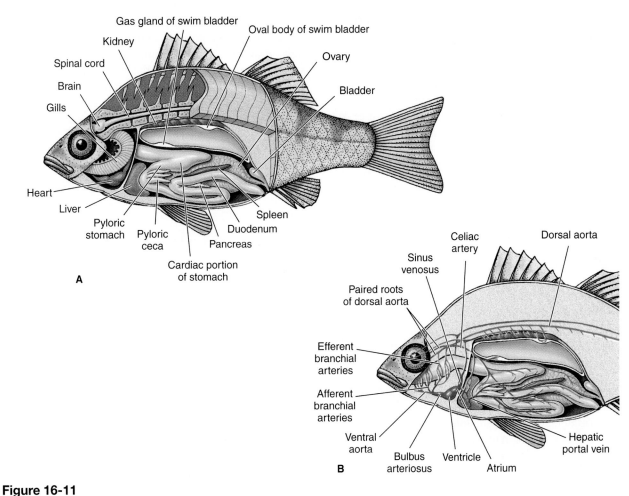

Figure 16-11

Yellow perch. **A,** Internal anatomy. **B,** Principal vessels of the circulatory system.

You have now exposed the abdominal cavity. This, together with the **pericardial cavity,** which contains the heart, makes up the **coelomic cavity.** What is the shiny lining of the coelom called? _____

Probably the first organ you will see will be the **intestine** encased in yellow fat. Carefully remove enough of the fat to trace the digestive tract anteriorly. Find the **stomach,** lying dorsal and somewhat to the left of the intestine. Anterior to the stomach is the **liver,** dark red in life but bleached to a cream color by preservative. The **spleen** (Figure 16-11A) is a dark, slender organ lying between the stomach and the intestine. The **gonads** are in the dorsoposterior part of the cavity. The **swim bladder** lies dorsal to these organs and to the peritoneal cavity. It is long and thin walled. Do not injure its walls or any of the blood vessels lying among the viscera. The **kidneys** are located dorsal to the swim bladder and will be seen later.

Digestive System. Run your probe through the mouth and into the opening of the **esophagus** at the end of the pharynx. Now lift up the liver and trace the esophagus from the pharynx to the large **cardiac portion** of the stomach. This ends posteriorly as a blind pouch. The

short **pyloric portion** of the stomach, opening off the side of the cardiac pouch, empties by way of a **pyloric valve** into the **duodenum,** the S-shaped proximal part of the intestine. Three intestinal diverticula, the **pyloric ceca,** open off the proximal end of the duodenum near the pyloric valve (Figure 16-11A). Follow the intestine to the **anus.** Note the supply of blood vessels in the **mesentery.** The **pancreas,** a rather indistinct organ, lies in the fold of the duodenum. The **liver** is large and lobed, with the **gallbladder** located under the right lobe. Bile from the liver is drained by tubules into the gallbladder, which in turn opens by several ducts into the duodenum posterior to the pyloric ceca.

Cut open the stomach and place its contents in a glass dish with water. Compare with your neighbor's findings. Examine the stomach lining. How is its surface increased? _____ Cut open and examine the pyloric valve. Open a piece of the intestine, wash, and examine the lining with a dissecting microscope. What type of muscle would you expect to find in the intestine? _____

Swim Bladder. The swim bladder is a long, shiny, thin, but tough-walled sac that fills most of the body cavity dorsal to the visceral organs. In some fishes (not the perch) it connects with the alimentary canal. Cut a slit in it and observe its internal structure. In its anterior ventral wall, look for the **gas gland,** which contains a network of capillaries, and the **rete mirabile** (rē′tē muh-rab′uh-lē), (L., wonderful net), which assists in the secretion of gases, especially oxygen, into the bladder at high pressure. Another capillary bed, the **oval,** lies in the dorsal body wall. Gases are resorbed from the swim bladder in this area. The swim bladder is a "buoyancy tank," or

hydrostatic organ, that adjusts the specific gravity of fish to varying depths of water so that the fish is always neutrally buoyant. Can you think of disadvantages to this type of buoyancy system? What will happen to the volume of the bladder and to the perch's specific gravity if the perch swims upward from some depth toward the surface? _____

When you dissected the dogfish shark did you find a swim bladder? _____

What does a shark use for buoyancy? _____

EXPERIMENTING IN ZOOLOGY
Aggression in Paradise Fish,
Macropodus opercularis

Many organisms compete for resources that are limited in the environment. These resources can be food, shelter or mates. The males of many species of animals compete for females. Competition for females often leads to aggressive encounters between conspecific males, and these encounters can escalate into fights that involve physical contact. Thus, natural selection has favored aggressive behavior in the males of some species so much that their aggression is innate (instinctive). Males in these species interact aggressively even when females are not present. Scientists that have observed aggressive interactions between males have suggested that the intensities of the encounters can depend on how closely matched the males are in size and strength. Small males can become injured when fighting larger males and may choose not to fight. However, when males are closely matched in size and strength aggressive encounters tend to be intense as each male attempts to establish its dominance and ultimately its access to the resource (females in this case).

Paradise fish, *Macropodus opercularis* (Figure 16-12), are tropical fish from Southeast Asia. They have special organs in their mouth that allow them to breathe air as well as use their gills. Males are more colorful than females and have red and blue bars that run vertically down their body. Several paradise fish can be kept together in a single tank; however, when two males are isolated from the other fish they begin to display aggressive interactions.

Getting Ready

Work with a partner for this exercise. Try to work in a room that can be darkened. You should have access to a fish tank (approximately 3 to 10 gallons) that has an overhead tank light. If an overhead light is not available,

illuminate the tank from the side with a small desk lamp. Turning off the room light will minimize the ability of the fish to see you and will minimize their distractions. Have a stopwatch available for timing fish behavior during your experiment.

How to Proceed

Select two male paradise fish from a communal tank. Try to select two males that appear very similar in size and other physical traits. Make sure to add the male fish to the experimental tank at exactly the same time. For the next 15 minutes you should quantify the interactions between the males. The easiest way to measure aggressive interactions is to record the amount of time the two males stay very close to each other. You might elect to record all time that the fish spend within one body length of one another. If the two males display aggression to each other you will notice several different types of behaviors. For example, males might chase each other, raise their dorsal fins, swim parallel very closely together or even come side to side and vibrate, thereby sending mechanical waves toward each other. If you decide to use categories of behaviors as well as total time spent near each other, you might want to consider observing two "practice" males before starting your experiment so that you can become familiar enough with the various behaviors to be able to count them during the experiment.

After recording your observations and data from the two males that are closely matched in size, select two new males. This time, however, choose two males that differ conspicuously in size (although make sure you have two males and are not placing a male and female together). Repeat the experiment, keeping track of time the males spend together and the number of aggressive behaviors displayed.

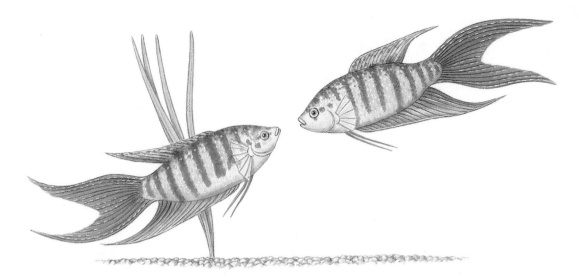

Figure 16-12
Two male paradise fish interacting.

Compile the data from the entire class and calculate the overall mean times males spend together in closely matched contests versus contests where males differed in size.

Do males that are closely matched in size interact more than males that differ in size? Do the closely matched males have encounters that escalate into more aggressive behaviors (e.g., biting, nipping) than males that differ in size?

Questions for Independent Investigations

1. Keeping the fish matched in size, how might light levels affect aggressive interactions?

2. Scientists that study fish aggression have found that individuals that are tank residents are more aggressive than fish that have been recently added to a tank. How could you test this with paradise fish?

3. Many kinds of fish have chemical alarm pheromones that are released when a fish is injured. Gently remove a couple of scales from a paradise fish and rinse his body with filtered water. This rinse water may now contain alarm pheromones. Add these chemicals to a tank with two interacting males. How might these chemicals affect the aggression levels of the males? Why?

4. What might happen to aggressive interactions if paradise fish are exposed to a potential predator?

References

Csanyi, V., J. Haller, and A. Miklosi. 1995. The influence of opponent-related and outcome-related memory on repeated aggressive encounters in paradise fish. Bio. Bull. **188:**83–86

Francis, R. C. 1983. Experiential effects on agonistic behavior in the paradise fish, *Macropodus opercularis*. Behaviour **85:**292–313.

Gerlai, R. 1993. Can paradise fish recognize a natural predator? An ethological analysis. Ethology **94:**127–136.

Jakobsson, S., O. Brick, and C. Kullberg. 1995. Escalated fighting behaviour incurs increased predation risk. Animal Behav. **49:**235–249.

Analysis of the Multiple Hemoglobin System in *Carassius auratus,* the Common Goldfish

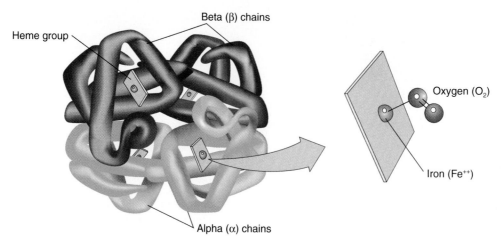

Figure 16-13

Human hemoglobin molecule showing the four polypeptide chains (two α and two β) each associated with a heme group to which an oxygen molecule will bind.

Carassius auratus, the Common Goldfish

Subphylum Vertebrata
  Class Actinopterygii
    Order Cypriniformes
      Family Cyprinidae (carps and minnows)
        Genus *Carassius*
          Species *Carassius auratus*

Oxygen is required by cells to utilize aerobic pathways in the production of ATP. For large multicellular animals, diffusion of oxygen from the environment into the animal is insufficient to supply their metabolic needs. This problem led to the development of complex oxygen exchange surfaces and organs in animals and also to the development of oxygen transport molecules within circulatory systems. In vertebrates, most oxygen is transported in the blood bound to hemoglobin within red blood cells. Oxygen is picked up by the hemoglobin in the gills or lungs at the exchange surfaces and delivered to tissues by circulation of the blood. The hemoglobin molecule in vertebrates consists of four noncovalently linked subunits that form a tetrameric molecule. Typically this tetramer is assembled from two α (alpha) polypeptide chains and two β (beta) polypeptide chains. Each polypeptide subunit of hemoglobin binds a heme group containing a ferrous iron responsible for binding the oxygen molecule (Figure 16-13).

Oxygen availability is relatively constant in the terrestrial environment; however, in the aquatic environment oxygen availability is highly variable because oxygen solubility in water is dependent upon temperature, ionic concentration, pH, and the amount of biomatter consuming or producing oxygen. Cyprinid fishes, which include carp, minnows, and goldfish, often inhabit shallow ponds in which temperature may vary by as much as 10° or 20° C each day. Typically, as the temperature of water increases, the oxygen solubility decreases. A change in water temperature from 0° C to 40° C causes oxygen available to drop by about one-half. An additional confounding factor is that animal tissue metabolism increases as temperature increases.

The decrease in oxygen availability combined with increased oxygen demand as water temperature increases poses a problem for maintaining aerobic respiration in fishes. One adaptation that fish have evolved to deal with variable oxygen availability is to produce multiple forms of hemoglobin. These multiple forms of hemoglobin in fishes may be distinguished electrophoretically (Riggs, 1970; di Prisco and Tamburrini, 1992) and may differ in physiological properties (Binotti et al., 1971; di Prisco and Tamburrini, 1992). The differences in physiological properties may allow for enhanced oxygen transport when oxygen availability is low in the environment. Trout are known to have up to nine distinctly different hemoglobin forms (I-IX) while goldfish exhibit three different hemoglobin forms

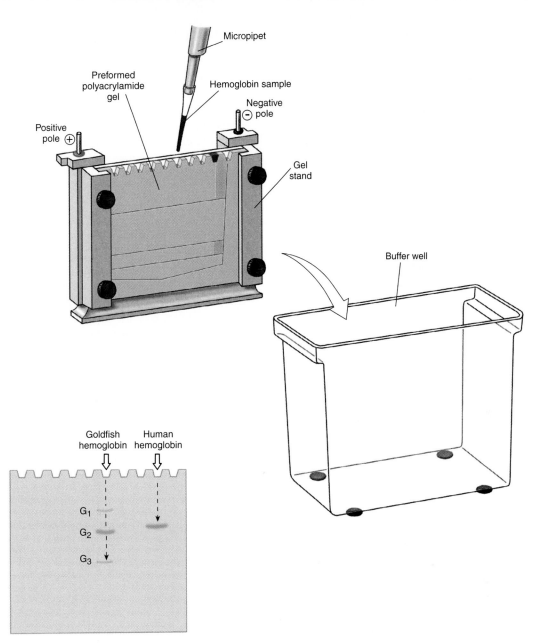

Figure 16-14

Diagram outlining the preparation, loading, and electrophoreses of a vertical polyacrylamide gel to separate the isoforms (G_1, G_2, and G_3) of hemoglobin from goldfish.

called G_1, G_2, and G_3. These three forms in goldfish vary in concentration as temperature varies, presumably brought about by rapid changes in aggregation of different subunits (Houston and Cyr, 1974; Houston et al., 1976; Houston and Rupert, 1976).

How to Proceed

We can visualize the multiple hemoglobin systems found in common goldfish using simple native polyacrylamide gel electrophoresis. In this procedure, we will remove red blood cells from goldfish that have been maintained in the laboratory at about 20° to 25° C

using a heparinized syringe. (Heparin is a compound that prevents coagulation of blood removed from the circulatory system.) The cells will be washed in isotonic saline (0.9% NaCl) to remove serum proteins and lysed in distilled water to release the hemoglobins. (Distilled water is hypotonic to the red blood cells, therefore the cells will swell and burst releasing hemoglobin and other cellular components.) The hemoglobin lysate will be mixed 1:1 with a loading buffer and a 20 μl sample applied to a well on an 8% native polyacrylamide gel (Figure 16-14). A sample of human hemoglobin prepared in the same way as the goldfish hemoglobin will serve as a marker on the gel. Electrophoretic separation will proceed for 30 to 45 minutes at 80 to 90 volts. Following electrophoresis, you should be able to see three faint reddish-orange bands in the goldfish lane indicating the presence of the three isoforms of hemoglobin. Typically the G_2 band (middle band) predominates while the G_1 (upper) and G_3 (lower) bands are fainter. Only one band will appear in the human hemoglobin lane. To assist in visualization of the hemoglobin bands, the gel will be removed from its case, stained with Coomassie Blue dye for a few minutes and then destained to remove dye not bound to protein. A picture of the gel may be taken if appropriate equipment is available.

Questions for Thought

How might possession of multiple hemoglobin isoforms allow for adaptation to different environments?

Why do you think trout have so many different forms of hemoglobin?

Can you suggest any other species of fish that might show multiple forms of hemoglobin?

References

di Prisco, G., and M. Tamburrini. 1992. The hemoglobins of marine and freshwater fish: the search for correlations with physiological adaptation. Comp. Biochem. Physiol. **102B:**661–671.

Houston, A. H., and D. Cyr. 1974. Thermoacclimatory variation in the haemoglobin systems of goldfish (*Carassius auratus*) and rainbow trout (*Salmo gairdneri*). J. Exp. Biol. **61:**455–461.

Houston, A. H., K. M. Mearow, and J. S. Smeda. 1976. Further observations upon the hemoglobin systems of thermally acclimated freshwater teleosts: pumpkinseed (*Lepomis gibbosus*), white sucher (*Catostomus commersoni*), carp (*Cyprinus carpio*), goldfish (*Carassius auratus*) and carp-goldfish hybrids. Comp. Biochem. Physiol. **54A:**267–273.

Houston, A. H., and R. Rupert. 1976. Immediate response of the hemoglobin system of the goldfish, *Carassius auratus,* to temperature change. Can. J. Zool. **54:**1737–1741.

Riggs, A. 1970. Properties of fish hemoglobins. In Hoar, W. S., and D. J. Randall (eds.). Fish physiology. Vol. 4. New York, Academic Press.

Class Amphibia
The Frog

The amphibians are a transition group between aquatic and the strictly land vertebrates. They still have the soft, moist epidermis of aquatic forms and therefore cannot stray too far from water or moist surroundings. Amphibian eggs lack the tough protective shell and specialized extra embryonic membranes characteristic of eggs of terrestrial vertebrates and so remain adapted to an aquatic habitat. Many amphibians have developed lungs for air breathing but still have aquatic larvae with external gills. The moist skin is also a respiratory organ. The evolution of the lung has brought a change in circulation that includes a pulmonary circuit as well as a systemic circuit. Amphibians have a three-chambered heart (two atria and one ventricle). They are usually four limbed for walking or jumping but often have webbed feet for swimming.

Frogs and toads belong to the order Anura (Gr. *an,* without, + *oura,* tail), the largest and most diverse group of living amphibians. The anurans differ from the salamanders (order Caudata) and tropical caecilians (order Gymnophiona) in several distinctive ways that are associated with a specialized jumping mode of locomotion. The frog's body is extremely shortened and virtually fused with the head, an adaptation that provides rigidity to the skeletal framework. It lacks true ribs. The caudal vertebrae (which would normally form the tail) are fused into a single pillarlike bone, the urostyle. The hindlegs are much larger and more powerful than the forelegs. Because of these specializations, and several others not obviously related to the frog's specialized locomotion, the frog is not an ideal generalized amphibian type. Nevertheless, the anurans are the most diverse of the amphibians and are commonly used in the general zoology laboratory because of their ready availability. The large bullfrog *Rana catesbeiana* and the small leopard frog *Rana pipiens* are commonly used for dissection.

The Frog

Phylum Chordata
    Subphylum Vertebrata
        Class Amphibia
            Order Anura (= Salientia)
                Family Ranidae
                    Genus *Rana*

EXERCISE 17A
The frog
Behavior and Adaptations

EXERCISE 17B
The Skeleton

EXERCISE 17C
The Skeletal Muscles
Directions for study of frog muscles

EXERCISE 17D
The Digestive, Respiratory, and Urogenital Systems
Mouthparts
Dissection of the frog

EXERCISE 17E
The Circulatory System

Where Found

Ranid frogs (family Ranidae) are almost worldwide in distribution. Their favorite habitats are swamps, low meadows, brooks, and ponds, where they feed on flies and other insects. Their young, the tadpoles, develop in water and are herbivorous. It should be noted that ranids, like many species of frogs, appear to be declining in numbers around the world. It is not entirely known what might be causing the declines but many scientists are now investigating the conservation biology of amphibians.

EXERCISE 17A
Behavior and Adaptations

☞ Place a live frog on a piece of wet paper toweling in a jar large enough not to cramp it. Do not let its skin become dry. Do not excite it unnecessarily. Observe its adaptations. Make notes of your observations. Have a mounted frog skeleton at your table for comparison.

Is the skin smooth or rough? Moist or dry? Examine the frog's feet. How are they adapted for jumping? For swimming? For landing on a slippery rock or log? Note the sitting position and compare with the mounted skeleton. In what ways is the body form of the frog adapted as a lever system that can catapult the animal into the air? _____

Compare the color of the dorsal and ventral sides. Imagine a predator approaching a swimming frog from above or from below. What is the protective advantage in this dorsal-ventral difference in coloration? _____

Knowing that a frog captures prey by striking at it with its protrusable tongue, how is the position of the eyes advantageous to the animal? _____ _____ How do the eyes close? _____ Examine the skeleton and see how this is possible. Note the transparent nictitating membrane. How is it used? _____

Feeding Reactions. Place some live fruit flies in the jar with the frog. If the frog is hungry and is not excited, it may feed. Describe how it seizes prey.

Breathing. Observe the movements of the throat and nostrils. Air is drawn into the mouth; then the nostrils close and throat muscles contract to force the air into the lungs. This form of breathing is called positive pressure breathing. The nostrils (nares) can be opened or closed at will. Record the number of movements per minute of the throat and then of the nostrils. _____ _____ Do their rates coincide? _____ Do the sides of the body move in breathing? _____ Excite the animal by prodding; then as soon as it becomes quiescent, count the rate of breathing again.

Righting Reaction. Place the frog on its back, release it, and note how it rights itself. Repeat this experiment but hold the frog down gently with your hand until it ceases to struggle. When you release it, the frog may remain in this so-called hypnotic state for some time.

Locomotion. Observe a frog in an aquarium. What is the floating position? How does it use its limbs in swimming? How does it dive from a floating position? Note how it jumps.

Written Report

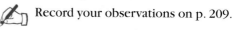 Record your observations on p. 209.

External Structure

☞ Study a preserved frog and compare it with a live frog and a mounted skeleton.

Note the **head** and **trunk.** Note the **sacral hump** produced by the protrusion of the pelvic girdle. Find this hump on the mounted skeleton. The **cloacal opening** is at the posterior end of the body.

On the **forelimbs,** identify the arm, forearm, wrist, hand, and digits. On the **hindlimbs,** find the thigh, shank, ankle, foot, and digits. How does the number of digits compare with your own? _____ Can you find the rudimentary thumb (prepollux) and the rudimentary sixth toe (prehallux)? During the breeding season the inner (thumb) digit of the male is enlarged into a nuptial pad for clasping the female. Observe the **webbed** toes of the hindfoot. How is the long ankle advantageous to the frog? _____ _____

The **eyes** are protected by **eyelids** and by a transparent **nictitating membrane** (L. *nictare,* to wink). Look in the corner of your neighbor's eye for a vestige of this membrane, the semilunar fold.

The **tympanic membrane** (eardrum) is a circular region of tightly drawn skin just back of the eye. The frog has no external ear—only the middle and internal ears.

Name_____

Date_____

Section_____

Observations on Behavior of Frog

Feeding reaction _____
_____
_____
_____

Breathing _____
_____
_____
_____

Reaction to touch _____
_____
_____
_____

Righting reaction _____
_____
_____
_____

Locomotion _____
_____
_____
_____

Other observations _____
_____
_____
_____
_____

EXERCISE 17B
The Skeleton

Shaping the Body for Life on Land

In amphibians, as in fishes, the well-developed skeleton provides a framework for the muscles in movement and protection for the viscera and nervous systems. The movement onto land and the necessity of transforming paddlelike fins into tetrapod legs capable of supporting the body's weight introduced a new set of stress and leverage problems. These changes are most noticeable in frogs and toads, where the entire musculoskeletal system is specialized for jumping and swimming by simultaneous extensor thrusts of the hindlimbs. Frogs no longer move in the typical sinuous or fishlike motion of their aquatic ancestors. Consequently the vertebral column has lost its flexibility and, together with the enlarged pelvic girdle, has become a rigid frame for transmitting force from the hindlimbs to the body.

☞ Prepared skeletons are brittle and delicate. Handle them with care and do not deface the bones in any way. Use a probe or dissecting needle, not a pencil, in pointing.

The **axial** (L., axis) **skeleton** includes the skull, the vertebral column, and the sternum. The **appendicular** (L. *ad*, to, + *pendare*, to hang) **skeleton** includes the pectoral and pelvic girdles and the forelimbs and hindlimbs.

For a discussion of the structure and growth of bones and their articulations, see Exercise 20A.

Axial Skeleton

Skull. The skull and jaws of a frog serve as protection for the brain and special sense organs. As with the rest of the frog skeleton, it is vastly altered as compared to early amphibian ancestors. It is much lighter in weight, is more flattened in profile, and contains fewer bones and less ossification. The front part of the skull, wherein are located the eyes, nose, and brain, is better developed, whereas the back of the skull, which in fishes contained the gill apparatus, is much reduced. Lightening of the skull was essential to mobility on land, and the other changes fitted the frog for its improved senses and means of feeding and breathing.

The skull includes the **cranium,** or braincase, the **visceral skeleton,** made up of the bones and cartilage of the jaws, the hyoid apparatus, and the little bones of the ears. All these elements of the visceral skeleton* are derived from the jaws and gill apparatus of fish ancestors. The **orbital fossae** and the **nasal fossae** are the dorsal openings where the eyes and external nares are located.

Vertebral Column. The backbone of the frog consists of only nine vertebrae and a **urostyle** (Gr. *oura*, tail, + *stylos*, pillar), the latter representing the fusion of several caudal vertebrae. The first vertebra, the **atlas** (Gr. *Atlas*, a Titan of Greek mythology, who bore the heavens on his shoulders), articulates with the skull. The ninth, or **sacral*** (L. *sacer*, sacred), vertebra has transverse processes for articulation with the ilia of the pelvic girdle.

No frogs of the large family Ranidae have ribs either as larvae or adults; ribs do appear, however, in two other frog families.

The sternum (L., breastbone) provides ventral protection for the heart and lungs and a center for muscular attachment. Its four parts, beginning at the anterior end, are the **episternum** (Gr. *epi*, upon), a cartilaginous rounded end often not seen in prepared skeletons; **omosternum** (Gr. *omos*, shoulder); **mesosternum** (Gr. *mesos*, middle), the section located posterior of the coracoid bone of the pectoral girdle; and **xiphisternum** (Gr. *xiphos*, sword), a cartilaginous heart-shaped end often not present on prepared skeletons. The episternum and omosternum are not visible in Figure 17-1.

Appendicular Skeleton

Pectoral Girdle and Forelimbs. The pectoral girdle serves as support for the forelimbs, which, in frogs, are used mainly to absorb the shock of landing after a jump. The pectoral girdle articulates with the sternum ventrally. Each half of the girdle includes a suprascapula and scapula (L., shoulder blade) and a clavicle (collar bone) (L., *clavicula*, small key) lying anterior to the coracoid (Gr. *korax*, crow, + *eidos*, form). Consult Figure 17-1 for the bones of the forelimb.

Pelvic Girdle and Hindlimbs. The pelvic girdle supports the hindlimbs. Each half is made up of the long **ilium** (L., flank), the anterior **pubis** (L. *pubes*, mature), and the posterior **ischium** (Gr. *ischion*, hip). Consult Figure 17-1 for the bones of the hindlimb.

Oral Report

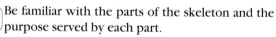 Be familiar with the parts of the skeleton and the purpose served by each part.

---

*The term "visceral" (from the Latin *viscera*, bowels) refers to elements and structures of the body associated with the gut tube. The visceral skeleton represents specialized parts of the primitive gut of jawed vertebrates.

*The origin of the term sacrum, "holy bone," is unknown. It is speculated that the curved appearance of the human sacrum suggested to Renaissance anatomists a resemblance to an obsidian knife used in ancient sacrifice.

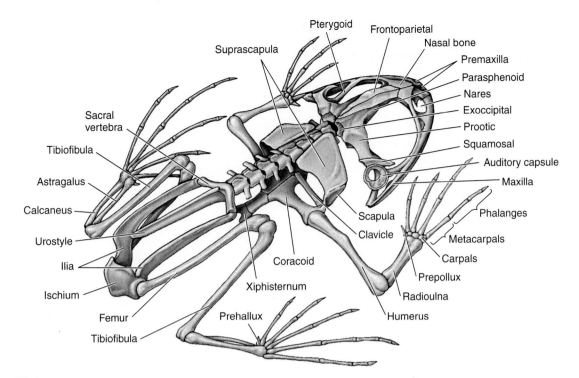

Figure 17-1
Skeleton of a frog.

EXERCISE 17C
The Skeletal Muscles

The frog has hundreds of muscles, but this exercise describes only the most important ones (more correctly, the most conspicuous ones, since they are all important to a frog).

If you learn a few of the Greek and Latin roots of some of the muscle names, they will give you clues to their orientation or action. For example, **rectus** means "straight," so the fibers of rectus muscles generally run along the long axis of the body. **Gracilis** means "slender." **Triceps** means "three heads," and this muscle has three tendons of origin; the **biceps** has two heads of origin. Long muscles are called **longus** and short muscles are called **brevis;** large muscles are termed **magnus** or **major. Anticus** means "anterior." Other muscles are named for specific movements. The **sartorius** is derived from the Latin word for "tailor" and is homologous to a human muscle of the same name that is active in crossing the legs (tailors, before the days of sewing machines, sat on the floor with crossed legs). The **gastrocnemius** (Gr. *gaster,* stomach, + *kneme,* tibia) is named for its fat "belly."

NOTE: You will also need some familiarity with the terminology relating to muscle connections and their actions. Refer to the general discussion of skeletal muscles in Exercise 20B.

Directions for Study of Frog Muscles

 To skin the frog, slit the skin midventrally from anal region to chin, keeping the scissors point up to prevent injuring the underlying muscles. Make a transverse cut completely around the body just above the hindlegs and another one anterior to the forelegs. A middorsal cut the length of the back will divide the skin into portions that can be peeled off easily. Loosen the skin with a blunt instrument and carefully pull off over a leg wrong side out. Be careful not to tear thin muscle attached to the skin. The skin can be pulled over the head and eyes in the same way.

The large spaces between skin and muscle where the skin is not attached are **subcutaneous lymph sacs.**

 In separating the muscles from each other, first observe the direction of the muscle fibers and the extent of the muscle; then use your fingers, a blunt probe, or the handle of a scalpel to loosen the tissues. *Never use scissors, scalpel blade, or needle for dissecting muscles.* Never cut a muscle unless instructed to do so. If it is necessary to cut superficial muscles to find deep muscles, cut squarely across the belly (middle fleshy portion) of the muscle, leaving the origin and insertion in place.

TABLE 17.1

Major Trunk and Leg Muscles of the Frog

Muscle	Origin	Insertion	Action
Ventral Trunk Muscles			
Pectoralis	Sternum and fascia of body wall	Humerus	Flex, adduct, and rotate the arm
Rectus abdominis	Pubic border	Sternum	Support abdominal viscera
External oblique	Dorsal fascia of vertebrae, also the ilium	Sternum, linea alba	Constrict abdomen and support viscera
Transversus	Ilium and vertebrae	Linea alba	Help support the abdominal viscera
Ventral Muscles of the Thigh			
Sartorius	Pubis	Tibiofibula	Flexes shank and adducts thigh
Adductor magnus	Pubic and ischial symphysis	Distal end of femur	Adducts and flexes thigh
Gracilis major	Ischium	Tibiofibula	Adducts thigh and flexes or extends shank, according to its position
Gracilis minor	Ischium	Tibiofibula	Same as gracilis major
Adductor longus	Ilium	Femur	Adducts thigh
Dorsal Muscles of the Thigh			
Triceps femoris	Three divisions, one head on the acetabulum, two on the ilium	Tibiofibula	Abducts thigh and extends shank
Biceps femoris	Ilium	Tibiofibula	Flexes shank
Semimembranosus	Ischium	Tibiofibula	Adducts thigh and flexes or extends shank, according to its position
Gluteus	Ilium	Femur	Rotates thigh forward
Muscles of the Shank			
Gastrocnemius	Two heads, distal end of femur and tendon from triceps femoris	By tendon of Achilles to sole of foot	Flexes shank and extends ankle and foot
Peroneus	Femur	Distal end of tibiofibula; head of the calcaneus	Extends shank; when foot is extended, extends it farther; when flexed, flexes it farther
Tibialis anterior longus	Femur; divides into two bellies	By two tendons on the ankle bones	Extends shank; flexes ankle
Extensor cruris	Femur	Ventral surface of tibiofibula	Extends shank
Tibialis posterior	Side of tibiofibula	Ankle	Extends foot when flexed; flexes foot when fully extended

Trunk Muscles

The trunk muscles of the frog, no longer required to produce the lateral flexion movements of the amphibians' swimming ancestors, have been modified to brace the back and to support the viscera in air. The ventral trunk muscles (Table 17-1) are arranged in layers that run in different directions. The **rectus abdominis** runs longitudinally, forming a sling from pubis to sternum (Figure 17-2). In the midventral line is a thin but tough band of connective tissue, the **linea alba** (literally, "white line"). Inserting on the linea alba are two oblique muscle bands, the **external oblique** and the **transversus,** the latter lying beneath the external oblique and rectus abdominis. (The transversus is not shown in Figures 17-2 or 17-3.) These assist the rectus abdominis in supporting the viscera.

Muscles of the Thigh

The frog's limb muscles are derived from muscles that raised and lowered the fins of fishes, now much modified to brace and move the limbs for walking and thrust-swimming. Not surprisingly, the limb muscles are complex since they perform several actions. Nevertheless we

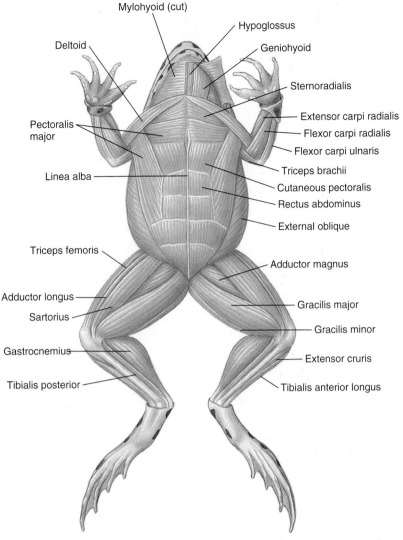

Mylohyoid (cut)

Hypoglossus

Geniohyoid

Deltoid

Sternoradialis

Extensor carpi radialis

Flexor carpi radialis

Pectoralis
major

Flexor carpi ulnaris

Triceps brachii

Linea alba

Cutaneous pectoralis

Rectus abdominus

External oblique

Triceps femoris

Adductor magnus

Adductor longus

Gracilis major

Sartorius

Gracilis minor

Gastrocnemius

Extensor cruris

Tibialis posterior

Tibialis anterior longus

Figure 17-2

Muscles of a frog, ventral view.

Reprinted by permission of Medical and Scientific Illustration/William C. Ober.

can recognize two major groups of muscles on any limb: an anterior and ventral group that pulls the limb forward (protraction) and toward the midline (adduction) and a second set of posterior and dorsal muscles that draws the limb back (retraction) and away from the body (abduction). Only the more conspicuous thigh muscles are included in this study (Table 17-1).

In the first group (protractors and adductors) are the **sartorius, adductor magnus, gracilis major,** and **adductor longus** (visible on the ventral surface, Figure 17-2); and **biceps femoris** and **gracilis minor** (visible on dorsal surface, Figure 17-3). All of these serve to draw the limb forward and toward the midline, or to flex more distal parts of the limb. In the second group (retractors and abductors) is the **triceps femoris,** a large muscle of three divisions, all of which serve to abduct the thigh and extend the shank (Figure 17-3).

Muscles of the Shank

The most conspicuous shank muscle is the **gastrocnemius,** which extends from the femur to the foot and inserts by the Achilles tendon; it extends the ankle during jumping and swimming. Because the gastrocnemius is easily dissected out together with the sciatic nerve that innervates it, it is commonly used in physiological studies of skeletal muscle. The ankle is flexed by the **tibialis anticus longus** and other muscles not shown in Figure 17-3. The major action of the **peroneus** is to flex the ankle joint and extend the shank.

If the frog you are dissecting is in the *Rana* group, it likely has well-developed leg muscles for swimming and jumping. Toads, another large group of frogs, are toxic and have less developed hindlimb muscles. Why might a toxic animal have less developed leg muscles? _____

_____

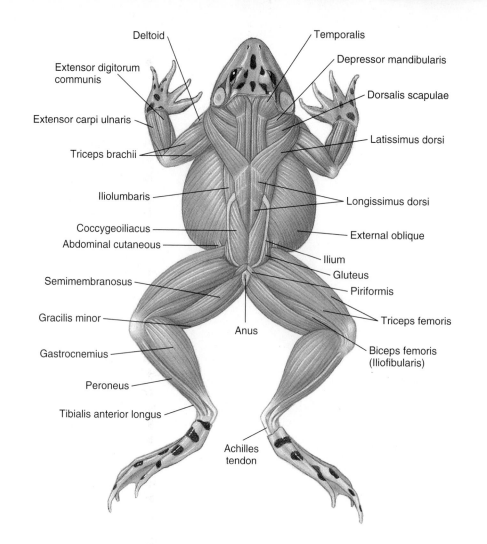

Figure 17-3

Muscles of a frog, dorsal view.

Reprinted by permission of Medical and Scientific Illustration William C. Ober.

Other Muscles

With the help of Figures 17-2 and 17-3 you can identify many of the muscles of the back, shoulder, head, and arm.

How the Muscles Act

A muscle has only one function—to contract. For effective action, muscles must be arranged in antagonistic pairs. The gastrocnemius and the tibialis anticus longus represent such an antagonistic pair. Loosen the body of each of these muscles, pull on the gastrocnemius, and see what happens. Now pull on the tibialis anticus longus. Which of these muscles would be used in jumping or diving? Which in sitting? See whether you can locate other antagonistic pairs.

For most movements, groups of muscles rather than single muscles are required. By varying the combi-

nation of these groups, many complicated movements are possible.

Oral Report

Be able to demonstrate a careful dissection of the muscles and to name the muscles and their actions.

Written Report

On p. 215 tell (1) what principal muscles of the hindleg are involved when a frog leaps, and (2) when the frog resumes a sitting position, what principal muscles contract.

Name_____

Date_____

Section_____

Principal Leg Muscles Involved in Leaping

Principal Leg Muscles Involved in Sitting

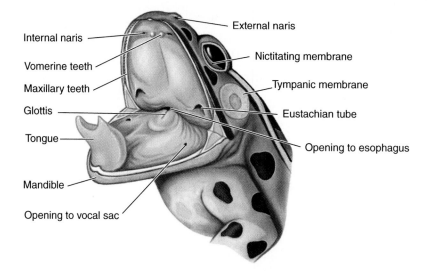

Figure 17-4
Mouthparts of a frog.

EXERCISE 17D
The Digestive, Respiratory, and Urogenital Systems

Mouthparts

☞ Pry open the mouth, cutting the angle of the jaw if necessary, and wash in running water.

The posterior portion of the mouth cavity is the **pharynx,** which connects with the **esophagus.** Feel the **maxillary teeth** along the upper jaw and the **vomerine teeth** in the roof of the mouth (Figure 17-4). Are these better adapted for biting and chewing or for holding the prey to prevent its escape? Find the **internal nares** (sing. **naris**) in the roof of the mouth and note how they connect with the external nares. Note how the ridge on the lower jaw fits into a groove in the upper jaw to make the mouth closure airtight. This is important in the frog's respiratory movements.

Eustachian tubes, which connect with and equalize the air pressure in the middle ear, open near the angle of the jaws. In male frogs, openings on the floor of the mouth slightly anterior to the eustachian tubes lead to **vocal sacs,** which, when inflated, serve as resonators to intensify the mating call. Examine the **tongue** and note where it is attached. Which end of the tongue is flipped out to catch insects? Feel the **sensory papillae** on the tongue surface. The free end of the tongue is highly glandular and produces a sticky secretion that adheres to the prey. Behind the tongue is a slight elevation in the floor of the mouth, containing the **glottis,** a slitlike opening into the **larynx.**

Dissection of the Frog

☞ Make an incision through the abdominal wall from the junction of the hindlegs to the lower jaw, cutting through the bones of the pectoral girdle as you go. Make transverse cuts anterior to the hindlegs and posterior to the forelegs and pin back the flaps of muscular tissue.

Note the three layers of the body wall: **skin, muscles** (with enclosed skeleton in some places), and **peritoneum,** which lines the large **coelom.**

In a mature female the ovaries with their dark masses of eggs may fill much of the coelomic cavity. In this case, remove the left ovary and its white, convoluted oviduct.

Note the **heart** enclosed in its **pericardial sac** and surrounded by lobes of the **liver.** Lift up the heart to find the **lungs.**

Digestive System

The digestive tract is relatively short in adult amphibians, a characteristic of most carnivores. The larval (tadpole) stages of the frog, however, are herbivorous, feeding on pond algae and other vegetation; they have a relatively long digestive tract, since their bulky food must be submitted to time-consuming fermentation before useful products can be absorbed.

You have seen the **mouth** and the **pharynx.** Lift the heart, liver, and lungs to see where the **esophagus** empties into the stomach (Figure 17-5). A **pyloric valve** controls movements of food into the **small intestine.** Note the blood vessels in the mesentery, which holds the stomach and small intestine in place.

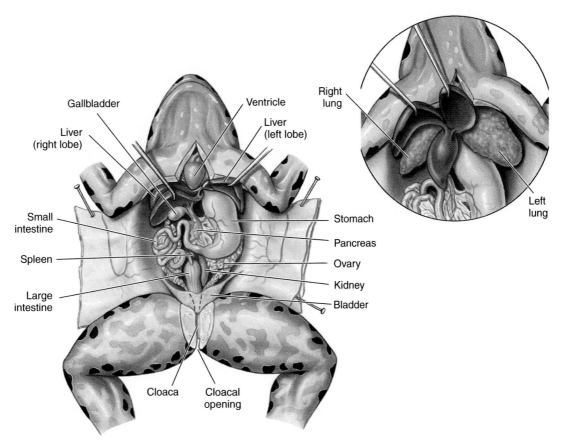

Figure 17-5
Abdominal cavity of a frog, ventral view. At **right,** the liver has been lifted up and turned back to expose the lungs.

Why must the digestive tract be so well supplied with blood? _____

The **liver,** the largest gland in the body, secretes bile, which is carried by a small duct to the **gallbladder** for storage. Find the gallbladder between the right and median lobes of the liver. The **pancreas** is thin and inconspicuous, lying in the mesentery between the stomach and duodenum.

The **large intestine** narrows down in the pelvic region to form the **cloaca,** which also receives urine from the kidneys and products from the reproductive organs. It empties through the **cloacal opening.** Not all vertebrate animals have cloacas; mammals do not.

Respiratory System

Amphibians breathe through their skin **(cutaneous respiration)** as well as with their lungs (although some adult amphibians do not have lungs and rely primarily on gas exchange through the skin). Some respiration also occurs through the lining of the mouth, which is highly vascular. Cutaneous respiration is very important for the frog, especially in winter, when it burrows into the bottom mud of ponds and ceases all lung breathing.

The frog has no diaphragm. Thus it draws air into its mouth cavity through nares by closing the **glottis** (the opening into the windpipe) and depressing the floor of its mouth. Then, by closing the nares and raising the floor of its mouth cavity, it forces air from the mouth through the glottis into the lungs (positive pressure breathing). Air is expelled from the lungs by the contraction of the muscles of the body wall and the elastic recoil of the stretched lung.

☞ Probe through the **glottis** into the **larynx.** Find the short **bronchus,** connecting each **lung** to the larynx. Slit open a lung and observe its internal structure.

Note the little pockets, or **alveoli** (sing. **alveolus**), in the lining. What is the purpose of this arrangement?

Did you find any parasites in the lungs of your specimen?

Urogenital System

Functionally, the urogenital system is two systems, the **urinary,** or **excretory system,** and the **reproductive system.** However, because some structures function in both systems, they are usually considered together.

Be careful not to injure the blood vessels as you study this system.

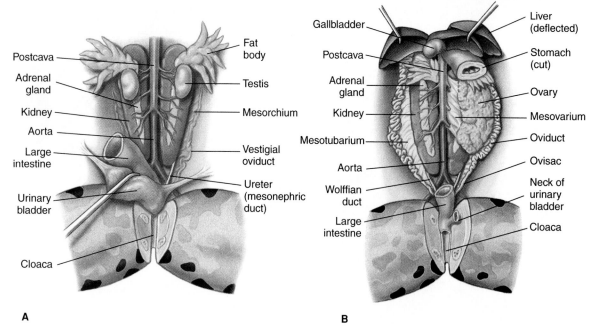

Figure 17-6
Urogenital system of a frog, ventral views. **A,** Male. **B,** Female.

Excretory System. The **mesonephric kidneys** separate the urine from the blood. They rid the body of metabolic wastes (aided by the lungs and skin), and they maintain a proper water balance in the body and a general constancy of content in the blood.

The kidneys (Figure 17-6) lie close to the dorsal body wall, separated from the coelom by a thin peritoneum. The **urinary bladder,** when collapsed, appears as a soft mass of thin tissue just ventral to the large intestine. It is bilobed and empties into the cloaca. The **ureters** connect the kidneys with the cloaca.

☞ To expose the cloaca, cut through the ischiopubic symphysis with a scalpel and push the pelvic bones aside. If you wish, you may open the cloaca just left of the midventral line, separate the cut edges, and find the bladder opening on the ventral wall of the cloaca and the openings of the ureters on the dorsal wall.

The **adrenal glands,** a light stripe on the ventral surface of each kidney, are endocrine glands, not urogenital organs. **Fat bodies** attached to the kidneys, but lying in the coelom, are for fat storage. They may be large in the fall and small or absent in the spring. Why? _____

Male Reproductive System (Figure 17-6A). A small, pale **testis** lies on the ventral side of each kidney. Sperm pass from the testis into some of the kidney tubules and then are carried by the ureter to the cloaca and hence to the outside. Thus the male ureters serve also as genital ducts. In leopard frogs a small **vestigial oviduct** runs parallel to the ureter (this is absent in some species of *Rana*).

Female Reproductive System. The **ovaries** are attached by mesenteries to the dorsal wall of the coelom. In winter and early spring the ovaries are distended with eggs. If the specimen was killed in summer or early fall, the ovaries will be small, pale, and fan shaped. Convoluted **oviducts** widen anteriorly (dorsal to the lungs) into funnel-like **ostia** and posteriorly into **uteri,** which empty into the cloaca. Eggs are released from the ovary into the coelom, carried in coelomic fluid to the ostia, and then down the oviducts by ciliary action to the outside. At **amplexus** (the courtship embrace) the male clasps the female and fertilizes the eggs externally as they are laid in the water.

During breeding season the thumbs of the male frogs of some species enlarge presumably to assist in clasping the female during amplexus. Why would the male need to clasp the female so firmly? _____ Who might be trying to displace the male? _____

Oral Report

Be able to demonstrate your dissection and give the functions of the structures you have studied.

Trace the route of food through the digestive tract and tell what happens to it at each stage.

Trace the route of eggs and spermatozoa and be able to explain the anatomy and physiology of excretion and reproduction.

Understand the mechanics and physiology of respiration and tell how the mechanics of respiration in the frog differ from those in a human.

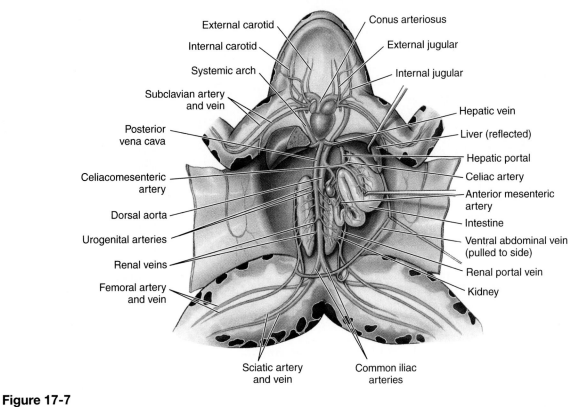

Figure 17-7
Circulatory system of the frog. Arterial, red; venous, blue.

EXERCISE 17E
The Circulatory System

The shift from gill to lung breathing during the evolution of amphibians required important changes in circulation. With the elimination of gills, a major resistance to blood flow was removed. But two new problems arose. The first was to provide a blood circuit to the lungs. This was solved by converting the last pair of aortic arches (which in fishes carried blood through the last gill arch) into **pulmonary arteries** to serve the lungs. New **pulmonary veins** then developed to return oxygenated blood to the heart. The second problem was to separate this new **pulmonary circuit** from the rest of the body's circulation such that oxygenated blood from the lungs would be selectively sent to the body, and deoxygenated blood from the body would be sent to the lungs. This was achieved by partitioning the heart into a double pump with each side serving each circuit. In this way a **double circulation** comprising separate **pulmonary** and **systemic** circuits was formed. As you will see in your dissection of the frog, amphibians approached this necessary advance only part way: the atrium, which was single in fish, is completely divided into two atria, but the ventricle remains undivided. The task of complete separation had to await the evolution of the birds and mammals, which have completely divided hearts of two atria and two ventricles.

To assist your study of the circulation, you will dissect a frog that has had the arterial system injected with a red or yellow latex and the venous system with blue latex. If only the arteries are injected, the veins may be filled with dark, clotted blood.

☞ Dissect carefully and do not cut or injure the blood vessels. With a probe you may loosen the connective tissue that holds them in place. Cut away a midsection of the pectoral girdle and pin back the arms so that the heart is fully exposed. Carefully remove the **pericardium,** the sac that contains the heart.

Identify the thick-walled conical **ventricle** (Figures 17-7 and 17-8), the thin-walled **left and right atria;** the **conus arteriosus,** arising from the ventricle and dividing to form the **truncus arteriosus** on each side; and on the dorsal side of the heart the thin-walled **sinus venosus,** formed by the convergence of three large veins—two **precaval veins** and one **postcaval vein.**[*]

---

[*]"Caval" derives from the Latin *cavus*, meaning hollow, and refers to the sinus venosus into which both the precaval veins (also called the anterior vena cava) and postcaval vein (= posterior vena cava) drain. The caval veins of tetrapod vertebrates replace the cardinal veins of fishes (see pp. 200–201).

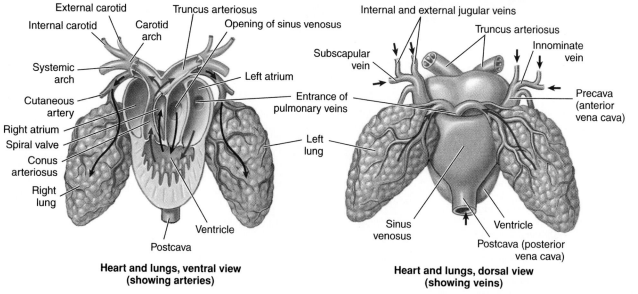

External carotid
Internal carotid
Carotid arch
Truncus arteriosus
Opening of sinus venosus
Systemic arch
Cutaneous artery
Right atrium
Spiral valve
Conus arteriosus
Right lung
Left atrium
Entrance of pulmonary veins
Left lung
Ventricle
Postcava

**Heart and lungs, ventral view
(showing arteries)**

Internal and external jugular veins
Truncus arteriosus
Innominate vein
Subscapular vein
Precava (anterior vena cava)
Sinus venosus
Ventricle
Postcava (posterior vena cava)

**Heart and lungs, dorsal view
(showing veins)**

Figure 17-8
Structure of the frog heart. **Left,** Ventral view of frontal section. **Right,** Dorsal view. What structure collects deoxygenated blood returning to the heart and is formed by the convergence of three large veins? _____ What structure may help prevent mixing of oxygenated and deoxygenated blood as it leaves the heart (use your textbook to help you if necessary)? _____ What artery carries deoxygenated blood to the skin where the blood can be oxygenated? _____

Venous System

The **precava (anterior vena cava)** (Figure 17-8) is formed by the union of (1) the **external jugular** from the tongue and floor of the mouth, (2) the **innominate** (L. *in,* not, + *nomen,* named), made up from the **subscapular vein** from the shoulder and the **internal jugular** (L. *jugulum,* collarbone) from the brain, and (3) the **subclavian** (L. *sub,* under, + *clavus,* key, i.e., below the clavicle), which receives blood from the arm and dorsal body wall.

The **postcava (posterior vena cava)** extends from the sinus venosus through the liver to the region between the kidneys. It receives **hepatic veins** from the liver, **renal veins** from each kidney, and the **ovarian** or **spermatic veins** from the gonads.

Pulmonary veins bring blood from the lungs to the left atrium. How does the oxygen content of the blood in these veins differ from that in any other vein? _____

Portal Systems

Ordinarily, veins carry blood directly from a capillary bed to the heart. This plan is interrupted in amphibians by capillary beds in two portal systems—the hepatic portal and the renal portal systems.

Hepatic Portal System. In the hepatic portal system, blood is carried to the capillaries of the liver by two veins. (1) The **ventral abdominal vein** in the ventral body wall collects from the pelvic veins, which are branches of the femoral veins; it empties into the liver. (2) The **hepatic portal vein** receives the splenic, pancreatic, intestinal, and gastric veins. From the capillary

bed in the liver the blood is picked up by the **hepatic veins,** carried to the postcava, and so to the sinus venosus. The hepatic portal system, present in all vertebrates, is of great importance because it delivers nutrients absorbed from the gut directly to the liver. This guarantees that the liver will have first opportunity to store and process food materials before they are released into the general circulation. In this way the blood leaving the liver remains relatively uniform regardless of digestive activities underway.

Renal Portal System. The amphibians also inherited a *renal* portal system from their fish ancestors. Most of the blood returning from the hindlegs is interrupted in its journey toward the heart to be diverted into a network of capillaries in the kidneys. The **renal portal vein,** found along the margin of each kidney, is formed by the union of the **sciatic** and **femoral veins** (See Figure 17-7). Then, from the kidney, blood is collected by the **renal veins** and carried to the **postcava.** All vertebrates except the mammals have a renal portal system. You will see then that both hepatic and renal portal systems fit the definition of a portal system as one that begins and ends in capillaries.

Arterial System

The **carotid, systemic,** and **pulmocutaneous arches** (known collectively as the **aortic arches**) arise from the **truncus arteriosus** (Figure 17-8).

Carotid Arch

The **common carotid artery** divides into (1) the **internal carotid** to the roof of the mouth, eye, brain, and spinal cord, and (2) the **external carotid (lingual)** to the floor of the mouth, tongue, and thyroid gland (See Figure 17-7).

Pulmocutaneous Arch

The third arch divides into the short **pulmonary artery** to the lungs and the longer **cutaneous artery** to the skin. Unlike the carotid and systemic arches, which carry oxygenated blood to the body, the pulmocutaneous arch carries *deoxygenated* blood to the lungs and skin, where the blood can be oxygenated.

Systemic Arch

Each systemic arch (one from each side) passes along the side of the esophagus to join middorsally to form the **dorsal aorta** (See Figure 17-7). Each gives off several arteries before joining, among these the **subclavian artery** to the shoulder.

Find the **dorsal aorta** by lifting the kidneys to see it. The dorsal aorta supplies all the body posterior of the head except the lungs and skin. Major branches of the dorsal aorta are:

1. The large **celiacomesenteric artery,** which in turn gives rise to the **celiac** (Gr. *koilia,* belly) **artery** to the stomach, pancreas, and liver; and the **anterior mesenteric artery** to the spleen and intestines.

2. Six pairs of **urogenital arteries** to the kidneys, fat bodies, and gonads.

3. **Lumbar arteries** (not shown in Figure 17-7) to the muscles of the back.

4. The **common iliac** (L. *ilia,* flanks) **arteries,** formed by the division of the dorsal aorta. Each iliac gives off the **femoral artery** to the thigh, as well as branches to the urinary bladder, abdominal wall, and rectum. The iliac then continues into the hindleg as the **sciatic artery.**

Heart

☞ Make a frontal section of the heart (Figure 17-8), dividing it into dorsal and ventral valves.

Find the opening from the sinus venosus into the **right atrium** and the opening from the pulmonary veins into the **left atrium.** Why is the ventricle more muscular than the atria? _____

_____

The **conus arteriosus,** which receives blood from the **ventricle,** divides to form a left and right truncus arteriosus. Valves to prevent backflow of blood guard the entrances to the atria and the conus.

Even though the amphibian heart is three-chambered with two atria and one undivided ventricle, there is nevertheless an effective separation of oxygenated and deoxygenated blood in the heart. Oxygenated blood from the lungs is sent preferentially to the body, whereas deoxygenated blood from the body is directed toward the pulmocutaneous arch. This partitioning is aided by a spiral fold inside the conus arteriosus and by pressure changes within the heart with each heart contraction.

Written Report

✍ Fill in the report on blood circulation on pp. 222-223.

Oral Report

🗨 Demonstrate your dissection of the circulatory system to your instructor and be able to explain orally any phase of its anatomy or its functions. Be able to trace the flow of blood into the heart from body tissues, through the pulmonary circuit, and back to the body tissues.

Observing the Heartbeat

☞ Open a pithed frog as you did the preserved frog, but avoid cutting the abdominal vein. Cut carefully through the pectoral girdle, keeping the scissors well up to avoid injuring the heart. Pin back the forelimbs and keep the heart well moistened with frog Ringer's solution.

Identify the ventricle, left and right atria, truncus arteriosus, and aortic arches. Watch the heartbeat and note the series of alternating contractions—first the two atria, then the ventricle, and finally the arterial trunk. Raise the ventricle carefully to view the contraction of the sinus venosus immediately before the contraction of the atria.

The frog is an **ectothermic** animal; that is, its temperature is governed by the environmental temperature, rather than by internal means (endothermic). You can examine the effect of temperature changes on the heart rate by a simple experiment.

☞ Count and record the number of beats per minute at room temperature. Flood the abdominal cavity with ice-cold frog saline solution. Count the beat again immediately, then wait a little while until the maximum effect of the cold is achieved, and then count and record again.

Replace the cold saline solution with frog saline solution warmed to about 40° C. Count the beat immediately and record. Allow the warmth to take effect; then count and record again. Replace the warmed saline with saline at room temperature and count the heart rate again. Record your results on p. 223.

What effect does temperature change have on the heart rate? What advantage would this reaction to cold have for the frog? Why should the solution not be warmed to more than 40° C? What is the effect of fever on human heart rate?

Name _____

Date _____

Section _____

Frog Circulation

1. Trace the shortest route a corpuscle could take on each of the following trips, underscoring each place where it would go through a **capillary bed.**

   a. Ventricle to lung and return _____

   _____

   _____

   b. Ventricle to brain and return _____

   _____

   _____

   _____

   c. Systemic arch to intestine to right arm _____

   _____

   _____

   _____

   d. Hindleg to left atrium by way of renal portal _____

   _____

   _____

   _____

2. What are the chief gains and losses that take place in the blood in these organs?

   a. Lung _____

   _____

   b. Kidney _____

   _____

   c. Intestinal wall _____

   _____

d. Liver _____

_____

e. Muscles _____

_____

3. Describe the shortest route from the **left atrium** to an **arm** and back to the **left atrium.**

_____

_____

_____

_____

_____

4. Effect of temperature on heart rate.

Rate of contraction

at room temperature _____ /min (first count)

of cooled heart _____ /min (first count)

of cooled heart _____ /min (later count)

of warmed heart _____ /min (first count)

of warmed heart _____ /min (later count)

at room temperature _____ /min (later count)

EXERCISE 18
The Painted Turtle

Reptiles, the first truly terrestrial vertebrates, show several important changes over their progenitors, the amphibians. Without doubt the most important of these was the development of a shelled, "amniotic" egg that could be laid on land, thus freeing reptiles from the aquatic environment. The reptiles further adapted their fitness for life on land by developing a tough, protective skin; a more efficient, high-pressure circulation; more efficient lungs; internal fertilization; and a somewhat more modified nervous system.

Living reptiles belong to two of three amniote lineages that appeared in the late Paleozoic. Most modern reptiles—lizards, snakes, and crocodilians—belong to the **diapsid** lineage, a group characterized by having a skull with two pairs of windowlike openings in the cheek (temporal) region. This is a derived condition that lightened the skull, furnished edges for jaw muscle attachment, and provided space that allowed the jaw muscles to bulge when the jaw was closed. The turtles, however, belong to a separate lineage called **anapsid.** Anapsids have a type of skull with no temporal openings at all. Turtles have changed little over the last 200 million years, and although they are highly specialized and much modified from the earliest known amniote fossils, they reveal several features that distinguish the reptilian lineages. The turtle's shell, the anatomical feature that makes it instantly recognizable, is undoubtedly one secret of their success, providing protection to an otherwise ungainly animal.

The Turtle

Phylum Chordata
    Subphylum Vertebrata
        Class Reptilia
            Order Testudines (Chelonia)
                Genus *Chrysemys*
                    Species *Chrysemys picta*

Where Found

The painted turtle, *Chrysemys picta,* is a familiar aquatic turtle of ponds, marshes, lake edges, and slow streams.

EXERCISE 18

The Painted Turtle
The turtle

It is widely distributed in the central and northern United States, as far south as Georgia and Louisiana, and north along the southern edge of the Canadian provinces. Painted turtles feed on aquatic vegetation, crayfish, snails, and insects.

External Structure

☞ If living turtles are available, study the locomotion and external features for adaptations that distinguish this group.

The anatomical form of the turtle is unusually broad and flattened as compared to that of other reptiles, with the limbs extending laterally from the sides and bent downward at elbow and knee. As a result, the turtle's gait is slow and seemingly awkward, although not as inefficient as it might appear—it has, after all, served this successful group for some 200 million years while numerous more agile reptiles have disappeared. The protective shell of the turtle, its exoskeleton, includes the dorsal **carapace** (Sp. *carapacho,* shell) and the ventral **plastron** (Fr. breastplate). Some turtles have hinged plastrons and can close their plastron tightly. Other species rely more on their snapping jaws for defense and have greatly reduced plastrons. The bones of the carapace are covered with horny **scutes** (homologous to the epidermal scales of other reptiles). Although the turtle's shell is confining, it is not the explanation for the awkward gait since a sprawling posture with the body dragging on the ground was characteristic of primitive land dwellers in general. The subsequent evolution of the shell, however, did prevent any further refinement in locomotion.

The head is stoutly protected with bone. A peculiarity of turtles is their lack of teeth. Instead, the edges of the jaws are formed into sharp ridges covered with strong, horny beaks. Locate the **external nares** (nostrils). Note that they are set close together at the tip of

Class Aves

EXERCISE 19
The Pigeon

EXERCISE 19
The Pigeon
The pigeon (rock dove)

Birds, described by the English zoologist Thomas Huxley as "glorified reptiles," do indeed bear the stamp of their reptilian heritage in many subtle ways. But birds have become so highly specialized into flying machines with all the constraints that design for flight requires that the truth of Huxley's comment is not instantly evident to one gazing casually at a bird.

The demands of flight, far more than anything else, have shaped the form and function of birds—and disguised their reptilian past. Virtually every adaptation found in flying birds focuses on two features: more power and less weight. This will be the central theme of this exercise.

The Pigeon (Rock Dove)

Phylum Chordata
   Subphylum Vertebrata
      Class Aves
         Order Columbiformes
            Family Columbidae
               Genus *Columba*
                  Species *Columba livia*, rock dove

Where Found

The common pigeon (rock dove), so familiar to city dwellers, is of Old World origin but has been introduced throughout the world. It is not migratory, yet it has excellent navigational proficiency; it is the homing pigeon used in carrying messages in wars from the time of Caesar's conquest of Gaul through World War II. It is one of the swiftest birds in flight. Many domesticated varieties have been developed over centuries of breeding; Charles Darwin was himself a pigeon fancier. The "city pigeon" is highly variable in coloration.

Feathers

☞ Examine a flight (contour) feather.

Identify the central **shaft** (also called the **rachis** [Gr. *rhachis,* spine]), which is a continuation of the **quill** that is thrust into the feather follicle of the living bird. The shaft bears numerous **barbs,** which spread laterally to form the feather's expansive webbed surface, the **vane.**

Flex the feather in your hands, noting its resilience and toughness, despite its remarkable light weight. If you run your fingers down the vane toward the quill, you will separate some of the barbs that are normally linked together by tiny **barbules.** Note that considerable force is needed to separate the barbs. Because this happens in the course of the bird's daily activities, the bird spends time each day preening: zipping the barbs back together by drawing the feather through its bill. You can do the same with your fingers. Why would preening be so important for a bird? _____

☞ Examine a prepared slide of a contour feather, using a dissecting microscope or low power of a compound microscope.

There are several types of feathers that may have been placed on display. The **contour feathers** that you have just examined give the bird its outward form; contour feathers used in flight are called **flight feathers.** **Down feathers** are soft tufts, lacking hooks, that are found on young birds or beneath the contour feathers of adult birds. Down feathers have excellent insulative value and function mainly to conserve heat. **Filoplume feathers** are hairlike feathers thought to play a sensory role. These feathers, consisting of a weak shaft with a tuft of short barbs at the tip, are the "hairs" visible on plucked fowl.

The Skeleton

☞ Study the mounted pigeon skeleton and unmounted bird bones on display.

The pigeon skeleton (Figure 19-1), like that of other birds, is a marvel of lightness combined with strength. It

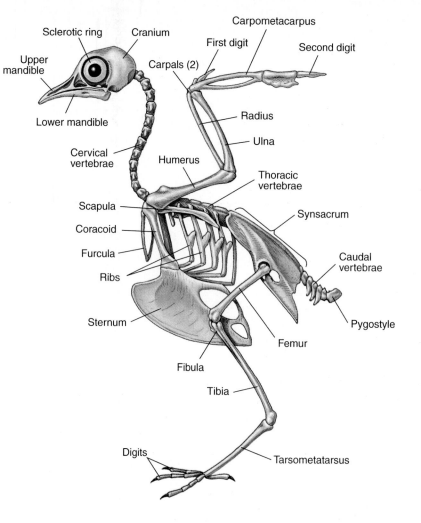

Figure 19-1

Skeleton of pigeon. Why is the keel of the sternum so large? _____ What bone is formed by the fusion of 13 vertebrae? _____ What bone is sometimes called the "wishbone"? _____ How would you describe the fibula of the bird when compared to other vertebrates (about the same, reduced, enlarged)? _____

is even lighter than it looks because many of the limb and girdle bones are hollow. If a long bone such as the humerus has been broken or cut open for examination, note its tubular form and the internal struts that have developed where stresses must be borne. Many of the bird's bones are "pneumatized": that is, they are penetrated by extensions of the air sac system and thus contain buoyant warm air rather than bone marrow typical of mammalian bones (some bird bones do contain marrow, but much of a bird's red and white blood cell production occurs in the spleen and liver).

A striking feature of the bird skeleton is its rigidity. Of the axial skeleton, only the neck remains flexible. The remaining vertebrae are fused together and with the pelvic girdle to form a stiff, boxlike framework to support the legs and provide rigidity for flight. The double-headed **ribs,** too, are mostly fused with the **thoracic vertebrae** and with the **sternum.** Unlike in mammals, nearly all of which have seven cervical vertebrae, the number of **cervical vertebrae** in birds varies

from 8 to 24, depending on the species (long-necked birds have the most). Note the complex articulations of the cervical vertebrae of this most flexible part of the pigeon's body. How many cervical vertebrae has the pigeon? _____

The last rib-bearing thoracic vertebra is fused with five **lumbar,** two **sacral,** and five **caudal** vertebrae to form a thin, platelike structure, the **synsacrum.** The **ilium** of the pelvic girdle is also fused with the synsacrum. This very light, but stout arrangement provides further rigidity to the body frame. Finally, note the short bony tail, consisting of five free **caudal vertebrae** that carry four caudal vertebrae fused into a **pygostyle,** which supports the tail feathers.

The pigeon skull is composed of individual bones that are completely united in the adult to form a single thin-walled and lightweight structure. Birds are descended from archosaurian reptiles, which belong to the diapsid lineage of amniotes; this lineage is characterized by skulls having two openings, or fenestra, in the

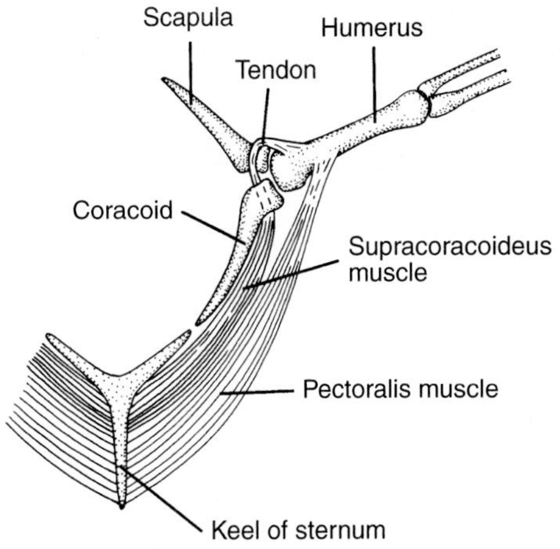

Figure 19-2

The major flight muscles of a bird are arranged to keep the center of gravity low in the body. Both the supracoracoideus and the pectoralis are anchored on the sternum keel. Contraction of the pectoralis muscle pulls the wing downward. As the pectoralis relaxes, the supracoracoideus muscle contracts and, acting as a pulley system, pulls the wing upward.

temporal region. Birds are so highly specialized, however, that it is difficult to see any trace of diapsid origin in their skulls. The large, bulging **cranium** encloses the brain, which is much larger, relative to body size, than the brain of a turtle; the complex coordinating problems of bird flight require far more central nervous coordination than does the plodding locomotion of a turtle. The pigeon's large eyes are housed in sockets and encircled in front by a protective ring of shingle-like bony plates, the **sclerotic ring.** The **beak** consists of a **lower mandible** hinged to the skull in a way that provides wide-gaping action. The **upper mandible** (also called the maxilla) is fused to the skull in pigeons, but some birds, parrots, for example, have kinetic skulls with movable bony elements that allow the upper mandible to tilt upward when the bird opens its mouth.

We will turn our attention now to the appendicular skeleton. Examine the pectoral girdle. It is a tripod of paired bones, the **scapula, coracoid,** and **furcula** ("wishbone"). The scapula is a thin, bladelike bone tied to the ribs by ligaments. The stout coracoid bone unites the scapula and **sternum.** Describe the sternum _____ Why is it so large? _____ Both the muscles that depress the wing (pectoralis) and those that raise the wing (supracoracoideus) are attached to the sternum. Where scapula and coracoid unite, there is a hollow depression into which the ball of the chief bone of the wing, the **humerus,** fits. The supracoracoideus is attached by a tendon to the upper side of the humerus so that it pulls from below by an ingenious "rope-and-pulley" kind of arrangement (Figure 19-2). In this way, muscle weight is kept below the center of gravity, providing greater flight stability.

The pelvic girdle, as we have seen, is a fused structure that is almost paper thin but is strengthened by bony ridges that can be seen by looking at the underside of the girdle. The **femur** (thighbone) is directed forward and is virtually buried in the flesh of the living bird. The **tibia** is the main bone of the shank ("drumstick"); the **fibula** is reduced to a thin splint. The ankle is greatly modified. Some of the pebblelike tarsal bones of the tetrapod limb are united with the tibia, and the others are fused with the metatarsals to form a single elongate **tarsometatarsus.** The pigeon, like most other birds, has four digits, three directed forward and the fourth directed backward.

The bones of the forelimbs are highly modified for flight. Note how the wing folds into a compact Z shape when the bird is at rest. Identify the **humerus** and locate the expanded dorsal surface for the attachment of the pectoral muscles. The **radius** and **ulna** are longer than the humerus, and the ulna, the larger of the two, carries the secondary flight feathers. Most modified are the wrist and digits, which carry the primary flight feathers. Identify the two **carpals** (wrist bones) and the two elongate **carpometacarpals** (palm bones), so called because they are formed by the fusion of three carpal and three metacarpal bones. There are only three **digits** (fingers). The first digit, or "thumb," carries the feathers of the **alula.** The second finger is by far the largest; this, together with the palm bones, carries the primary flight feathers. The third digit, like the first, is reduced to a small bone (not shown in Figure 19-1) that carries a single outermost flight feather.

Class Mammalia
The Fetal Pig

Fetal Pig

Class Mammalia
  Subclass Theria
    Infraclass Eutheria
      Order Artiodactyla
        Genus *Sus*
          Species *Sus domesticus*

The Mammalia are those animals whose young are nourished by milk from the breasts of the mother. Mammals have a muscular diaphragm, a structure found in no other class, and a four-chambered heart. Most of them are covered with hair. Their nervous system is especially well developed. Their eggs develop in a uterus, with placental attachment for nourishment (except the monotremes, which lay eggs, and the marsupials, in which the placental attachment is only weakly and briefly developed).

The order Artiodactyla includes the even-toed, hoofed mammals such as deer, sheep, cattle, and camels. These usually have two toes, but some, such as hippopotamuses and pigs, have four toes.

The embryo depends on maternal blood to bring it nutrients and oxygen and to carry off the waste products of metabolism because its own organs cannot serve these functions until birth. This exchange of materials between fetal blood and maternal blood takes place within the placenta of the mother's uterus. The difference between the fetus and the adult is largely physiological, but there are also a few morphological differences, especially in the circulatory system, which you will observe in your specimen.

The period of gestation in a pig is 16 to 17 weeks, as compared with 20 days in a rat, 8 weeks in a cat, 9 months in a human being, 11 months in the horse, and 22 months in an elephant. Pig litters average 7 to 12 but may have as many as 18 piglets. The pigs are about 30 cm long at birth and weigh from 1 to 1.5 kg (about 2 to 3 pounds). The age of a fetus may be estimated from the length of its body:

At 3 weeks the fetus is about 1.3 cm long.
At 7 weeks the fetus is about 3.8 cm long.
At 14 weeks the fetus is about 23 cm long.
At full term the fetus is about 30 cm long.

EXERCISE 20A

The Skeleton
The skeleton

EXERCISE 20B

The Muscular System
Organization of skeletal muscles
Muscles of the forequarter
Muscles of the hindquarter

EXERCISE 20C

The Digestive System
Head and throat
The abdominal cavity
Thoracic cavity and neck region
The digestive tract

EXERCISE 20D

The Urogenital System
Urinary system
Male reproductive system
Female reproductive system

EXERCISE 20E

The Circulatory System
Heart

External Structure

☞ Before proceeding with the regular exercises, look at the external structure of your pig.

On the head, locate the **mouth** with fleshy lips and the **nostrils** at the tip of the snout. The snout has a tough rim for rooting and bears **vibrissae,** stiff sensory hairs (whiskers). Each eye has two lids and a small membrane in the medial corner, which represents the nictitating membrane. The fleshy pinnae, or ear flaps, contain the external auditory opening.

On the trunk, locate the **thorax,** supported by ribs, sternum, and shoulder girdle with forelimbs attached; the **abdomen,** supported by a vertebral column and muscular walls; the **sacral** region, comprising the pelvic girdle with hindlimbs attached; the **umbilical cord;** five to eight pairs of **mammae,** or nipples, on the abdomen; and the **anus** at the base of the tail.

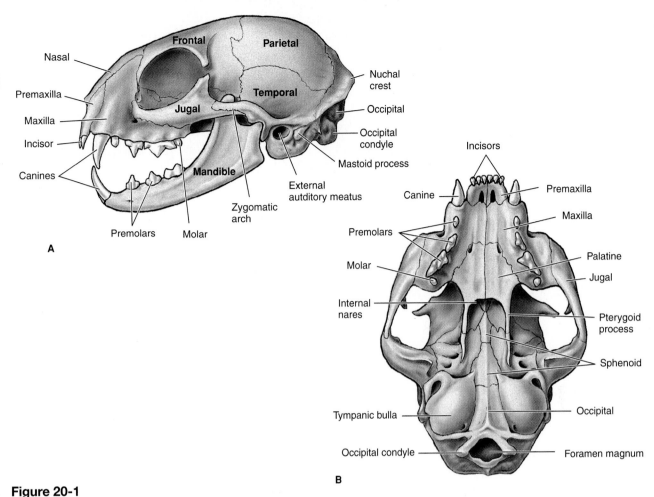

Figure 20-1
A, Skull of a cat, lateral view. **B,** Skull of a cat, ventral view.

What is the **sex** of your specimen? In the **male** the **urogenital opening** is just posterior to the umbilical cord; the **scrotal sacs** form two swellings at the posterior end of the body (the **penis** can sometimes be felt under the skin as a long thin cord passing from the urogenital opening back between the hindlegs). In the **female** the urogenital opening is just ventral to the anus and has a fleshy tubercle projecting from it.

Examine the cut end of the umbilical cord. Note the ends of four tubes in the cord. These represent an umbilical vein, two umbilical arteries, and an allantoic duct, all of which during fetal life are concerned with transporting food, oxygen, and waste products to or from the placenta of the mother's uterus.

Notice that the entire body of the fetal pig is covered with a thin cuticle called the **periderm.**

EXERCISE 20A

The Skeleton

The skeleton of an adult pig, although different in size and proportion from the skeletons of other mammals such as a dog, cat, or human, is, nonetheless, quite similar. The bones are homologous, and the origins and insertions of muscles as described for one can usually be traced on the others.

Because of its immature condition, the skeleton of a fetal pig is unsuitable for classroom study. However, if you are planning to dissect the muscles of a fetal pig, it is essential to be familiar with the bones. Skeletons of a cat or dog can be used quite satisfactorily for this purpose. Figures 20-1 and 20-2 of a cat skeleton will help you in your identification.

☞ As you study the mounted skeleton of the dog or cat, compare the parts with those of the fetal pig skeleton (Figure 20-3). Then try to locate these parts and visualize their relationships within the flesh of the preserved pig. As you do, notice the similarities with the human skeleton (Figure 20-4).

The Skeleton

As in a frog, the pig skeleton consists of the **axial skeleton** (skull, vertebral column, ribs, and sternum) and the **appendicular skeleton** (pectoral and pelvic girdles and their appendages).

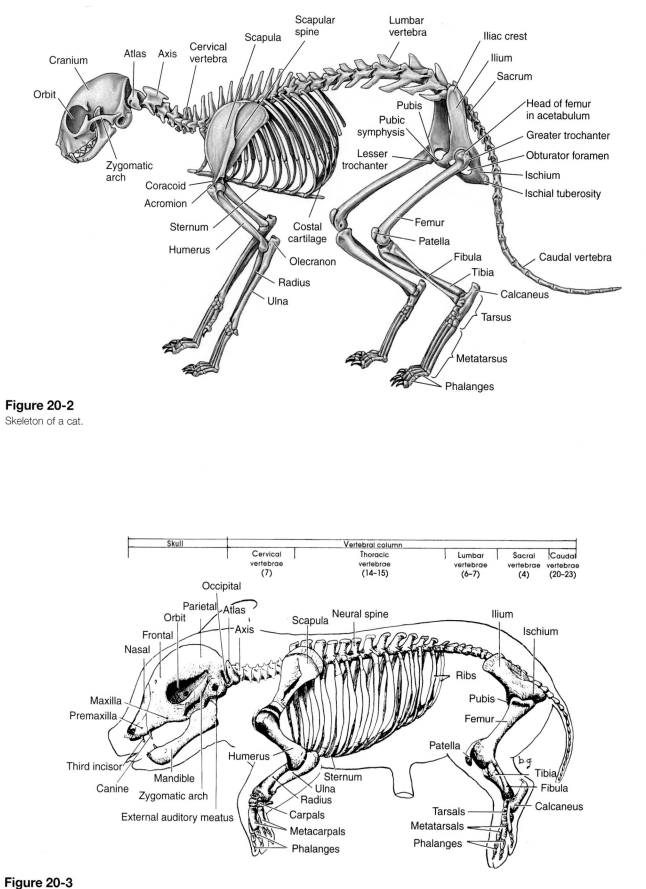

Figure 20-2

Skeleton of a cat.

Figure 20-3

Skeleton of a fetal pig.

20-3

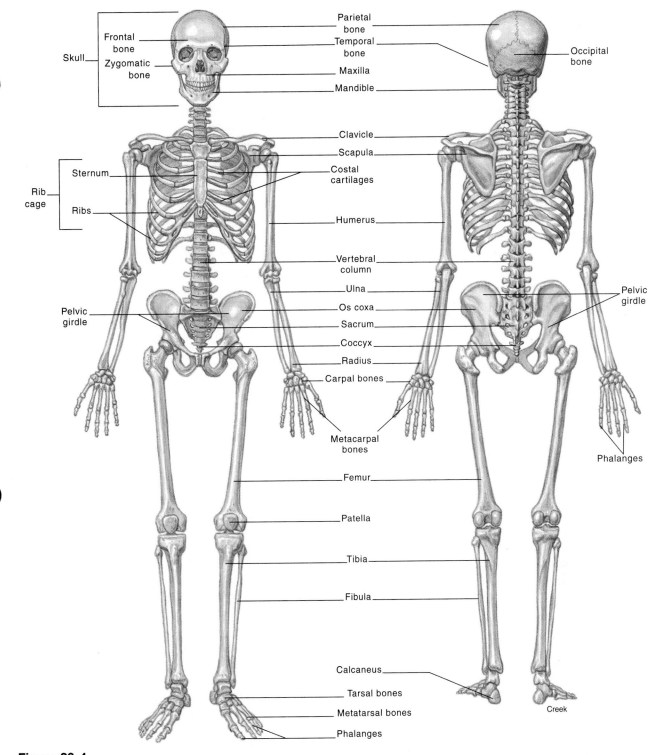

Figure 20-4

Human skeleton.

From Kent M. Van De Graaff, Human Anatomy, *4th ed. Copyright © 1995 Times Mirror Higher Education Group, Inc., Dubuque, Iowa. Reprinted by permission. All Rights Reserved.*

Axial Skeleton

Skull. The skull can be divided into a **facial region,** containing the bones of the eyes, nose, and jaws, and a **cranial region,** which houses the brain and ears. The smooth, rounded occipital condyles (Gr. *kondylos,* knuckle) of the skull articulate with the ring-shaped first cervical vertebra (called the **atlas,** named for Atlas of Greek mythology, condemned to hold the heavens on his shoulders for all eternity). The foramen magnum ("great opening") is the opening at the posterior end of the braincase for the emergence of the spinal cord. Many of the important bones of the skull can be identified with the help of Figure 20-1.

Vertebral Column. Note the five types of vertebrae—**cervical,** in the neck; **thoracic,** bearing the ribs; **lumbar,** without ribs but with large transverse processes; **sacral,** fused together to form a point of attachment for the pelvic girdle; and **caudal,** the tail vertebrae. In humans, three to five vestigial caudal vertebrae are fused to form the coccyx of the tail bone. All of the vertebrae are built on the same general plan with recognizable individual differences (Figures 20-2 and 20-3).

Ribs. Observe the structure of a rib and its articulation with a vertebra. Each rib articulates with both the body of the vertebra and a transverse process. The **shaft** of the rib ends in a **costal** (L. *costa,* rib) **cartilage,** which attaches to the sternum or to another costal cartilage and so indirectly to the sternum. The cat has one pair of free, or floating, ribs. The pig has 14 or 15 pairs of ribs, of which 7 pairs attach directly to the sternum and 7 or 8 attach indirectly. How many does the human have? _____

Sternum. The sternum is composed of a number of ossified segments, the first of which is called the **manubrium,** and the last of which is called the **xiphisternum.** Those in between are called the **body** of the sternum.

Appendicular Skeleton

Pectoral Girdle and Its Appendages. The pectoral girdle comprises a pair of triangular **scapulae,** each with a lateral **spine** and a **glenoid fossa** at the ventral point for attachment with the head of the humerus. The **forelimb** (Figures 20-2 to 20-4) includes (1) the **humerus;** (2) the two forearm bones—a shorter, more medial **radius** and a longer, more lateral **ulna,** with an **olecranon** (ol-ek′re-non) **process** at the proximal, or elbow, end; (3) the **carpus,** consisting of two rows of small bones; (4) the **metacarpals** (five in the cat and human and four in the pig); and (5) the **digits,** or toes, made up of **phalanges.**

Pelvic Girdle and Its Appendages. The pelvic girdle in adult mammals consists of a pair of **innominate** (L., without name) **bones,** each formed by the fusion of the **ilium** (L., flank), **ischium** (Gr. *ischion,*

hip), and **pubis** (L., mature). A lateral cavity, the **acetabulum** (L., vinegar cup), accepts the head of the femur. The pubic bones and the ischial bones of opposite sides unite at **symphyses,** and the ilia articulate with the sacrum so that the innominates and the sacrum together form a complete ring, or **pelvic canal.** Each pelvic appendage includes (1) the **femur,** or thigh bone; (2) the larger **tibia** and more slender **fibula** of the shank; (3) the ankle, or **tarsus,** comprising seven bones, and the fibular tarsal (**calcaneus;** cal-ka′nee-us; L. *calx,* heel), forming the projecting heel bone; (4) the **metatarsals** (five in the cat, of which the first is very small, and four in the pig); and (5) the **digits,** or toes (four in both the cat and the pig), composed of **phalanges.**

Structure of a Long Bone

Longitudinal and transverse sections through a long bone (Figure 20-5) show it to consist of a shell of **compact bone,** within which is a type of **spongy bone (cancellous bone),** the spaces of which are filled with **marrow.** The **shaft** (also called the **diaphysis;** di-af′-uh-sis; Gr. *dia,* through; + *phyein,* to grow) is usually hollowed to form a **marrow cavity.** In the young animal there is only red marrow, a blood-forming substance, but this is gradually replaced in the adult with yellow marrow, which is much like adipose tissue. The extremities (**epiphyses;** ee-pif′uh-sees; Gr. *epi,* upon; + *phyein,* to grow) of the bone usually bear a layer of **articular cartilage.** The rest of the bone is covered with a membrane, the **periosteum** (pear-ee-os′te-um; Gr. *peri,* around; + *ostrakon,* shell). Arteries, veins, nerves, and lymphatics pass through the compact bone to supply the marrow and cells responsible for bone formation and maintenance.

Growth of a Bone

The primitive skeleton of the embryo consists of cartilage and fibrous tissue, in which the bones develop by a process of ossification. Bones that develop in fibrous tissue, namely, some of the bones of the cranium and face, are called membranous bones. Most of the bones of the body develop from cartilage and are designated as endochondral ("within cartilage") bones.

In a typical long bone there are usually three primary centers of ossification—one for the diaphysis, or shaft, and one for each epiphysis, or extremity (Figure 20-5). As long as this cartilage persists and grows, new bone may form, and the length may increase.

Articulations

An articulation, or joint, is the union of two or more bones or cartilages by another tissue, usually fibrous tissue or cartilage or a combination of the two.

Three types of joints are recognized:

Synarthrosis. A synarthrosis (sin-ar-thro′sis; Gr. *syn,* with; + *orthron,* joint) is an immovable joint. Interlocking

TABLE 20.1

Muscles of the Forequarter—cont'd.

Muscle	Origin	Insertion	Action
Muscles of the Face, Neck, Chest, and Shoulder—cont'd.			
Supraspinatus	Anterior and dorsal portion of the scapula and the scapular spine	Proximal end of humerus	Extends the humerus
Infraspinatus	Lateral surface and spine of scapula	Lateral surface of the proximal end of the humerus	Abducts and rotates the forelimb
Muscles of the Foreleg			
Triceps brachii	Long head: posterior border of scapula Lateral head: lateral side of the proximal end of the humerus Medial head: medial surface of the proximal end of the humerus, covering the insertion of the teres major	All three heads insert on the medial and lateral surfaces of the olecranon process of the ulna	Extend the forearm
Brachialis	Proximal third of the humerus, ventral to the lateral head of the triceps	Medial surface of the distal end of the radius and ulna	Flexes the elbow
Biceps brachii	Ventral surface of the scapula near the glenoid fossa	Proximal ends of the radius and ulna	Flexes the elbow

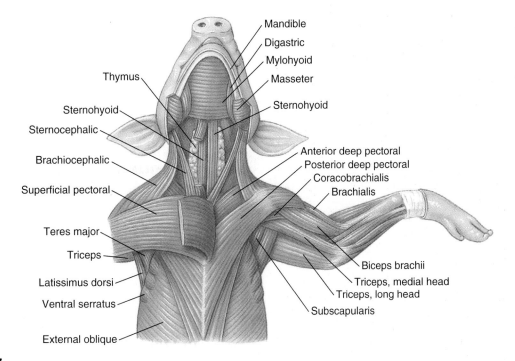

Figure 20-7

Muscles of the ventral thoracic region of the fetal pig. The pectoral muscle has been removed from the pig's left side to reveal the underlying musculature.

Reprinted by permission of Medical and Scientific Illustration/William C. Ober.

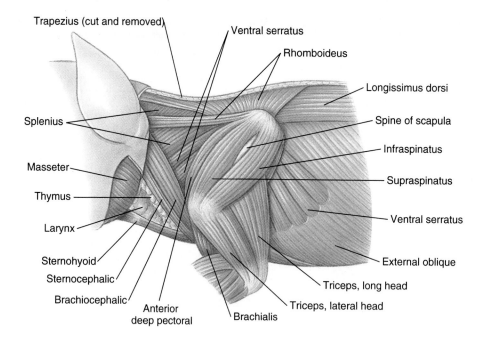

Figure 20-8

Muscles of neck and shoulder of the fetal pig, lateral view. The trapezius muscle has been removed to reveal underlying musculature.

Reprinted by permission of Medical and Scientific Illustration/William C. Ober.

the small, dark, compact **thyroid gland** lying on the ventral side of the trachea.

The **digastric** muscle is the major depressor of the mandible: that is, it acts to open the jaw. It originates by a strong tendon from the base of the skull. Stretching ventrally between the mandibles as a thin transverse sheet is the **mylohyoid;** it compresses the floor of the mouth and assists in swallowing. The large muscle of the cheek is the **masseter.** This muscle and another (the temporal, not shown in Figure 20-8) are the major muscles that elevate the jaw and close the mouth.

Locate the **sternocephalic,** a flat muscle band that passes diagonally across the throat posterior to the submaxillary gland and beneath the parotid gland; it turns the head. When both sternocephalics contract together the head is depressed. Posterior to the sternocephalic is the **brachiocephalic,** a large band-shaped muscle that originates on the skull and inserts on the shoulder. This muscle raises the head, or, if the head is fixed in position by other muscles, it draws the forelimb forward. It originates on two different processes (nuchal crest and mastoid process) of the skull.

Examine now the muscles that position the shoulder girdle and move the forelimb (Figure 20-7). The **superficial pectoral** (equivalent to the pectoralis major of humans) adducts the humerus. The **posterior deep pectoral** and **anterior deep pectoral** both retract and adduct the forelimb. The most superficial muscle of the back (after removal of the cutaneous maximus) is the thin and triangular **trapezius**

(Figure 20-8). This muscle elevates the shoulder and moves the scapula. The **latissimus dorsi** (See Figure 20-6) is a large, broad muscle that fans out across the back; it flexes the shoulder and is a major retractor of the humerus.

Now carefully cut through the trapezius at its insertion on the spine of the scapula and separate it from the underlying muscles. A triangular muscle just beneath the trapezius is the **rhomboideus.** Anterior to this is the straplike **rhomboideus capitis.** Both of these muscles act upon the scapula to rotate it or draw it forward or dorsally. Beneath the rhomboideus capitis locate the triangular **splenius,** which helps to elevate and turn the head.

Cut the latissimus dorsi at its origin along the spine and remove most of the muscle, leaving only a centimeter or two at its insertion on the humerus. Locate the **ventral serratus** (Figure 20-8), an extensive fan-shaped chest muscle that originates on the cervical vertebrae and several ribs and inserts on the scapula beneath the insertion of the rhomboideus muscles. It acts to shift the scapula forward and backward and serves as a muscular support to sling the weight of the trunk.

Arising near the insertion of the trapezius on the scapula is the **deltoid.** It inserts on the humerus and acts to protract (move forward) the upper foreleg. Carefully trim away the deltoid muscle at both origin and insertion. Beneath it lies the **infraspinatus,** which abducts and rotates the forelimb. Anterior to this is the **supraspinatus,** a fleshy muscle on the anterior surface of the scapula; it acts to extend the humerus.

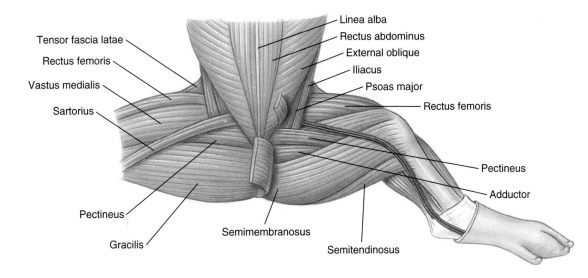

Figure 20-9

Muscles of the hindlimb of the fetal pig, ventral view. The gracilis and sartorius have been cut and removed from the pig's left leg.

Reprinted by permission of Medical and Scientific Illustration/William C. Ober.

Muscles of the Foreleg (Table 20.1)

The largest muscle is the **triceps brachii,** an extensor of the forearm (Figures 20-7 and 20-8). It arises from three heads: a triangular-shaped long head at the posterior border of the scapula, a lateral head from the lateral surface of the humerus, and a medial head from the medial surface of the humerus. Two smaller muscles lying on the anterior and ventral surfaces of the humerus are the **brachialis** and **biceps brachii;** both act upon the elbow to flex the forearm.

Muscles of the Hindquarter

Muscles of the Abdomen, Back, and Hip (Table 20.2)

Three thin sheets of muscle lie in the lateral abdominal wall. Most superficial is the **external oblique** lying immediately beneath the cutaneous (See Figure 20-6). Beneath this is the **internal oblique;** this is revealed by cutting a window high up on the external oblique. Then, by separating fibers of the internal oblique you will see the **transverse abdominal,** the deepest layer and the thinnest of the three muscles (and the most difficult to see). All three muscles insert on the **linea alba,** a tendinous band that extends from the pubis to the sternum (Figure 20-9). Together, these muscles support the abdominal wall and compress the viscera during expiration and defecation. Beneath the external oblique and extending between the pelvic girdle and ribs on each side of the midventral line is a longitudinal band of muscle, the **rectus abdominus** (Figure 20-9). It also supports and constricts the abdomen.

Dorsal and lateral to the vertebral column locate the **longissimus dorsi,** a very long muscle extending from the sacrum to the neck (See Figure 20-8). Acting together these muscles extend the back and neck; acting singly each will flex the spine laterally.

The most anterior superficial thigh muscle is the **tensor fasciae latae** (See Figures 20-6 and 20-10); it acts to flex the hip joint and extend the knee joint. Posteriorly, the most superficial thigh muscle is the **biceps femoris** (See Figures 20-6 and 20-10). Its action is complex, acting across both the hip and knee joints to retract the thigh and flex the shank. Between the tensor fasciae latae and biceps femoris, and partially covered by them, is the **gluteus medius** (See Figures 20-6 and 20-10). It acts to abduct the thigh.

Muscles of the Hindleg (Table 20.2)

The **quadriceps femoris** is a large muscle group covering the anterior and lateral sides of the femur that comprises four muscles: **rectus femoris,** a thick muscle on the anterior side of the femur; **vastus lateralis** (Figure 20-10), lateral to the rectus femoris and partly covering it; **vastus medialis** (See Figure 20-9), on the medial surface of the rectus femoris; and **vastus intermedialis,** a deep muscle lying beneath the rectus femoris. All four of these muscles converge on the patella (kneecap) and then continue as the patellar ligament to insert on the tibia. These are the extensors of the shank.

The posteromedial half of the thigh is covered with a thin, wide muscle, the **gracilis** (See Figure 20-9). It adducts the thigh and flexes the shank. Just anterior to the gracilis is the **sartorius,** a thin band of muscle that covers the

TABLE 20.2

Muscles of the Hindquarter

Muscle	Origin	Insertion	Action
Muscles of the Abdomen, Back, and Hip			
External oblique	Lateral surface of the last 9 or 10 ribs and the lumbodorsal fascia	Linea alba, ilium, and femoral fascia	Compress the abdomen, arch the back; singly, it flexes the trunk laterally
Internal oblique	Similar to external oblique	Similar to external oblique	Similar to external oblique
Transverse abdominal	Similar to external oblique	Similar to external oblique	Similar to external oblique
Rectus abdominus	Pubic symphysis	Sternum	Constricts the abdomen
Longissimus dorsi	Sacrum, ilium, and neural processes of the lumbar and thoracic vertebrae	Transverse processes of most of the vertebrae and lateral surfaces of the ribs except the first	Singly, flexes the spine laterally; together, extend the back and neck; the rib attachments may aid in expiration
Tensor fasciae latae	Crest of the ilium	Fascia over the knee, the patella, and the crest of the tibia	Flexes the hip joint and extends the knee joint
Biceps femoris	Lateral part of ischium and sacrum	By a wide aponeurosis to the patella and the fascia of the thigh and leg	Abducts and extends limb; may also flex the knee joint
Gluteus medius	Fascia of the longissimus dorsi, the ilium, and the sacroiliac and sacrosciatic ligaments	Proximal end of the femur	Abducts the thigh
Muscles of the Hindleg			
Quadriceps femoris, a large muscle group consisting of			
1. Rectus femoris	Ilium	Patella and its ligament	Extends the shank
2. Vastus lateralis	Proximal end of femur	Patella and its ligament	Extends the shank
3. Vastus medialis	Proximal end of femur	Patella and its ligament	Extends the shank
4. Vastus intermedialis	Proximal end of femur	Patella and its ligament	Extends the shank
Gracilis	Pubic symphysis and ventral surface of the pubis	Patellar ligament and proximal end of the tibia	Adducts the hindlimb
Sartorius	Iliac fascia and tendon of the psoas minor (external iliac vessels lie between the two heads)	Patellar ligament and proximal end of the tibia	Adducts hindlimb and flexes the hip joint
Semimembranosus	Ischium	Distal end of the femur and proximal end of the tibia, both on medial side	Extends the hip joint and adducts the hindlimb
Semitendinosus	First and second caudal vertebrae and the ischium	Proximal end of tibia and the calcaneus	Extends the hip and the tarsal joint and flexes the knee joint
Adductor	Ventral surface of the pubis and ischium and the tendon of the origin of the gracilis	Proximal end of the femur	Adducts the hindlimb and extends and rotates the femur inward
Pectineus	Anterior border of the pubis	Medial side of the shaft of the femur	Adducts the hindlimb and and flexes the hip
Iliacus	Ventral surface of the ilium and wing of the sacrum	Proximal end of the femur together with the psoas major	Flexes the hip and rotates the thigh outward
Psoas major	Ventral sides of the transverse processes of the lumbar vertebrae and the last two ribs	With the iliacus on the proximal end of the femur	Flexes the hip and rotates the thigh outward

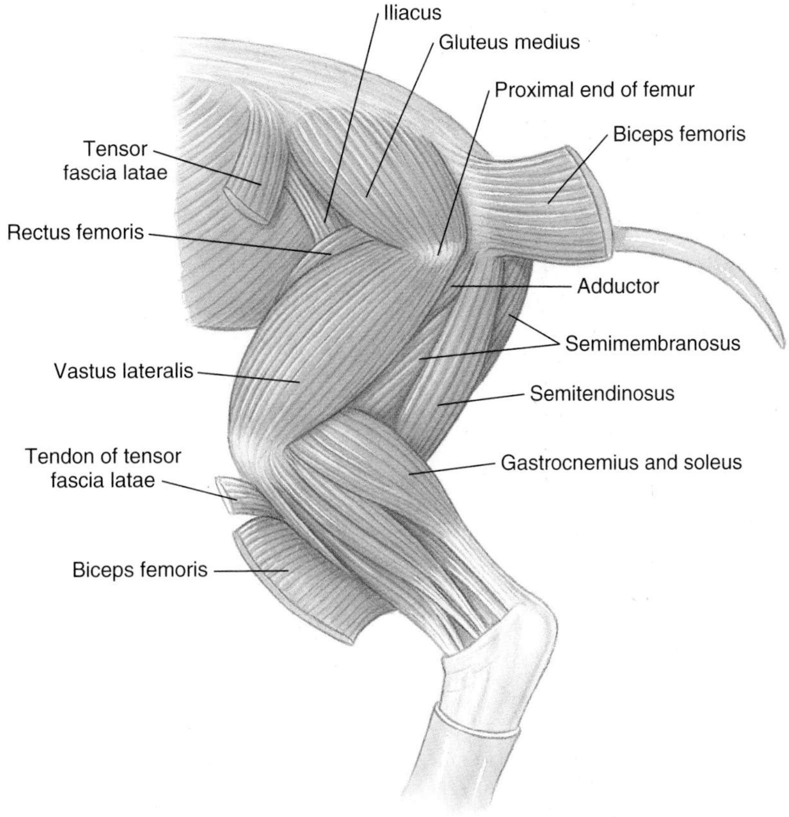

Labels on figure:
Iliacus
Gluteus medius
Proximal end of femur
Biceps femoris
Tensor fascia latae
Rectus femoris
Adductor
Semimembranosus
Semitendinosus
Vastus lateralis
Tendon of tensor fascia latae
Gastrocnemius and soleus
Biceps femoris

Figure 20-10
Muscles of the hindlimb, lateral view.

Reprinted by permission of Medical and Scientific Illustration/William C. Ober.

femoral blood vessels; it is delicate and easily destroyed if not identified. Cut through the gracilis and sartorius to reveal the large **semimembranosus** muscle in the medial portion of the thigh (Figure 20-9). It extends the hip joint and adducts the hindlimb. Just posterior to the semimembranosus is the thick, band-shaped **semitendinosus;** it acts mainly to extend the hip. The semimembranosus and semitendinosus, together with the biceps femoris, are the hamstring muscles of humans.

The **adductor,** lying anterior to the semimembranosus and covered by the gracilis, is a triangular-shaped muscle that, as its name suggests, adducts the femur, that is, draws it toward the midline.

Also on the medial side of the thigh are three smaller muscles: the triangular-shaped **pectineus,** an adductor of the thigh; the **iliacus,** which flexes the hip and rotates the thigh outward; and the **psoas major,** which acts the same as the iliacus.

Some of the shank muscles are shown in Figure 20-10 but are not described in this exercise.

Identification ◄

Be able to identify and to explain the origin, insertion, and action of as many of the foregoing muscles as your instructor has assigned.

EXERCISE 20C
The Digestive System
Head and Throat
Salivary Glands

Three pairs of salivary glands produce a continual background level of fluid secretions containing lysozymes and immunoglobulins that flush the teeth and mouth cavity and help to keep bacterial growth under control. During meals, much larger quantities of saliva are produced that contain lubricating glycoproteins called **mucins** and a large amount of **salivary amylase** (α-amylase) that begins the breakdown of complex carbohydrates such as starch.

☞ On the right side of the face, neck, and chin, carefully remove the skin if you have not already done so. A muscle layer will tend to adhere, but push this layer back into place gently so as not to destroy the glands beneath. Now carefully remove the thin muscles back of the angle of the jaw and beneath the ear to uncover the **parotid gland.** Do not destroy any large blood vessels.

The triangular parotid (pa-rot′id; Gr. *para,* beside; + *ous,* ear) gland is broad, thin, and rather diffused,

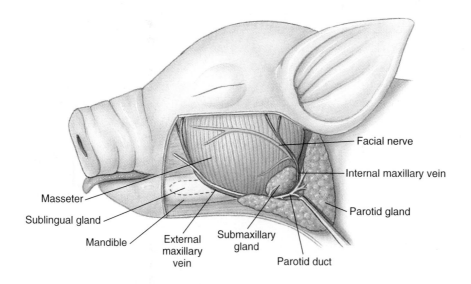

Figure 20-11

Dissection of the head and neck of a fetal pig to show some superficial veins, nerves, and salivary glands.

Reprinted by permission of Medical and Scientific Illustration/William C. Ober.

extending from almost the midline of the throat to the base of the ear (Figure 20-11). Do not confuse the salivary glands, which are choppy and lobed in appearance, with the lymph nodes, which are more smooth and shiny. The **parotid duct** comes from the deep surface of the gland and follows the ventral border of the masseter (cheek) muscle along the external maxillary vein to the corner of the mouth (Figure 20-11).

The **submaxillary (= mandibular) gland** lies under the parotid gland and just posterior to the angle of the jaw. It is darker, compact, and oval. Its duct comes from the anterior surface of the gland and passes anteriorly, medial to the mandible, and through the sublingual gland to empty into the floor of the mouth. This duct is very difficult to trace.

> To find the **sublingual glands,** remove the mylohyoid muscle and the slender pair of geniohyoid muscles immediately beneath it.

On each side a whitish, elongated sublingual gland is located between the diagastric muscle, which lies inside the mandible, and the genioglossus, which is one of the muscles of the base of the tongue. A sublingual artery and vein will be seen along the ventral side of each gland. The sublingual glands empty by way of several short ducts to the floor of the mouth.

Mouth Cavity and Pharynx

> Cut through the angle of the mouth on both sides with scissors. Cut posteriorly, pulling open the mouth as you proceed. Follow the angle of the tongue and do not cut into the roof of the mouth. Continue the cuts to the esophagus to expose the oral cavity and pharynx fully (Figure 20-12).

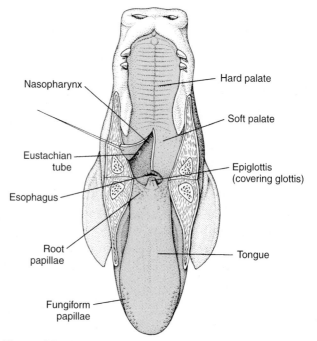

Figure 20-12

Oral cavity of the fetal pig.

The **teeth** may not be erupted yet, although the canines and third pair of incisors may be seen in older fetuses. The young pig will have three incisors, one canine, and four premolars on each side of each jaw.

> Remove the flesh along the right jaws and carefully cut away enough of the jawbone to expose the buds of the embryonic teeth.

The third incisors and the canines are the first to erupt; the second incisors are the last.

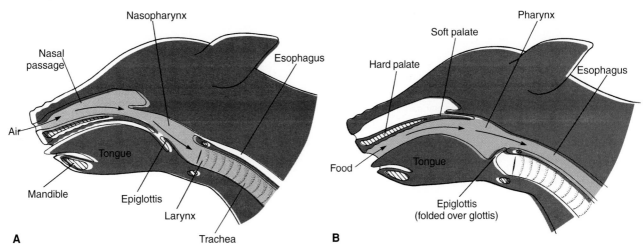

Figure 20-13

Relationship of respiratory passage to the mouth and esophagus, in breathing and swallowing. **A,** During breathing, the glottis is open to receive air from nostrils and is protected from food and saliva by the epiglottis. **B,** For swallowing, the larynx is pushed anteriorly, causing epiglottis to fold over the glottis, thus closing air passage to lungs. Feel your Adam's apple (larynx) as it moves up when you swallow.

The mouth cavity is roofed by a narrow, bony **hard palate,** sheathed ventrally with mucous membranes that are ridged into transverse folds. Extending posteriorly from the hard palate is the **soft palate** composed of thick membrane. The hard and soft palates of mammals completely separate the oral cavity from the air passages above, an innovation that allows a mammal to chew a mouthful of food at leisure while breathing freely through its nose. Only crocodilians among other vertebrates have a hard palate; the soft palate is unique to mammals.

Open the mouth wide, drawing down the tongue, to locate at the posterior end of the soft palate the opening into the **nasopharynx** (fair′inks; Gr. *pharyngx,* gullet), the space above the soft palate. It connects with the nasal passages from the nostrils.

☞ To expose the nasopharynx, make a midline incision of the soft palate. Locate the small openings on either side of the roof of the nasopharynx.

From these openings the **eustachian tubes** (named after B. Eustachio, an Italian physician) lead to the middle ear.

Posterior to the nasopharynx is the **laryngeal pharynx,** which connects the oral cavity with the **esophagus.** Both nasal and laryngeal pharynx are derived from the pharynx of ancestral chordates, that evolved as a filter-feeding apparatus. Pharyngeal (gill) pouches develop in this region in all vertebrate embryos. In fishes these pouches break through to develop into gill chambers, but in tetrapods they become transformed into other structures: middle ear cavity and glandular tissue (thyroid, parathyroid, and thymus).

The **larynx** lies in the floor of the laryngeal pharynx. Locate the flaplike **epiglottis,** which folds up over

the **glottis** (the open end of the larynx) to close it when food is being swallowed. Note that in the mouth the air passages are *dorsal* to the food passage. In the throat, however, the air is carried through the larynx and trachea, which are *ventral* to the food passage (esophagus). These passageways cross in the pharyngeal cavity (Figure 20-13). When the animal is respiring, the epiglottis fits up against the opening into the nasopharynx, allowing air into the larynx but preventing the entrance of saliva or food from the mouth. During swallowing the larynx is pushed forward, causing the epiglottis to fold over the glottis, thus opening the food passage while closing off the air passage.

☞ Continue your dissection of the neck region by making a midventral incision down the neck.

After parting the skin and clearing away some tissue around the larynx, you will expose the **thymus,** a large, soft, and irregular mass of glandular tissue lying lateral to the sternohyoid muscles (Figure 20-14). It is an extensive gland in the fetus and young animal but after puberty it decreases in size although continuing to function throughout life. The thymus extends caudally under the sternum with its posterior portion overlying the heart. The thymus is part of the body's lymphoid system and is filled with lymphocytes of all sizes and is especially rich in T cells, which are important in immunological responses.

As tissue is cleared from the larynx, find the **trachea** (windpipe) extending caudally from it. The trachea is stiffened by a series of **C**-shaped cartilage rings, which are incomplete dorsally where the trachea lies against the esophagus.

Tracing the trachea posteriorly, you will see the **thyroid gland,** a small, dark red, oval gland lying on the

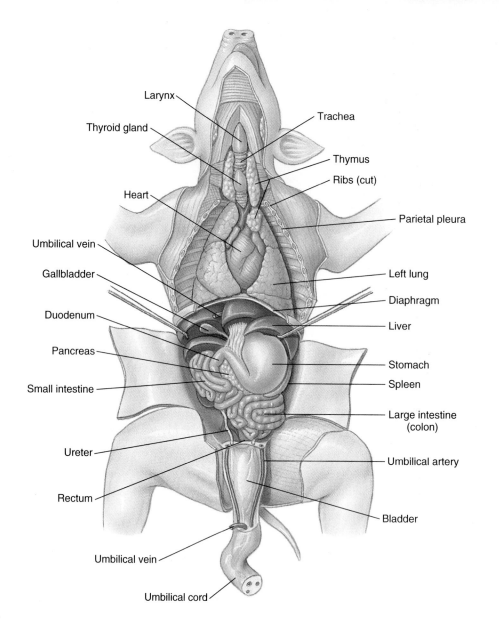

Figure 20-14

Internal anatomy of the fetal pig, ventral view. The first part of the small intestine (just past the stomach) is called the _____ . The liver has how many lobes? _____ What long reddish organ serves as both an important lymph organ and has important immunological function? _____

Reprinted by permission of Medical and Scientific Illustration/William C. Ober.

trachea beneath the sternothyroid muscles (Figure 20-14). The thyroid is an endocrine gland that produces thyroxin and triiodothyronine, two hormones that promote growth and development and regulate the metabolic rate.

The Abdominal Cavity

Directions for Dissection

To proceed further it is now necessary to expose the organs of the abdominal cavity. Place the pig ventral side up in the dissecting pan. Tie a cord or rubber band around one forelimb, loop the cord under the pan, and fasten it to the other forelimb. Do the same to the hindlegs.

☞ With a scalpel, make a midventral incision through the skin and muscles but not into the body cavity, continuing the incision already made in the neck posteriorly to within 1 cm of the umbilical cord (incision 1, Figure 20-15). Cut around each side of the cord (2). If your specimen is female, continue on down the midline from the cord to the anal region (3). If it is male, make two incisions, one on each side of the midline, to avoid cutting the penis, which lies underneath (3a). Now, in either sex, deepen the incisions you have made in the abdominal region through the muscle layer to reach the body cavity, taking care not to injure the underlying organs. With scissors,

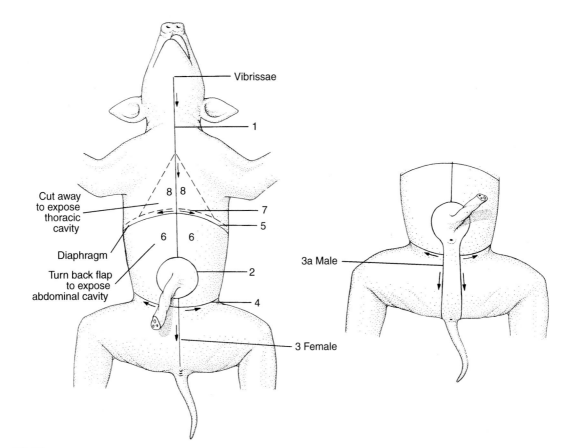

Figure 20-15

Cutting diagram. The numbers indicate the order in which each incision is to be made. *1* to *6* expose the abdominal cavity; *7* and *8* expose the thoracic cavity.

make two lateral cuts on each side, one just anterior to the hindlegs (4) and the other posterior to the ribs (5), and turn back the flaps of the body wall (6). Flush out the abdominal cavity with running water.

All visceral organs are invested in mesentery and held in place with connective tissue. Loosen this tissue carefully to separate organs, tubes, and vessels, being careful not to cut or tear them. *Do not remove any organs unless you are specifically directed to do so.* Be careful in all your preliminary dissection not to destroy blood vessels or nerves, to keep them intact for later dissection of the circulatory and nervous systems.

It is important to remember that instructions referring to the "right side" refer to the animal's right side, which will be on your left as the animal lies ventral side up in the dissection pan.

Notice that the umbilical cord is attached anteriorly by a tube, the **umbilical vein.**

☞ Tie a string around the umbilical vein in two places and sever the vein between the two strings.

The strings will identify this vein later. Lay the umbilical cord between the hindlegs and identify the following parts.

The **body wall** consists of several layers: (1) tough external **skin,** (2) two layers of **oblique muscle** and an inner layer of **transverse muscle** (try to separate the layers and determine the direction of the fibers), and (3) an inner lining of thin, transparent **peritoneum.**

The **diaphragm** is a muscular, dome-shaped partition separating the peritoneal cavity (abdominal cavity) from the thoracic cavity, which together constitute the coelom. *Do not remove the diaphragm.*

The peritoneum is the smooth, shiny membrane that lines the abdominal cavity and supports and covers the organs within it. That which lines the body walls is called the **parietal peritoneum.** It is reflected off the dorsal region of the body wall in a double layer to form the **mesenteries,** which suspend the internal organs, and then continues on around the organs as a cover, where it is called the **visceral peritoneum.**

The **liver** is a large, reddish gland with four main lobes lying just posterior to the diaphragm. The greenish, saclike **gallbladder** may be seen under one of the central lobes (See Figure 20-14).

The **stomach** is nearly covered by the left lobe of the liver. The **small intestine** is loosely coiled and held by mesenteries. Note the blood vessels in the mesentery that supports the digestive tract. The **large intestine** is compactly coiled on the left side posterior to the stomach.

The **spleen** is a long, reddish organ attached by a mesentery to the greater curvature of the stomach. The spleen contains one of the largest concentrations of lymphatic tissue in the body. It functions to phagocytize spent blood components and salvage the iron from hemoglobin for reuse. It is also immunologically important in initiating immune responses by B cells and T cells.

The **umbilical arteries** are two large arteries extending from the dorsal wall of the coelom to and through the umbilical cord (See Figure 20-14).

The **allantoic bladder,** the fetal urinary bladder, is a large sac lying between the umbilical arteries. It connects with the allantoic duct in the umbilical cord.

The **kidneys** are two large, bean-shaped organs attached to the dorsal wall dorsal to the intestines. They are *outside* the peritoneal cavity in the **cisterna magna** and are separated from the other abdominal organs by the peritoneum.

Thoracic Cavity and Neck Region

☞ With scissors, begin just anterior to the diaphragm and cut along the midventral line through the sternum, to a point midway between the forelegs. Keep the lower blade of the scissors up to prevent injuring the heart underneath. Now make a lateral cut on each side just anterior to the diaphragm. (See incision 7, Figure 20-15.) This exposes the thoracic cavity but leaves the diaphragm in place.

The **mediastinal septum,** which separates the right and left lung cavities, is a thin, transparent tissue attached to the sternal region of the thoracic wall (Figure 20-16).

☞ Separate the mediastinal septum carefully from the body wall. Now lift up one side of the thoracic wall and look for the small **internal thoracic artery** and **vein** (also called sternal or mammary) embedded in the musculature of the body wall. Carefully separate these vessels on each side and lay them down over the heart and lungs for future use. Now you may cut away some of the ventral thoracic wall (incision 8, Figure 20-15) to allow a better view of the thoracic cavity containing the left and right **lungs** and the **heart.**

The **pleura** (Gr., side) is the name given to the peritoneum that lines each half of the thoracic cavity and covers the lungs (Figure 20-16). The peritoneum lining the thoracic cavity is the **parietal** (L. *paries,* wall) **pleura;** the part applied to the lungs is the **visceral** (L., bowels) **pleura.** The small space between is the **pleural cavity,** in which lubricating **pleural fluid** prevents friction. The portions of the parietal pleurae on the medial side next to the heart are called the **mediastinal pleurae.** The **mediastinum** is the region between the

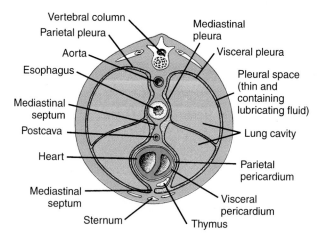

Figure 20-16
Diagrammatic transverse section through the thorax in the region of the ventricles to show the relations of the pleural and pericardial membranes, mediastinum, and lung cavities.

mediastinal pleurae. It contains the pericardium and heart and the roots of the big arteries and veins, as well as the trachea, esophagus, part of the thymus, and other parts.

The double-walled **pericardium** enclosing the heart is made up of an outer **parietal pericardium** and a **visceral pericardium** applied to the heart with pericardial fluid in the space between.

The Digestive Tract

The digestive system consists of the alimentary canal, extending from mouth to anus, and glands such as salivary glands, liver, and pancreas, that assist in its function of converting food into a form that can be assimilated for growth and energy requirements. You have already studied the anterior portions of the alimentary canal: mouth cavity and salivary glands. We will now consider the digestive tract proper, beginning with the esophagus.

The **esophagus** is a soft, muscular tube that leads from the pharynx to the stomach. Locate it in the neck region posterior to the larynx, where it is attached to the dorsal side of the trachea by connective tissue. Find the esophagus in the thoracic cavity posterior to the lungs, and in the abdominal cavity find where it emerges through the diaphragm at the cardiac end of the stomach. The muscles at the anterior end of the tube are striated (voluntary), gradually changing to smooth muscle. How does this affect swallowing?

The **stomach** (See Figure 20-14) is a large, muscular organ that breaks up food and thoroughly mixes it with gastric juice. Identify its **cardiac end** near the heart, its **pyloric end** that joins the duodenum, its **greater curvature** where the spleen is attached, and its **lesser curvature.** The **fundus** (L., bottom) is the anterior blind pouch. The contents of the fetal digestive tract, made green by pigments in the bile salts, are called **meconium** and contain epithelium sloughed from the mucosa lining, sebaceous secretions, and amniotic fluid swallowed by the fetus. Open the stomach longitudinally, rinse out, and find (1) the **rugae,** or folds, in its walls; (2) the opening from the esophagus; and (3) the **pyloric** (Gr. *pylōros,* gatekeeper) **valve,** which regulates the passage of food into the duodenum.

The **small intestine** includes the **duodenum** (doo-uh-dē′num; L., "twelve-each," referring to its length in humans, which equals about 12 finger-widths), or first portion, which lies next to the pancreas and receives the common bile duct and pancreatic duct, and the **jejunum** and **ileum,** indistinguishable in the fetal pig, which make up the remainder of the small intestine.

☞ Remove a piece of the intestine, open it, and examine it *under water* with a hand lens or dissecting microscope. Observe the minute, fingerlike **villi** (pl. of **villus,** L., shaggy hair), which greatly increase the absorptive surface of the intestine.

Most of the digestion and absorption take place in the small intestine.

The **large intestine** includes the long, tightly coiled **colon** and the straight, posterior **rectum,** which extends through the pelvic girdle to the **anus.** Its primary function is the absorption of water and minerals from the liquified chyme that enters it.

Find the **cecum,** which is a blind pouch of the colon at its junction with the ileum. In humans and the anthropoid apes the cecum has a narrow diverticulum (a tube, blind at distal end) called the **vermiform** (L., worm-shaped) **appendix.** Open the cecum (on its convex side opposite the ileum) and note how the entrance of the ileum forms a ring-shaped **ileocecal valve.** The posterior end of the rectum will be exposed in a later dissection.

Digestive Glands. The **liver,** a large, brownish gland posterior to the diaphragm, has four main lobes: the left and right lateral lobes and the left and right central lobes. One of the many important functions of the liver is the production of **bile,** a fluid containing bile salts, which are steroid derivatives responsible for the emulsification of fats. Bile is stored and concentrated in the **gallbladder,** a small greenish, oval sac embedded in the dorsal surface of the right central lobe of the liver. The liver is connected to the upper border of the stomach by a tough, transparent membrane, the **gastrohepatic ligament,** in which are embedded blood vessels (in the left side) and ducts (in the right side). Carefully loosen the gallbladder and note its tiny **cystic duct.** This unites with **hepatic ducts** from the liver to form the **common bile duct,** which carries bile to the duodenum. Probe the gastrohepatic ligament and adjoining liver tissue carefully to find these ducts. Do not injure the blood vessels lying beside them.

The **pancreas** is a mass of soft glandular tissue in the mesentery between the duodenum and the end of the stomach. Push the small intestine, except the duodenum, to the left to explore the gland. Its pancreatic juice is carried by a **pancreatic duct** to the pyloric end of the duodenum. The pancreas is a double gland having both endocrine and exocrine portions. Its endocrine portion produces two hormones, insulin and glucagon, that are of great importance in carbohydrate and fat metabolism. The exocrine portion secretes the pancreatic juice—a mixture of water, electrolytes, and enzymes. The enzymes include carbohydrases, which digest sugars and starches; lipases, which split lipids; and proteases, which break down proteins. The pancreas is the only source of lipases in the digestive system.

Other digestive juices are secreted by the **mucosa lining** of the stomach and of the small intestine.

Histological Study of the Intestine. Examine a cross section of human or other mammalian small intestine and compare it with the histological structure of amphibian intestine. Examine mammalian liver and pancreas slides.

Identification

Be able to locate and give the functions of the parts of the digestive system.

EXERCISE 20D
The Urogenital System

Urinary System

☞ Read the directions carefully and dissect cautiously. Do not tear or remove any organs, blood vessels, or ducts. Instead, separate them carefully from the surrounding tissues. You will dissect the urogenital system of only one sex, then exchange your dissected specimen with another student who has dissected the opposite sex. Therefore, prepare your specimen with the care you would give to a demonstration dissection.

The urinary system consists of a pair of kidneys, a pair of ureters, a urinary bladder, and a urethra (shared with the reproductive system in the male).

The fetal **urinary bladder** is the **allantoic** (Gr. *allas,* sausage) **bladder,** a long sac located between the umbilical arteries (Figure 20-17). It narrows ventrally to form the **allantoic duct,** which continues through the umbilical cord and is the fetal excretory canal. The bladder narrows dorsally to empty into the **urethra,** the adult excretory canal. The urethra will be dissected later. After birth the allantoic end of the bladder closes to form the urinary bladder.

The **kidneys** are dark and bean shaped. They lie outside the peritoneum on the lumbar region of the

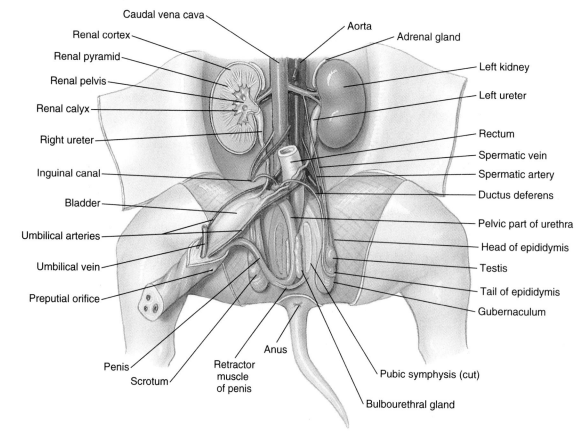

Figure 20-17

Male urogenital system of the fetal pig. The right kidney has been longitudinally sectioned to show internal structure.

Reprinted by permission of Medical and Scientific Illustration/William C. Ober.

dorsal body wall. Uncover the right kidney carefully. A depression on the median side of each kidney is called the **hilus.** Through it pass the renal blood vessels and the **ureter,** or excretory duct. Follow the left ureter posteriorly to its entrance into the bladder. Be careful of small ducts and vessels that cross the ureter. (Note the small **adrenal gland,** an endocrine gland lying close to the medial side of the anterior end of the kidney and embedded in fat and peritoneum.)

☞ Slit open the right kidney longitudinally, cutting in from the outer border. Remove the ventral half of the kidney and lay it in a dish of water.

Study with a hand lens or dissecting scope. Identify the **cortex** (L., bark), or outer layer, containing the microscopic renal corpuscles; the **medulla** (L., marrow, pith), or deeper layer, containing the radially arranged blood vessels and collecting tubules; and the **renal pyramids,** which contain groups of collecting tubules coming together to empty through **papillae** into the **pelvis.** The pelvis is a thin-walled chamber that connects with the ureter. Divisions of the pelvis into which the papillae empty are referred to as **calyces** (ka′luh-sez; sing. **calyx**) (Figure 20-17).

☞ For a description of the anatomy and physiology of excretion, read your textbook. Examine the demonstration specimens of sheep kidneys on display. You will be expected to understand the structure and functioning of the kidney and its functional unit, the nephron.

Male Reproductive System

The location of the testes in the fetus depends on the stage of development of the fetus. Each testis originates in the abdominal cavity near the kidney. During the development of the fetus a prolongation of the peritoneum, the **processus vaginalis,** grows down into each half of an external pouch, the **scrotum.** Later the testis "descends" into the scrotum through the **inguinal canal,** where it lies within the sac, or sheath, formed by the processus vaginalis. The inguinal canal is the tubular passage connecting the abdominal cavity with the scrotal sac (Figure 20-17). The descent of both testes into the scrotum is usually completed shortly before birth. Part of the cutaneous scrotum can be seen ventral to the anus.

Lay the umbilical cord and allantoic bladder between the legs and locate the **urethra** (from Gr.

ouron, urine) at the dorsal end of the bladder. The urethra bends dorsally and posteriorly to disappear into the pelvic region. Find the sperm ducts (pl., **vasa deferentia;** sing., **vas deferens**), two white tubes that emerge from the openings of the **inguinal** (L., groin) **canals,** cross over the umbilical arteries and ureters, and come together medially to enter the urethra. Also emerging from each inguinal ring are the spermatic artery, vein, and nerve. Together with the vas deferens these make up the **spermatic cord,** which leads to the testis.

Now lay the bladder up over the abdominal cavity and locate the thin, hard, cordlike **penis** under the strip of skin left posterior to the urogenital opening. The penis lies in a sheath in the ventral abdominal wall, ending at the **urogenital opening.**

☞ Carefully separate the penis from surrounding tissue, then cut away the skin and muscle that covered and surrounded it. Complete the removal of any skin remaining on the inside of the thigh and rump, and carefully separate away the underlying fascia. The thin-walled scrotal sac extends posteriorly across the ventral surface of the high muscles toward the cutaneous scrotum. Free the left scrotal sac from the surrounding tissues.

Pass a probe through the inguinal canal into the processus vaginalis, which houses the **testis.** The testis is a small, hard oval body containing hundreds of microscopic **seminiferous tubules** in which the sperm develop.

☞ Cut open the left scrotal sac to expose the testis.

The seminiferous tubules of the testis unite into a much-coiled **epididymis** (Gr. *epi,* upon; + *didymos,* testicle). The epididymis begins as a whitish lobe on the anterior surface of the testis and passes posteriorly around one side of the testis to its caudal end, where it unites with the **vas deferens** (sperm duct). The vas deferens (L. *vas,* vessel; + *deferre,* to carry off) passes cranially through the inguinal canal, loops over the ureter, and enters the urethra, as already seen. A fibrous cord attaches the testis and the epididymis to the posterior end of the processus vaginalis. It is called the **gubernaculum** (L., rudder). A narrow band of muscle (the cremaster) runs along the lateral and posterior part of the processus vaginalis parallel with the vas deferens.

☞ Separate the tissues on each side of the penis in the pelvic region; then, being careful not to injure the penis or cut too deeply, use a scalpel to cut through the cartilage of the pelvic girdle.
Spread the legs apart to expose the **urethra** and its connection with the penis.

The urethral canal extends throughout the length of the penis and serves as a common duct for both sperm and urine.

☞ Now, beginning at its juncture with the bladder, follow the urethra posteriorly and locate these **male glands:**

Seminal Vesicles. The seminal vesicles are a pair of small glands on the dorsal side of the urethra. These glands contribute a secretion to the semen that is rich in fructose, a six-carbon sugar that stimulates previously inactive but mature spermatozoa to become highly motile. In humans, the secretion of the seminal vesicles makes up more than 60% of the volume of the semen.

Prostate Gland. The prostate gland is poorly developed in the fetus; it lies between and often partly covered by the seminal vesicles, but may be difficult to find. The alkaline secretions of the prostate assist in neutralizing acids normally present in the urethra, as well as in the vagina of the female. In humans, the prostate secretion is known to contain a compound that may help to prevent urinary tract infections in males.

Bulbourethral Glands (= Cowper Glands). The bulbourethral glands are a pair of narrow glands about 1 cm long on each side of the urethra near its junction with the penis (Figure 20-17). These glands add a thick, sticky alkaline mucus to the semen that has lubricating properties.

Identification

Be able to follow the path of urine and of sperm to the outside. Compare the male urogenital system of the pig with that of a frog and with that of a human.

Female Reproductive System

The **ovaries** are small, pale organs lying just posterior to the kidneys (Figure 20-18). Each is suspended by a mesentery, the **mesovarium** (Gr. *mesos,* middle; + L. **ovarium,** ovary), that can be seen extending between the kidney and the ovary. The **uterus** (L., womb) is Y-shaped. The **horns of the uterus** (the arms of the Y) extend from the ovaries to unite medially at the **body of the uterus,** which leads to the **vagina** (L., sheath). It is in the horns of the uterus, not the body of the uterus, that the fetal pigs develop. (Most mammals have a similar Y-shaped uterus, called **bicornuate** ["double-horned"], but in the higher primates, including humans, the two uterine horns are completely fused, a condition called **simplex.**) Each uterine horn is suspended by a mesentery, the **broad ligament.** Follow the uterine horn anterior to the ovary where it gives rise to a highly convoluted **oviduct** (also called **fallopian tube** after G. Fallopio, Italian anatomist). The oviduct coils around the ovary and terminates at a wide, ciliated funnel, the **infundibulum** (L., funnel). In adult pigs, eggs released

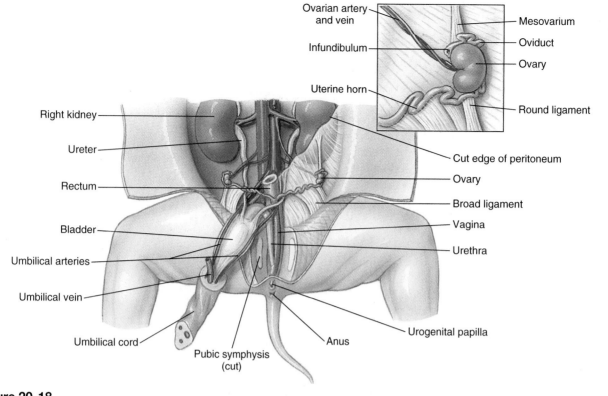

Figure 20-18

Female urogenital system of the fetal pig. What structures serve as excretory ducts for the kidneys? _____ What structure serves to drain the bladder? _____ Embryonic pigs develop in what paired structures? _____

Reprinted by permission of Medical and Scientific Illustration/William C. Ober.

from the ovary at ovulation are swept into the opening **(ostium)** of the infundibulum by ciliary currents.

To expose the rest of the reproductive system the pelvis must be cut open.

☞ Lay the allantoic bladder anteriorly over the abdominal viscera. Cut through the muscle medially between the legs and through the pelvic girdle. Be careful not to cut through the urethra or vagina. Spread the legs apart and separate the urethra from the surrounding tissue. (Note where the ureters join the dorsal end of the bladder.)

Lay the bladder and urethra to one side. Follow the body of the uterus posteriorly to a slight constriction called the **cervix.** From here the tube widens and is called the **vagina.** The vagina and urethra soon join to form a **urogenital sinus,** which is a short common passageway for the two systems.

The **vulva** is the external opening of the urogenital sinus, ventral to the anus. The ventral side of the vulva extends out to form a pointed **genital papilla.** A small, rounded **clitoris** may be seen extending from the ventral floor of the urogenital sinus. This is not always evident.

In adults, at copulation, the male penis places the spermatozoa, contained in a seminal fluid, into the vagina. The sperm must pass through the uterus to the oviduct to fertilize the egg. After fertilization the zygote passes down to the horn of the uterus to develop. Note how the horns are adapted for carrying a litter. How does this compare with the human uterus? _____ How is the developing fetus nourished? _____ How are waste products from the fetus disposed of? _____

The placenta of the pig is known as a **chorioallantoic,** or **diffuse, placenta** (Figure 20-19). In the uterus of the dog and some other carnivores the placenta is called a zonary placenta because the chorionic villi are located in a girdlelike zone, or band, around the middle of each pup rather than over the whole surface of the chorion as in the pig. The human placenta is disc shaped.

Identification ◀

Be able to trace the path of the unfertilized egg from the ovary to the uterus. Compare with the frog.

Be able to trace the path of urine in the female. Are any parts shared by both urinary and reproductive systems?

If you have not studied the male system, trade your specimen for a male and make a thorough study of the male system.

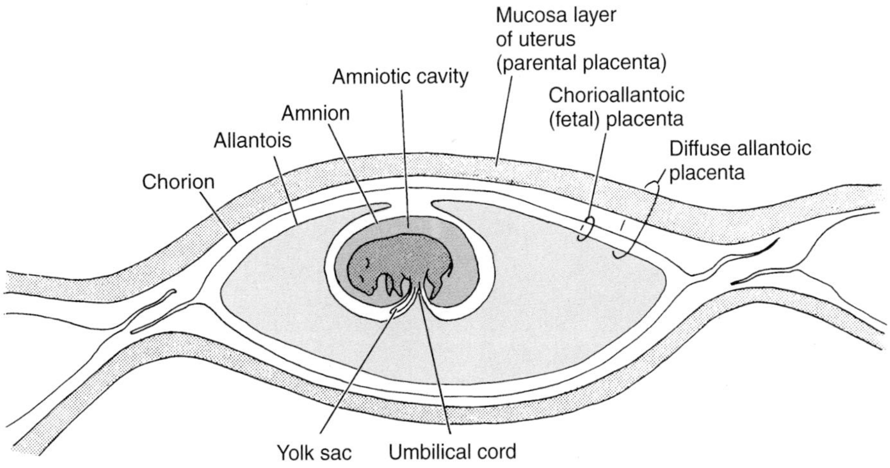

Figure 20-19
Diagram of pig fetus in utero, showing the relationship of fetal and parental membranes in a diffuse placenta.

Oral Report

Be able to identify and give the functions of the reproductive organs of both the male and female pig.

Written Report

On separate paper write a comparison of the mammalian and amphibian reproductive systems as illustrated by your study of the pig and frog. How is each adapted to its own type of reproduction?

EXERCISE 20E
The Circulatory System

The circulatory system of the pig is quite similar to that of other mammals, including humans. In contrast to the single-circuit system of the fish, with its two-chambered heart, and the incomplete double circuit of the three-chambered amphibian heart, the mammal has an effective four-chambered heart, which allows the blood two complete circuits: a **systemic** circuit through the body, followed by a **pulmonary** circuit to the lungs for oxygenation. The right side of the heart receives the oxygen-poor blood returning from the body tissues and pumps it to the lungs; the left side receives the oxygen-rich blood returning from the lungs and pumps it to the body tissues.

The study of fetal pig circulation has the added advantage of illustrating not only typical mammalian circulation but also typical **fetal circulation.** The changes in circulation necessary for the transition from a non-breathing, noneating fetus to an independent individual with a fully independent circulation are both crucially important and elegantly simple.

Uninjected vessels will contain only dried blood (or they may be empty) and therefore will be fragile, flat-tened, and either brown or colorless. If injected, the arteries will have been filled with latex or a starchy injection mass through one of the arteries in the cut umbilical cord and will be firm and pink or yellow. The veins will have been injected through an external jugular vein in the neck and will be blue. Sometimes, however, the injection medium does not fully penetrate the vasculature. The lymphatic system will not be studied.

You should uncover and separate the vessels with a blunt probe and trace them as far into the body as possible, but be careful not to break or remove them. Nerves often follow an artery and vein and will appear as tough, shiny, white cords. Do not remove them. You may in fact find it efficient to identify the major nerves at this time as you find them. As you identify a vessel, separate it and carefully remove investing muscle and connective tissue, taking care not to break the vessel or destroy other vessels that you have not yet identified. Do not remove any body organs unless specifically directed to do so.

As it is often difficult to trace the arterial system without damaging the venous system, which lies above it, you will study the veins first. However, since corresponding arteries and veins usually lie side by side, it is often convenient to study both systems at the same time.

Keep in mind that there is considerable variation among individual pigs in the points of vessel bifurcation, especially in the venous system of the neck and shoulder region. *The venous arrangement in your pig almost certainly will not look exactly like the manual illustrations.* Make notes or sketches of any variations that you find in your specimen. With careful dissection, both veins and arteries can be left intact.

Heart

Note carefully the shape and slope of the diaphragm and how it forms the posterior boundary of the thoracic cavity. Then cut the diaphragm

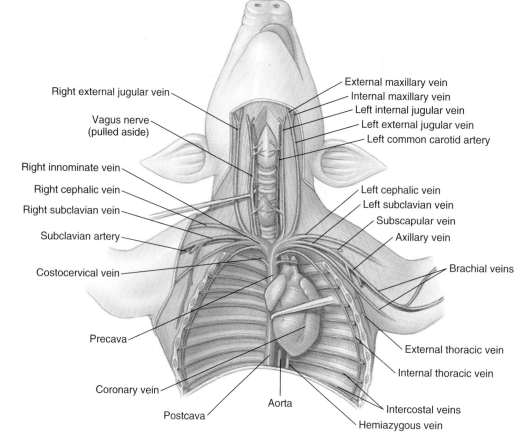

Figure 20-20

Veins of head, shoulders, and forelimbs of the the fetal pig. The internal thoracic veins, shown laterally on the opened thoracic cavity, actually lie ventral to the heart.

Reprinted by permission of Medical and Scientific Illustration/William C. Ober.

away from the body wall to make entrance into the thoracic cavity more convenient.

Open the pericardial sac and examine the heart. It has two small thin-walled atria and two larger muscular-walled ventricles.

Right Atrium (Anterior and Ventral). Lift the heart (Figure 20-20) and see the precaval and postcaval veins that empty into the right atrium. The **postcava** from the abdominal region emerges through the diaphragm; the **precava** comes through the space between the first ribs (Figure 20-20).

Left Atrium (Anterior and Dorsal). The left atrium receives the **pulmonary veins** on the dorsal side. Note the conspicuous earlike right and left auricles (L. *auricula,* ear) lying on each side of the heart. The term "auricle" is often incorrectly used as a synonym for "atrium"; however, the term should be reserved for the earlike flap that protrudes from each atrium.

Right Ventricle. The right ventricle is large and thick walled. The **pulmonary artery** leaves the right ventricle and passes over the anterior end of the heart

to the left, where it divides back of the heart to go to the lungs.

Left Ventricle. The left ventricle is larger and covers the apex of the heart (posterior). It gives off the large **aorta,** which rises anteriorly just behind (dorsal to) the pulmonary artery. The **coronary sulcus** is the groove on the surface of the heart between the right and left ventricles. It contains the **coronary artery** and **vein,** which supply the tissues of the heart itself.

General Plan of Circulation

In mammalian circulation (after birth) the systemic and pulmonary systems of circulation are separate. Blood from all parts of the body except the lungs returns by way of the large **precava** (anterior vena cava) and **postcava** (posterior vena cava) to the **right atrium.** From there it goes to the **right ventricle** to be pumped out through the **pulmonary arteries** to the lungs to be oxygenated. The blood returns through the **pulmonary veins** to the **left atrium** and then to the **left ventricle,** which sends it through the **aorta** to branches that carry the blood finally to the capillaries of all parts of the body. The venous system returns it again to the right side of the heart to begin another circuit. The mammal has a **hepatic portal system** but no renal portal system as in the amphibian.

Before birth, when the lungs are not yet functioning, the fetus depends on the placenta for nutrients and oxygen, which are brought to it through the **umbilical vein.** Therefore the general plan of circulation is modified in the fetus so that most of the pulmonary circulation is short-circuited directly into the systemic bloodstream to be carried to the placenta instead of the lungs. Only enough blood goes to the lungs to nourish the lung tissue until birth. These modifications will be mentioned as they arise in later descriptions.

Veins of the Head, Shoulders, and Forelimbs

☞ Carefully remove the remainder of the sternum and the first rib without damaging the veins beneath.

You will find considerable variation in the veins of individual specimens.

The precaval division of the venous circuit carries blood from the head, neck, thorax, and forelimbs to the heart.

Begin by locating the **internal jugular vein** found lying along the trachea and adjacent to the common carotid artery and the vagus nerve (Figure 20-20). It drains the brain, larynx, and thyroid.

Trace the internal jugular vein posteriorly to its confluence with the **external jugular vein** and **subclavian vein;** these join to form the **innominate vein.** Follow the two innominate veins (right and left) through the opening at the level of the first rib into the chest cavity. Here they join to form the **precaval vein** (anterior vena cava), which enters the right atrium. The precava also receives at this level a pair of **internal thoracic veins** lying on either side of the sternum. This is the vein that, along with the accompanying artery, you detached from the ventral muscle wall of the chest cavity (Exercise 20C). You will also find its continuation in the abdominal muscle wall.

Another vessel entering the precava dorsolaterally close to the heart is the **costocervical trunk.** It receives several veins from the neck and dorsal thorax.

Lift the heart and left lung and push them to the right. The unpaired **hemiazygous (= azygous) vein** lies in the chest cavity along the left side of the aorta and receives from the left and right intercostal veins. Follow the hemiazygous anteriorly to the point at which it crosses the aorta and goes under the heart. Lift the heart up to see the vein cross the pulmonary veins and enter the right atrium along with the postcava.

Return now to the neck region and locate the **external jugular vein.** This vein was probably used to inject the venous system and will be tied and severed on one side of the neck. The external jugular is formed by the union of the **internal maxillary** (dorsal) and the **external maxillary** (ventral) near the angle of the jaw. These vessels drain superficial parts of the head.

Joining the base of the external jugular (or sometimes the subclavian) is the **cephalic vein,** a large superficial vein from the arm and shoulder. It lies just beneath the skin.

The **subclavian vein** extends from the shoulder and forelimb. It is a deeper vein that follows the subclavian artery. (There may be from one to three subclavian or brachial veins. If there are more than one, there may be considerable anastomosing [networklike branching and rejoining of vessels] among them.) The subclavian, as it continues into the arm, is called the **axillary** (L. *axilla,* armpit) in the armpit and the **brachial** (Gr. *brachion,* upper arm) in the upper arm. Large nerves of the brachial nerve plexus overlie these veins.

Arteries of the Head, Shoulders, and Forelimbs

Arteries carry blood away from the heart, and, with the exception of the pulmonaries, all branch from the main artery, the aorta.

The **aorta** begins at the left ventricle (ascending aorta), curves dorsally (aortic arch) behind the left lung, and extends posteriorly along the middorsal line (descending, or dorsal, aorta). The first two branches are the **brachiocephalic trunk** and, just to the left of it, the **left subclavian artery** (Figure 20-21).

Brachiocephalic Trunk. The brachiocephalic is a large single artery that extends anteriorly a short distance and branches into the carotid trunk and the right subclavian artery.

The **carotid trunk** may extend anteriorly for as much as a centimeter or it may divide at once into the **left** and **right common carotid** arteries, which form a Y over the trachea and extend up each side of it. Each common carotid artery follows an internal jugular vein and the vagus nerve toward the head, giving off branches to the esophagus, thyroid, and larynx. Near the head each common carotid gives off an **internal carotid,** a deep artery passing dorsally to the skull and brain and an **external carotid,** which is a continuation of the common carotid that supplies the tongue and face.

The **right subclavian artery** arises from the brachiocephalic and continues into the right arm as the **axillary** in the armpit, the **brachial** in the upper arm, and the **radial** and **ulnar** in the lower arm. The **subscapular** extends to the deep muscles of the shoulder and branches off the axillary part of the artery.

Left Subclavian Artery. The left subclavian arises directly from the aorta. Otherwise it is similar to the right subclavian artery.

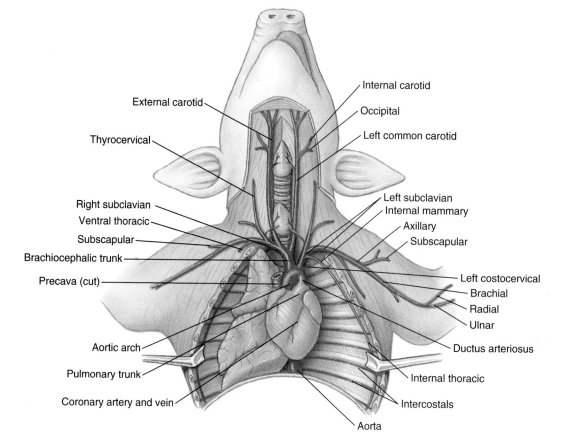

Figure 20-21

Arteries of head, shoulders, and forelimbs of the fetal pig. The internal thoracic arteries, shown laterally, actually lie ventral to the heart. What pair of arteries supply blood to the tongue and face? _____ What artery supplies blood to the thyroid and parotid glands? _____ Near the head the common carotid artery splits into what two arteries? _____

Reprinted by permission of Medical and Scientific Illustration/William C. Ober.

Each of the subclavian arteries gives off several branches, which include the **internal thoracic artery** and the **thyrocervical artery.** The internal thoracic artery (also called the mammary or sternal artery and, in humans, an artery used in coronary bypass operations) is the artery you detached earlier. It supplies the ventral muscular wall of the thorax and abdomen.

The thyrocervical artery (= internal cervical) arises from the subclavian at the same level as the internal thoracic. It supplies the thyroid and parotid glands and some of the pectoral muscles.

Pulmonary Circulation

Pulmonary Trunk. The pulmonary arises from the right ventricle, ventral to the aorta. Follow it as it branches into **right** and **left pulmonary arteries,** which carry oxygen-poor blood to the lungs. Scrape away some lung tissue to find branches of these vessels.

Pulmonary Veins. Pulmonary veins empty oxygen-rich blood into the left atrium.

☞ Probe gently under the left atrium to expose the veins.

You should be able to find the large trunk entering the heart just under the hemiazygous vein and a pair of vessels servicing each lobe of the lungs.

Fetal Shortcuts

Foramen Ovale. This is a fetal opening in the wall between the right and left atria. Part of the blood from the right atrium can pass through the foramen ovale directly to the left atrium, where it can go to the left ventricle, to the aorta, and back into the systemic circulation without going to the lungs at all (Figure 20-22). The remainder of the blood enters the pulmonary trunk.

Ductus Arteriosus. This is the short connection between the pulmonary trunk and the aorta. Trace this connection, which begins where the smaller pulmonary arteries branch off toward the lungs. Part of the blood from the right ventricle goes to the lungs, and part is shunted through the ductus arteriosus to the aorta (Figure 20-22).

Ductus Venosus. This is a third fetal shortcut that connects the umbilical vein with the postcava, passing through a channel in the liver tissue. During fetal life the anterior part of the postcaval vein carries a mixture

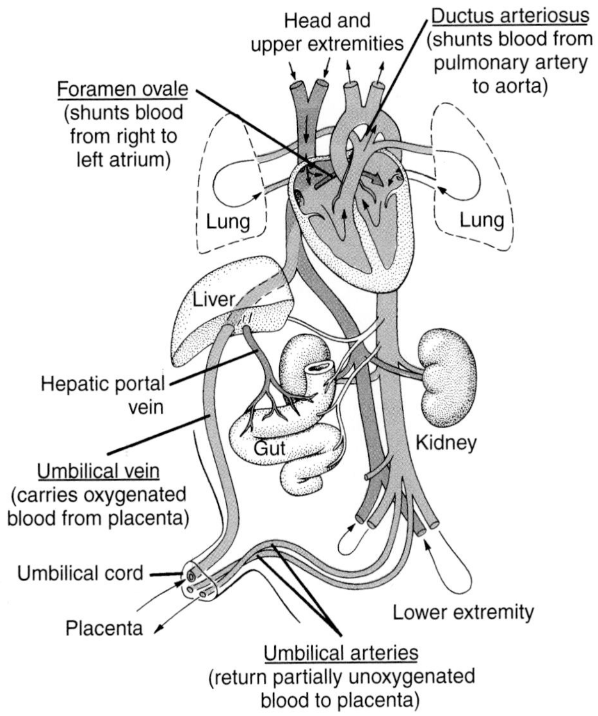

Figure 20-22
Scheme of the fetal circulation.

of oxygen-poor blood from body tissues and oxygen-rich blood (and nutrients) from the placenta by way of the umbilical vein (the vein tied and cut earlier). After birth both the umbilical vein and the ductus venosus degenerate.

☞ You will be able to see the arrangement of the vessels in the liver tissue by using the probe to scrape or "comb out" the liver tissue to separate it from the vessels. Wash out the loose tissue.

Some of the vessels you see will probably be hepatic ducts.

Changes in Circulation at Birth

Two crucially important events happen at the moment of birth: (1) the placental bloodstream upon which the fetus has depended is abruptly cut off, and (2) the pulmonary circulation immediately assumes the task of oxygenating the blood. One of the most extraordinary aspects of mammalian development is the perfect preparedness of the circulatory architecture for this event.

When the newborn piglet (or human) takes its first breath, the vascular resistance through the lungs is suddenly lowered as the lungs expand. The ductus arteriosus functionally constricts almost immediately, and the lungs receive full blood circulation. Blood now returns from the pulmonary veins to the left atrium, raising the pressure in this chamber enough to close the flaplike valve over the foramen ovale. With blood no longer

passing between the right and left atrium, the foramen ovale gradually grows permanently closed, leaving a scar called the **fossa ovalis.** The ductus arteriosus also closes permanently, becoming a ligament between the pulmonary artery and the aortic arch.

With severing of the umbilical cord, flow through the umbilical vein and arteries ceases immediately and these vessels are eventually reduced to fibrous cords.

Postcaval Venous Circulation

The **postcava** carries blood to the heart from the hindlimbs and trunk. Follow it from the right atrium through the intermediate lobe of the lung and through the diaphragm to the liver (Figure 20-23). Push the intestines to the left side, clean away most of the dorsal portion of the liver, and note where the **postcava** emerges and continues posteriorly. Clean away the peritoneum that covers the postcava, and find the following veins that enter it on each side.

One or two **renal veins** extend from each kidney. Do both right and left veins enter at the same level? The narrow, band-shaped adrenal gland (= suprarenal gland) lying against the anteromedial border of each kidney is drained by a small **adrenal vein** that may enter the renal vein or may empty directly into the postcava. It is accompanied by the adrenal artery. Just posterior to the adrenal vein and at about the same level as the renal veins, find the **parietal vein** extending to the body wall, accompanied by the parietal artery. The **genital vein** and **artery** serve the testes and ovaries. In the male the vein is called the **spermatic vein** and is incorporated together with the spermatic artery into the spermatic cord to the testis. In the female it is called the **ovarian vein;** it lies suspended in the mesovarium along with the ovarian artery.

Follow the postcava posteriorly to its bifurcation into the **common iliac veins.** These in turn divide into the **internal** and **external iliac veins.** The internal iliac vein (hypogastric) is the medial branch extending dorsally and posteriorly into the deep tissue and draining blood from the rectum, bladder, and gluteal muscles. The external iliac vein is the lateral branch and the largest vein entering the common iliac. It drains the foot and leg and is formed by the union of the large **femoral vein,** which extends ventrally toward the knee, and the **deep femoral vein,** which extends dorsally and posteriorly into the deep muscles of the thigh. It also receives a lateral branch, the **circumflex iliac vein,** which drains the muscles of the abdomen and the upper thigh.

Hepatic Portal System

The hepatic portal system is a series of veins that drain the digestive system and spleen. The blood is collected in the **hepatic portal vein,** which carries it to the liver capillaries. As the blood laden with nutrients from the

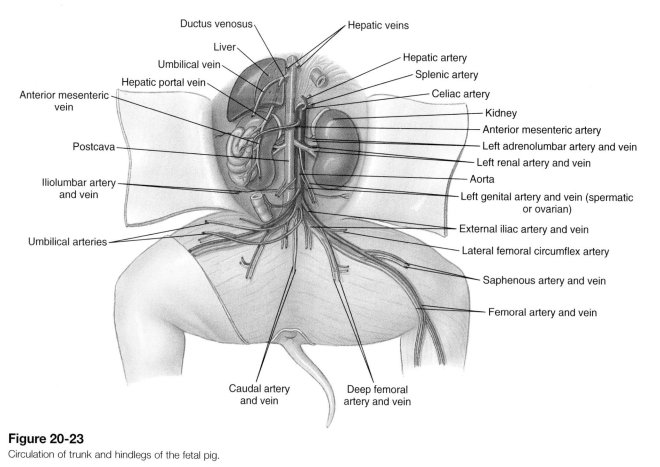

Figure 20-23

Circulation of trunk and hindlegs of the fetal pig.

Reprinted by permission of Medical and Scientific Illustration/William C. Ober.

intestine passes through the liver, the liver may store excess sugars in the form of glycogen, or it may give up sugar to the blood if sugar is needed. Some of the amino acids formed by protein digestion may also be modified here. Blood from the liver capillaries is collected by the **hepatic veins,** short veins that empty into the postcava.

The hepatic portal system in your specimen may not be injected. Unless these vessels are injected or are filled with dry blood, they may be difficult to locate. Try to examine an injected specimen.

Lift up the liver, stomach, and duodenum and draw the intestine posteriorly to expose the pancreas. Loosen the pancreatic tissue from the hepatic portal vein that runs through it. Now lay the duodenum over to your right and see how the vein enters the lobes of the liver near the common bile duct (in the gastrohepatic ligament). Lay the small intestine over to your left and fan out the mesenteries of the small intestine to see how the veins from the intestines are collected into the **anterior mesenteric vein** (Figure 20-23). This joins with the small **posterior mesenteric vein** from the large intestine and with veins from the stomach and spleen to form the **hepatic portal vein.**

Arteries of the Trunk and Hindlegs

Descending Aorta. The descending aorta follows the vertebral column posteriorly, first lying dorsal and then ventral to the postcava (Figure 20-23). After passing through the diaphragm, the aorta gives rise to a single large artery, the **celiac artery.** To find it, clip away the diaphragm and remove the tissue around the aorta at the anterior end of the abdominal cavity. The celiac artery divides almost immediately into arteries serving the stomach, spleen, pancreas, liver, and duodenum.

Just posterior to the celiac find another unpaired artery, the **anterior mesenteric artery.** This vessel must be dissected from the surrounding tissue. It sends branches to the pancreas, small intestine, and large intestine.

Locate next the paired **parietal arteries, renal arteries,** and **spermatic** or **ovarian arteries.** Posterior to these is the single **posterior mesenteric artery,** which divides and sends branches to the colon and rectum. It will probably have been broken off during earlier dissection of the abdominal cavity.

At its posterior end, the aorta now divides into two large lateral **external iliac arteries** to the legs, two medial **internal iliac arteries** to the sacral region, and a small continuation of the aorta into the tail, the **caudal artery.** Trace one of the external iliac arteries. It first gives off a lateral **circumflex artery** that supplies some of the pelvic muscles, then penetrates the peritoneal body wall to enter the leg as the **femoral artery.** Note that nearly all of the caudal arterial circulation is accompanied by corresponding venous circulation, already described. Follow the femoral artery to the point where, on its median side, it gives off a **deep femoral artery.** This artery serves the deep muscles of the thigh.

Locate now the origin of the **internal iliac arteries.** Each internal iliac gives off at once a large **umbilical artery** and then, as a smaller vessel, continues dorsally and posteriorly into the sacral region beside the internal iliac vein, giving off branches to the bladder, rectum, and gluteal muscles.

In the fetus the umbilical arteries are major vessels that return blood from the fetus to the placenta by way of the umbilical cord. After birth they become smaller and serve the urinary bladder.

Structure of the Heart

External View.
☞ Use fresh or preserved pig or sheep hearts and compare with the heart of the fetal pig.

Locate the right and left **atria** and the right and left **ventricles.** Find the **coronary sulcus** and the **coronary artery** and **vein,** which supply the muscles of the heart. Now identify on the ventral side the large, thick-walled **pulmonary trunk** leaving the right ventricle and the large **aorta** leaving the left ventricle. On the dorsal side, find the large, thin-walled **precaval** and **postcaval veins** entering the right atrium and the **pulmonary veins** entering the left atrium.

Frontal Section.
☞ Now make a frontal section, dividing the heart into dorsal and ventral halves. Start at the apex and direct the cut between the origins of the pulmonary and aortic trunks. Leave the two halves of the heart attached at the top. Wash out the heart cavities (or, if injected, carefully pull out the latex filling).

The cavities of the heart are lined with a shiny membrane, the **endocardium.** Identify the chambers and the valves at the entrance to each chamber that prevent a backflow of blood.

Right Atrium. The right atrium is thin walled, with openings from the precava, postcava, and hemiazygous veins. Find the entrance of these veins. Now, in the dorsal half of the fetal pig heart, probe for an opening between the two atria. This is the **foramen ovale,** one of the fetal shortcuts.

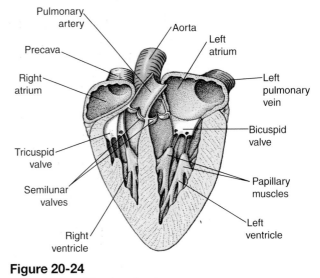

Figure 20-24
Frontal section of the sheep heart.

Right Ventricle. The right ventricle is thick walled. An atrioventricular valve, called the **tricuspid valve,** prevents backflow of blood to the atrium (Figure 20-24). The valve is composed of three **cusps,** or flaps of tissue, that extend from the floor of the atrium and are connected by fibrous cords, the **chordae tendineae** (ten-din′ee-ee), to **papillary muscles** projecting from the walls of the ventricle.

In the ventral half of the heart the opening into the pulmonary artery is guarded by **semilunar valves.**

☞ Cut the pulmonary artery close to the heart and remove the latex. Slit the vessel for a short distance into the ventricle and look into it to see the three cusps, or pockets of tissue. Determine how they would work to allow passage into the vessel but prevent return of blood into the ventricle.

Left Atrium. The left atrium is thin walled. Find the entrance of the pulmonary veins that return oxygen-rich blood to the heart.

Left Ventricle. The left ventricle is the most muscular chamber of the heart, for it must send the freshly oxygenated blood at high pressure to all the tissues of the body. An atrioventricular valve, the **bicuspid valve** (also called the mitral valve), guards the entrance from the left atrium. Find the cusps of this valve. Push the valve open and closed to see how it works.

Find the three-cusped **semilunar valve,** which prevents backflow from the aorta (Figure 20-24).

Notice the difference in the thickness of the walls of arteries and veins. Of what are these walls composed? _____ Would you find the same kind of muscle in the heart as in the vessels? _____ What is the pacemaker of the heart? _____ Consult your text.

Identification

Be familiar with the direction of blood flow throughout the body. Be able to describe how fetal circulation differs from postnatal circulation. Be able to trace blood from any part of the heart to any part of the body and back to the heart.

Written Report

On separate paper list the two chief differences between mammalian and amphibian circulatory systems as illustrated by your study of the pig and the frog. How are these differences adaptive to an all-terrestrial or half-aquatic life?

APPENDIX

Sources of Living Material and Prepared Microslides

Sources of Living Material

United States

Carolina Biological Supply Co.
Burlington, NC 27215
800-334-5551, 800-547-1733, 800-547-1733 (orders)
Fax: 503-656-4208
www.carolina.com

Connecticut Valley Biological Supply Co., Inc.
P.O. Box 326
Southampton, MA 01073
800-628-7748

Delta Biologicals
P.O. Box 2666
Tucson, AZ 85726
800-821-2502
www.deltabio.com

Fisher Scientific Co.
485 S. Frontage Road
Burr Ridge, IL 60521
800-766-7000 (orders)
Fax: 800-926-1166
www.fisherscientific.com

Frey Scientific Co.
P.O. Box 8101
Mansfield, OH 44901-8101
800-225-FREY
www.freyscientific.com

Gulf Specimen Co., Inc.
P.O. Box 237
Panacea, FL 32346
850-984-5297
www.gulfspecimen.org

Lapine Scientific Co.
P.O. Box 780
Blue Island, IL 60406
708-388-4030

NASCO
P.O. Box 222
Germantown, WI 53022
800-767-4288, 800-558-9595
sales@nascoinc.com

Nebraska Scientific
3823 Leavenworth
Omaha, NE 68105
800-228-7117
Fax: 402-346-2216
www.nebraskascientific.com

Parco Scientific Co.
P.O. Box 189
Vienna, OH 44473
800-247-2726
www.parcoscientific.com

Sargent-Welch
P.O. Box 5229
Buffalo Grove, IL 60089-5229
800-SARGENT (orders)
www.sargentwelch.com

Science Kit and Boreal Laboratories
P.O. Box 5003
Tonawanda, NY 14151-5003
800-828-7777
www.sciencekit.com

Ward's Natural Science Establishment, Inc.
P.O. Box 92912
Rochester, NY 14692-9019
800-962-2660 (orders)
Fax: 800-635-8439
www.wardsci.com

Canada

Northwest Laboratories, Ltd., Eastern Division
90 Monarch Road
P.O. Box 1356
Guelph, Ont. N1H 6N8
519-836-7720, 800-265-7250
Fax: 519-836-4105

Northwest Laboratories, Ltd., Western Division
P.O. Box 6100, LCD1
Victoria, B.C., VP 5L4
604-592-2438, 800-663-5890
Fax: 604-592-1341

Ward's Natural Science, Ltd.
397 Vansickle Road
Saint Catherines, Ont. L2S 3T5
905-984-8900, 800-387-7822 (only from Canada)
Fax: 905-984-5952

Sources of Living Marine Material

United States

AquaScience Research Group, Inc.
1100 Gentry Street
North Kansas City, MO 64116
816-842-5936
www.petsforum.com/aquascience

Carolina Biological Supply Co.—
See p. 263

Connecticut Valley Biological Supply Co.—See p. 263

Gulf Specimen Marine Labs—
See p. 263

Marine Bait Co.
Academy Street
Newcastle, ME 04553
207-563-3000

Marine Biological Laboratory
Supply Department
Woods Hole, MA 02543
508-548-3705
www.mbl.edu

Pacific Bio Marine Laboratories, Inc.
124 North Ash
Inglewood, CA 90301
310-677-1056

Sea Life Supply
740 Tioga Avenue
Sand City, CA 93955
831-394-0828 (orders)
Fax: 831-899-3399
www.sealifesupply.com

Ward's Natural Science—See p. 263

Canada

Northwest Laboratories, Ltd., Eastern
Division—See p. 263

Northwest Laboratories, Ltd., Western
Division—See p. 263

Ward's Natural Science, Ltd.—
See p. 263

Sources of Prepared Microslides

United States

R. P. Cargill Laboratories, Inc.
Scientific Division
55 Commerce Road.
Cedar Grove, NJ 07009
973-239-6633
Fax: 973-239-6096

Carolina Biological Supply Co.
Burlington, NC 27215
800-334-5551
Fax: 800-222-7112
www.carolina.com

Connecticut Valley Biological
P.O. Box 326
Southampton, MA 01073
413-527-4030, 800-628-7748

Dako Corp.
6392 Via Real
Carpinteria, CA 93013
805-566-6655, 800-235-5743
www.dakousa.com

Fisher Scientific Co.
Science Education Division
485 South Frontage Road
Burr Ridge, IL 60521
www.fishersci.com

Frey Scientific Co.
P.O. Box 8101
Mansfield, OH 44901-8101
800-225-FREY
www.freyscientific.com

Insect Lore Products
P.O. Box 1535
Shafter, CA 93263
800-548-3284 (orders)
Fax: 805-746-6047
www.insectlore.com

NASCO
P.O. Box 222
Germantown, WI 53022
800-767-4288, 800-558-9595
mailto:sales@nascoinc.com

Triarch, Inc.
P.O. Box 98
Ripon, WI 54971
414-748-5125, 800-848-0810

Ward's Natural Science Establishment,
Inc.
P.O. Box 92912
Rochester, NY 14692-9019
800-962-2660 (orders)
Fax: 800-635-8439
www.wardsci.com

Canada

Northwest Laboratories, Ltd., Eastern
Division
90 Monarch Road
P.O. Box 1356
Guelph, Ont. N1H 6N8
519-836-7720, 800-265-7250
Fax: 519-836-4105

Northwest Laboratories, Ltd., Western
Division
P.O. Box 6100, LCD1
Victoria, B.C., VP 5L4
604-592-2438, 800-663-5890
Fax: 604-592-1341

Ward's Natural Science, Ltd.
397 Vansickle Road
Saint Catherines, Ont. L2S 3T5
905-984-8900, 800-387-7822 (only from
Canada)
Fax: 905-984-5952

CREDITS

Photos

Part Openers

1: C. P. Hickman; **2:** C. P. Hickman

Chapter 3

3.1, 3.5: Courtesy of Leica Microsystems Inc., Bannockburn, IL

Chapter 4

4.1a: © Carolina Biological Supply Co./ Phototake; **4.8:** © David M. Phillips/Visuals Unlimited; **4.10:** Eric V. Grave/Photo Researchers Inc.; **4.13:** C. P. Hickman; **4.16:** © Karl Aufderheide/Visuals Unlimited; **4.19:** © Phil A. Harrington/Peter Arnold, Inc.; **4.20, 4.21:** © M. I. Walker/ Photo Researchers, Inc.

Chapter 5

5.3: C. P. Hickman; **5.6:** © Stanley Flegler/ Visuals Unlimited; **5.7a, 5.7b:** Frances M. Hickman.

Chapter 6

6.3: C. P. Hickman; **6.5a, 6.5b:** © Dr. D. P. Wilson/FPLA; **6.6:** © Robert Brons/Biological Photo Service; **6.9:** © William C. Ober

Chapter 7

7.1: © Carolina Biological Supply Co./ Phototake; **7.6:** © CABISCO/Visuals Unlimited; **7.7:** C. P. Hickman.

Chapter 8

8.3: © Robert Calentine/Visuals Unlimited; **8.5:** Courtesy Indiana University School of Medicine

Chapter 9

9.1: Frances M. Hickman.

Chapter 10

10.2a: © Daniel W. Gotshall/Visuals Unlimited; **10.2b:** William C. Ober; **10.7:** C. P. Hickman

Chapter 11

11.3: C. P. Hickman

Chapter 14

14.1: © Dr. D. P. Wilson/FPLA; **14.2:** C. P. Hickman; **14.5a, 14.5b, 14.7:** William C. Ober; **14.9a, 14.9b:** © R. Harbo

Chapter 15

15.2, 15.4: C. P. Hickman

Chapter 16

16.5: © C. Stehr/Visuals Unlimited

INDEX

Page numbers followed by *f* and *t* indicate figures and tables, respectively.

A

Abdomen
    of crayfish, 136
    of cricket, 155
    of grasshopper, 149–150
    of honey bee, 151
    of horseshoe crab, 130
    of pig, 232, 243, 243*f*, 244*t*
    of spider, 131–132
Abdominal cavity, of yellow
        perch, 200–202
Abduction, 238
Abductor muscles, 238
Aboral surface
    of sea star, 167
    of sea urchin, 176
Acanthobdella, 113
Acanthocephala, 10, 95, 95*f*
Accessory gland, 158
Acetabulum, 78, 80, 236
Acheta domesticus, 155–158,
        156*f*, 157*f*
Aciculum, 114
Acoelomate animals, 72–87
Acontia, 69, 70
Acorn barnacle, 142
Acorn worms, 11
Acrania, 180
Acromial process, 226
Actin, 27
Actinophrys, 25, 29, 29*f*
Actinopod amebas, 28–29
Actinopodans, 25
Actinopterygii, 11, 181
Action, of muscles, 238
Adduction, 238
Adductor longus, 212*t*, 213
Adductor magnus, 212*t*, 213
Adductor muscles, 101, 103,
        212*t*, 213, 238, 244*t*, 245
Adhesive papillae, 184
Adrenal glands, 218, 252
Adrenal vein, 259
Aedeagus, 158
Afferent branchial arteries, 186
Aggression, in paradise fish,
        202–203
Agnatha, 11, 181, 188–192
Alcyonium, 58

Alimentary canal, 120
Allantoic bladder, 250, 251
Allantoic duct, 251
Alula, 231
Alveolar pockets, 228
Alveoli, 217
Ambulacral groove, 171
Ambulacral ossicles, 169, 171
Ambulacral plates, 175
Ambulacral pores, 169
Ambulacral regions, 174
Ambulacral ridge, 169
Ambulacral spines, 167
Amebas, 25–28
    actinopod, 28–29
    naked, 25–28
    parasitic, 28, 28*f*
    shelled, 28, 29*f*
Amebocytes, 52, 52*f*
"American killer," 92
Ammocoete larva,
        188–191, 189*f*
Amoeba, 25–28
Amoeba proteus, 26
Amphiarthrosis, 237
Amphibia, 11, 181, 207–223
Amphiblastula larvae, 51, 52
Amphioxus, 11, 180, 184–187,
        185*f*, 187*f*
Amphipoda, 142
Amplexus, 218
Ampulla, 166, 169, 177
Amylase, 245
Anadromous fish, 188
Anal fin, 198
Anal plates, 176
Anal pore, 45
Anapsid lineage, 224
Ancylostoma caninum,
        92, 93*f*
Ancylostoma duodenale, 92
"Animal–flagellates," 33
Annelida, 10, 112–126
Annuli, 123
Anodonta, 99
Anopheles mosquito, 42, 43
Anoplura, 161, 161*f*
Anostraca, 141, 142*f*, 143
Antedon, 166
Antenna cleaner, 151
Antenna comb, 151
Antennae, 134
    of centipedes, 146
    of crayfish, 135, 136,
        138, 141

of grasshopper, 149
    of honey bee, 150
Antennal arteries, 140
Antennal glands, 141
Antennules, 138, 141
Anterior adductor muscles, 101
Anterior aorta, 105
Anterior cardinal sinuses, 196
Anterior cardinal veins, 191
Anterior deep pectoral,
        240*t*, 242
Anterior foot retractor
        muscles, 101
Anterior mesenteric artery,
        221, 260
Anterior mesenteric vein, 260
Anterior trunk, 75
Anterior vena cava, 220
Anthozoa, 9, 58, 68–71
Anticus, 211
Antipathes, 58
Anura, 207
Anus
    of ammocoete larva, 189
    of amphioxus, 185, 186
    of *Ascaris*, 89
    of centipedes, 146
    of clam, 106
    of clamworm, 113
    of crayfish, 136, 140
    of earthworm, 116, 118
    of horseshoe crab, 130
    of leech, 123
    of pig, 232, 251
    of sea cucumber, 176, 177
    of sea squirt, 183
    of sea star, 167
    of sea urchin, 176
    of spider, 132
    of squid, 111
    of yellow perch, 198, 201
Aorta, 256, 257, 261. *See also*
        Dorsal aorta
    anterior, 105
    descending, 260
    left, 228
    posterior, 105
    right, 228
    ventral, 186, 187, 191, 197
Aortic arches, 117, 118
Apertures, 99–100, 107
    excurrent, 102
    genital, 108
    incurrent, 102
Apex, 107

Apicomplexa, 24, 25, 41–43
Apis, 150–152, 150*f*–152*f*
Aplacophora, 98
Aplysia, 99
Aponeurosis, 238
Apopyles, of *Sycon*, 50, 51, 51*f*
Appendages
    of crayfish, 136–138, 137*f*
    of horseshoe crabs,
        129–130
    jointed, 127
Appendicularians, 181
Appendicular skeleton, 199,
        210, 225, 231, 236
Appendix, vermiform, 251
Aquatic snail, 108
Arachnida, 11, 127–128,
        130–132
Arbacia, 166, 173–176,
        174*f*, 175*f*
Arbacia punctulata, 173
Arcella, 25, 28, 29*f*
Architeuthis, 109
Argiope, 128, 130–132
Argiope aurantia, 130, 131*f*
Argiope trifasciata, 130
Aristotle's lantern, 174
Arms
    of sea star, 166, 167,
        170–171
    of squid, 111
Arolium, 149
Arrow worms, 10
Artemia salina, 143
Arterial system
    of dogfish shark, 197, 197*f*
    of frog, 220–221
Arthropoda, 10, 127–163
Articular cartilage, 236
Articulations, 236–237
Artiodactyla, 232
Ascaris, 88–92, 90*f*, 91*f*
Ascaris lumbricoides, 88
Ascaris megalocephala, 89
Ascaris suum, 88
Ascidiacea, 181
Asconoid sponge, 50*f*, 53–54
Asexual reproduction
    in amebas, 28
    in *Clonorchis*, 78
    in cnidaria, 57
    in euglenoids, 35
    in *Hydra*, 61
    in *Obelia*, 63
    in *Paramecium*, 46

in planarians, 75
in *Plasmodium,* 42–43
in *Schistosoma,* 79
in sea squirt, 184
in *Sycon,* 52
in *Volvox,* 33, 34*f*
in *Vorticella,* 47
Asterias, 166–171, 167*f*–169*f*
Asteroidea, 9, 166–171,
  167*f*–169*f*
Astrangia, 58, 70–71, 71*f*
Astrangia danae, 71*f*
Astrophyton muricatum, 172*f*
Atlas, 210, 236
Atrial siphon, 182
Atriopore, 185, 186
Atrium, 182, 186, 187, 196, 219,
  221, 228, 256, 261
Aurelia, 58, 66–68, 67*f*
Aurelia aurita, 66
Auricles, 73, 74*f,* 105
Aves, 11, 181, 229–231
Axial skeleton, 199, 210, 225,
  230–231, 236
Axillary artery, 257
Axillary vein, 257
Axopodia, 25
Azygous vein, 257

B

Back, muscles of, 243,
  243*f,* 244*t*
Balanus, 142
Barbs, 60, 229
Barbules, 229
Barnacle, 142, 142*f*
Basal disc, 59, 69
Basal plate, 71
Basket stars, 166, 172*f*
Bath sponges, 49
Bdelloura, 75
Beak, 231
Biceps, 211
Biceps brachii, 241*t,* 243
Biceps femoris, 212*t,* 213,
  243, 244*t*
Bicornuate, 253
Bicuspid valve, 261
Bilateral symmetry, 72, 127
Bile, 194, 227, 251
Bile duct, 251
"Bilharzia," 79
Binary fission
  in amebas, 28
  in *Paramecium,* 46
  in *Vorticella,* 47
Binocular microscope, 19
Binomial nomenclature, 7
Biramous appendages, 136
Birds, 11, 181, 229–231
Bivalvia, 10, 99–107, 100*f*–106*f*
Bivium, 167
Black corals, 58
Bladder, of *Clonorchis,* 78
Blastostyle, 63

Blepharisma, 25
Blood fluke, 78–81
Body cavity, 88
Body covering, in euglenoids,
  34
Body of sternum, 236
Body of uterus, 253
Body tube, 16, 17*f*
Body wall
  of *Ascaris,* 89
  of *Clonorchis,* 78
  of *Metridium,* 70
  of pig, 249
Bone
  growth of, 236
  structure of, 236
Bony fishes, 181, 198–202,
  199*f*–201*f*
Book gills, 130
Book lungs, 131
Brachial artery, 257
Brachialis, 241*t,* 243
Brachial vein, 257
Brachiocephalic artery,
  228, 242
Brachiocephalic muscle, 240*t*
Brachiocephalic trunk, 257
Brachiopoda, 10
Brain
  of ammocoete larva,
    189, 190
  of crayfish, 141
Branchial arteries, 197
Branchial chambers, 197
Branchial sac, 182
Branchial system, of
  crayfish, 138
Branchinecta, 141, 142*f*
Branchiopoda, 141–142
Branchiostegites, 135
Branchiostoma, 180, 184–187,
  185*f,* 187*f*
Brevis, 211
Brine shrimp, 143
Brittle stars, 9, 166, 171–173
Broad ligament, 253
Bronchus, 217, 227
Brood chamber, 103
Brook lamprey, 188
Buccal cavity, 186, 189
Buccal chamber, 190
Buccal funnel, 191
Buccal membrane, 111
Buccal podia, 174
Buccal shields, 171
Budding
  in *Astrangia,* 71
  in *Hydra,* 61
  in *Obelia,* 63
  in sea squirt, 184
  in *Sycon,* 52
  in tapeworms, 81
  in *Vorticella,* 47
Bulbourethral glands, 253
Bullfrog, 207
Bursa, 158, 171
Busycon, 99

C

Caecilians, 11
Calcarea, 49–55
Calciferous glands, 118
Calyces, 252
Cambarus, 128, 134
Canal systems, 49
  asconoid, 50*f,* 53–54
  leuconoid, 50*f,* 53
  syconoid, 50, 50*f*
Cancellous bone, 236
Canines, 246
Carapace, 129, 135, 224
Carassius auratus, 204–206
Cardiac chamber, 140
Cardiac end, 250
Cardiac portion of stomach, 201
Cardiac stomach, 169
Carotid arch, 220–221
Carotid arteries, 220, 228
Carotid trunk, 257
Carpals, 231
Carpometacarpals, 231
Carpus, 236
Carrying capacity, 3
Cartilaginous fishes, 192–198,
  193*f*–195*f,* 197*f*
Carybdea, 58
Cassiopeia, 58
Castes, 152
Cat
  skeleton of, 234*f*
  skull of, 233*f*
Caudal artery, 197, 261
Caudal fin, 185, 189, 191,
  193, 198
Caudal sucker, 123
Caudal vein, 196
Caudal vertebrae, 226, 230, 236
Caudofoveata, 98
Cecum, of pig, 251
Celiac artery, 197, 221, 260
Celiacomesenteric artery, 221
Cell differentiation, 33
"Cell drinking." *See* Pinocytosis
Cells
  division of labor among, 33
  epitheliomuscular, 60, 61*f*
  flagellated flame, 75
  gland, 60, 61*f*
  interstitial, 60, 61*f*
  longitudinal muscle, 89
  nutritive–muscular, 60, 61*f*
  pigment, 110
  reproductive, 33
  sensory, 61, 61*f*
  sex, 33
  somatic, 33
Cellular level of organization,
  49
Centipedes, 11, 128, 146, 147*f*
Central axis, 138
Central canal, 187
Central disc, 167, 173*f*
Cephalaspidomorphi, 11, 181,
  188–192

Cephalic vein, 257
Cephalization, 72, 127
Cephalobus, 92
Cephalochordata, 11, 180,
  184–187, 185*f,* 187*f*
Cephalopoda, 10, 99,
  109–111, 110*f*
Cephalothorax
  of crayfish, 135
  of horseshoe crabs, 129–130
  of spider, 131
Ceratium, 37, 37*f*
Cercariae, 78, 79
Cercus, 150
Cerebral ganglia, 75, 119, 182
Cerebropleural ganglia, 106
Cerianthus, 58
Ceriantipatharia, 58
Cervical groove, 135
Cervical vertebrae, 226, 230, 236
Cervix, 254
Cestoda, 10, 73, 81–84, 81*f*–83*f*
Chaetoderma, 98
Chaetognatha, 10
Chaetonotus, 94, 95*f*
Chaetopleura, 98
Chagas' disease, 37
Chelae, 129, 135, 138
Chelicerae, 129, 131
Chelicerata, 127, 128–133
Chelipeds, 135, 138
Chemoreceptors, 186
Chemotaxis
  in *Paramecium,* 46
  in planarians, 73, 74
Chest
  muscles of, 239–242
Chilaria, 130
Chilopoda, 11, 128, 146
Chimaeras, 181
Chironex, 58
Chironex fleckeri, 66
Chitin, 127, 129, 135
Chitinous teeth, 141
Chitons, 10, 98
Chlamydomonas, 25
Chlamydophrys, 25
Chloragogue, 118, 120
Chlorohydra viridissima, 58, 59
Chlorophyll, 34, 35
Chlorophyta, 25, 33
Chloroplasts, 33, 34, 35
Choanocytes, 49, 51–52, 52*f*
Chondrichthyes, 11, 181,
  192–198, 193*f*–195*f,* 197*f*
Chordae tendineae, 261
Chordata, 180–187
Chorioallantoic placenta, 254
Christmas tree worms, 115*f*
Chromatophores, 110, 111, 189
Chrysemys picta, 224–228,
  225*f*–227*f*
Cilia, 44, 47
Ciliary action, 44, 45*f*
Ciliated epidermis, 175
Ciliated ridges, 191
Ciliophora, 9, 24, 25, 43–48

Ciona, 181–184, 182*f*–184*f*
Ciona intestinalis, 181, 182*f*
Circular muscle layer, of
    earthworm, 120
Circular muscles, 76
Circulation/circulatory system
    in amphioxus, 186
    in annelids, 112
    in clam, 105, 105*f*
    in crayfish, 140
    in dogfish shark,
        196–197, 197*f*
    in earthworm, 118, 120
    in frog, 219–223, 219*f,* 220*f*
    in horseshoe crab, 130
    in molluscs, 98
    in pig, 255–262, 256*f,*
        258*f*–261*f*
    in sea squirt, 182–183
    in turtle, 228
Circumflex artery, 261
Circumflex iliac vein, 259
Circumpharyngeal connectives,
    119
Cirri, 113, 114
Cirripedia, 142, 142*f*
Cisterna magna, 250
Cladocera, 141–142, 142*f*
Clamworm, 112, 113–114, 114*f*
Clasper, 196
Class, 7
Classification, 7–13
    of annelida, 112–113
    of arthropoda, 127–128
    of chordata, 180–181
    of cnidaria, 57–58
    of echinodermata, 165–166
    of insects, 159–163
    of mollusca, 98–99
    of platyhelminthes, 72–73
    of protozoa, 24–25
    of sponges, 49
    using taxonomic keys for,
        7–11
Clathrulina, 25, 29, 29*f*
Claws, 149
Cliona, 49, 53
Clitellum, 116, 118, 123
Clitoris, 254
Cloaca
    of ammocoete larva, 189
    of *Ascaris,* 90
    of dogfish shark, 196
    of frog, 217
    of sea cucumber, 177
    of turtle, 227
Cloacal opening, 189, 193,
    208, 217
Clonorchis, 73, 78, 79*f*
Closed circulatory system
    in annelids, 112
    in earthworm, 118
Cnidaria, 9, 57–71
Cnidocils, of *Hydra,* 59, 60*f*
Cnidocytes, 59, 63
Coarse–adjustment knob, 17–18,
    17*f,* 19

Cobb, N. A., 92
Coccidea, 25, 42–43
Coccyx, 236
Cockroaches, *Gregarina* in guts
    of, 41
Cocoon, 118*f,* 119
Coelom, 88, 120, 187. *See also*
    Pseudocoel; True coelom
Coelomic cavity, 105, 169, 170,
    177, 190, 201
Coelomic fluid, 115, 116,
    119, 169
Coenosarc, 63
Cog muscles, 174
Coleoptera, 163, 163*f*
Collagen, 89
Collar, 108, 111
"Collar cells." *See* Choanocytes
Collecting insects, 158
Collembola, 162, 162*f*
Colloblasts, 57
Colon
    of pig, 251
    of turtle, 227
Colonies
    of *Astrangia,* 71
    daughter, 33, 34*f*
    hydroid, 58, 61–63
    of sea squirt, 184
    of *Volvox,* 33, 34*f*
Color change, in squids, 110
Colpoda, 25
Columba livia, 229–231
Columella, 108
Comb jellies, 9
Commercial bath sponges, 49
Common cardinal veins, 196
Common carotid artery, 220
Common goldfish, 204–206
Common green hydra, 59
Common iliac arteries, 221, 259
Compact bone, 236
Compound eyes
    of crayfish, 141
    of grasshopper, 149
    of honey bee, 150
Compound light microscope,
    16–22
Concave surface of mirror, 17
Conchs, 99
Condenser, 16
Confirmation, in scientific
    method, 2
Conjecture, in scientific
    method, 2
Conjugation, 46
Contour feathers, 229
Contractile vacuole, 27, 30,
    35, 44
Conus arteriosus, 196, 219, 221
Convergent evolution, 111
Copepoda, 142, 142*f*
Copulatory organs, 138
Coracoid, 210, 231
Coracoid process, 226
Corals, 9, 57
Corona, 94

Coronary arteries, 228, 256, 261
Coronary sulcus, 256, 261
Coronary vein, 256, 261
Cortex, 252
Costal cartilage, 236
Costocervical trunk, 257
Countershading, 194
Cowper glands, 253
Coxa, 131, 149
Cranial region, 236
Craniata, 181
Cranium, 210, 231
Craspedacusta, 63
Crawfish. *See* Crayfish
Crayfish, 134–141, 135*f*–137*f,*
    139*f,* 143
Cricket, 155–158, 156*f,* 157*f*
Crinoidea, 9, 165–166
Crocodilians, 11, 181
Crop, 118
Crossed reflexes, 182
Crustacea, 11, 128, 134–145
Ctenoid scales, 199
Ctenophora, 9, 57–71
Cubozoa, 58
Cucumaria, 166, 176, 177
Cucumaria miniata, 176
Cusps, 261
Cutaneous artery, 221
Cutaneous maximus, 239
Cutaneous muscle, 238–239
Cutaneous respiration, 217
Cuticle
    of *Ascaris,* 89, 91
    of earthworm, 119
    of rotifers, 94
Cuttlefishes, 10
Cuvier, ducts of, 186, 196
Cyanea capillata, 66
Cyclops, 142, 142*f*
Cyclosis, 45
Cyclostomata, 181
Cylindrical body, 69, 88, 95
Cylindrical column, 69
Cypris, 142*f*
Cyst, 28, 28*f*
Cystic duct, 251
Cytopharynx
    of *Paramecium,* 44, 45
    of *Vorticella,* 47
Cytoproct, 45
Cytostoma, 47
Cytostome, of *Paramecium,*
    44, 45

D

Dactylozooids, 64
Daphnia, 141–142, 142*f,* 143,
    144–145
Darwin, Charles, 37
Daughter colonies, 33, 34*f*
Decapoda, 142
Deep femoral artery, 261
Deep femoral vein, 259
Deltoid, 240*t,* 242

Demibranchs, 198
Demospongiae, 49, 53, 54*f*
Dendrocoelopsis vaginata, 73
Dentalium, 99
Depressor muscles, 238
Dermal branchiae, 165, 167, 171
Dermal endoskeleton, 165
Dermal ostia, 50, 51, 51*f*
Dermaptera, 163, 163*f*
Dermis
    of amphioxus, 186
    of dogfish shark, 193
    of sea star, 168, 170
Descending aorta, 260
Deuterostomes, 165
Deutomerite, 41
Dextral shells, 108
Diaphragm, 249
Diaphysis, 236, 237*f*
Diapsid lineage, 224
Diarrhea, 37
Diarthrosis, 237
Dichotomous key, 8
Didinium, 25
Difflugia, 25, 28, 29*f*
Diffuse placenta, 254
Digastric muscle, 240*t,* 242
Digenetic flukes, 10, 72–73,
    78–81
Digestion/digestive system
    in amebas, 27
    in ammocoete larva, 189
    in amphioxus, 186
    in annelids, 112
    in *Ascaris,* 89–90
    in *Aurelia,* 68
    in clam, 106
    in *Clonorchis,* 78
    in crayfish, 140–141
    in cricket, 158
    in dogfish shark,
        194–195, 194*f*
    in earthworm, 117–118, 117*f*
    extracellular, 59, 75
    in frog, 216–217, 217*f*
    in *Hydra,* 59
    intracellular, 59, 75
    in land snail, 108
    in molluscs, 98
    in *Paramecium,* 45
    in pig, 245–251, 246*f*–249*f*
    in planarians, 75
    in sea cucumber, 177
    in sea squirt, 183
    in sea star, 169–170
    in turtle, 226–227, 227*f*
    in yellow perch, 201
Digestive glands, 106, 140
Digits, 231, 236
Dinoflagellates, 37
Dioecism
    in *Hydra,* 61
    in spiny–headed worms, 95
Diphyllobothrium, 73
Diphyllobothrium latum, 83
Diploblastic structure, 60
Diplogaster, 92

Diplopoda, 11, 128, 148
Diptera, 159, 159f
Dipylidium, 81–84
Dipylidium caninum, 81, 82f, 84
Directional illumination, 74
Direct reflexes, 182
Dirt
    on ocular, 19
    on slide, 19
Dissection
    of crayfish, 136–138
    of dogfish shark, 194
    of earthworm, 116–117
    of freshwater clam,
        100, 101f
    of frog, 216
    of pig, 238–239,
        248–249, 249f
    of sea star, 169–170
Division of labor, among
    cells, 33
Dogfish shark, 192–198,
    193f–195f, 197f
Dog hookworm, 92, 93f
Dog tapeworm, 81, 81f
Dorsal abdominal artery, 140
Dorsal aorta, 186, 187, 190, 191,
    197, 221, 228
Dorsal blood vessel, 116, 117,
    118, 120
Dorsal cirrus, 114
Dorsal fin, 185, 186, 189, 191,
    193, 198
Dorsal lines, 89
Dorsal nerve cords, 91
Dorsal pore, 116
Dorsal tubular nerve cord, 180,
    185, 186
Dorsoventral muscle fibers, 76
Dorylaimus, 92
Double circulation, 219
Down feathers, 229
Drones, 152
Ducts of Cuvier, 186, 196
Ductus arteriosus, 258
Ductus venosus, 258–259
Dugesia, 73–76, 74f, 76f
Dugesia dorotocephala, 73
Dugesia tigrina, 72
Duodenum
    of dogfish shark, 195
    of pig, 251
    of turtle, 227
    of yellow perch, 201
Dysentery, 28

E

Ears, of yellow perch, 198
Earthworms, 10, 112, 115–122,
    117f–120f
Ear vesicles, 190
Ecdysis, 129, 135, 155
Echinodermata, 165–179
Echinoidea, 9, 166, 173–176,
    174f, 175f

Ecological relationships of
    animals, 2–5
Ectoplasm, 26, 34, 44
Ectothermic animals, 221
Efferent branchial artery,
    186, 197
Egg(s)
    of *Hydra,* 61
    of tapeworms, 84
Ejaculatory duct, of *Ascaris,* 90
Ejection reflex, 183
Elasmobranchii, 192
Elliptio, 99
Endamoeba, 25, 28
Endocardium, 261
Endoplasm, 26, 34, 44
Endopod, 138
Endopterygota, 155
Endoskeleton
    of echinodermata, 165
    of sea cucumber, 178
    of sea star, 168–169
    of sea urchin, 175
    of turtle, 225
Endostyle, 183, 186, 189
Endothermic animals, 221
End sac, 141
Entamoeba, 28
Entamoeba gingivalis, 28
Entamoeba histolytica, 28, 28f
Enterobius vermicularis,
    93–94, 94f
Epaxial muscles, 199
Ephemeroptera, 159, 159f
Ephyrae, 68
Epidermal nerve plexus, 170
Epidermis
    of ammocoete larva, 190
    of amphioxus, 186
    of *Ascaris,* 91
    of crayfish, 138–140
    of dogfish shark, 193
    of earthworm, 120
    of *Gonionemus,* 64
    of *Hydra,* 60, 61f
    of *Metridium,* 70
    of *Obelia,* 63
    of planarians, 76
    of sea star, 167, 168, 170
    of sea urchin, 175
Epididymis, 253
Epidinium, 25
Epiglottis, 247
Epigynum, 131
Epipharyngeal groove. *See*
    Hyperbranchial groove
Epiphyses, 236
Episternum, 210
Epistylis, 46f, 47
Epitheliomuscular cells, 60, 61f
Errantia, 114
Esophagus
    of ammocoete larva,
        189, 190
    of clam, 106
    of *Clonorchis,* 78
    of crayfish, 140–141

of cricket, 158
of dogfish shark, 194
of earthworm, 117
of frog, 216
of pig, 247, 250
of sea squirt, 183
of turtle, 226, 227
of yellow perch, 201
Eubranchipus, 141, 143
Euchlanis, 95f
Eucoelomate body plan, 112
Eudistyllia polymorpha, 115f
Eudorina, 37f
Euglena, 25
Euglena gracilis, 34
Euglena viridis, 34
Euglenida, 25
Euglenoidea, 25
Euglenoid movement, 34, 35f
Euglenozoa, 25, 34–35, 35f,
    36–37
Euplotes, 25, 46f
Eurycea longicauda
    longicauda, 7
Eurypterida, 127
Eusnpongia, 50f
Eustachian tubes, 216, 226, 247
Evolution, convergent, 111
Excretion/excretory system
    in acoelomates, 72
    in ammocoete larva, 189
    in annelids, 112
    in *Ascaris,* 89, 91
    in clam, 105
    in *Clonorchis,* 78
    in crayfish, 141
    in dogfish shark,
        195–196, 195f
    in earthworm, 119
    in frog, 218, 218f
    in pig, 251–252, 252f
    in planarians, 75
    in sea cucumber, 177
    in sea squirt, 184
    in tapeworms, 83
Excretory canals
    of *Ascaris,* 89, 91
    of flatworms, 75
    of tapeworms, 83, 84
Excretory pore, 78, 89
Excurrent aperture, 102
Excurrent siphon, 182
Exopod, 138
Exopterygota, 155
Exoskeleton
    of arthropods, 127
    of crayfish, 135
    of grasshopper, 149
    of horseshoe crab, 129
    of molluscs, 98, 99
    of spider, 131
    of turtle, 224
Exponential population growth, 3
Extension, 238
Extensor cruris, 212t
Extensor muscles, 140, 238
External carotid artery, 220, 257

External gill slits, 189, 191, 197
External iliac arteries, 261
External iliac veins, 259
External jugular vein, 220, 257
External maxillary vein, 257
External nares, 224–225
External oblique, 212, 212t,
    243, 244t
Extracellular digestion, 59, 75
Exumbrella, 63
Eyelids
    of dogfish shark, 193
    of frogs, 208
    of turtle, 225
Eyepieces, of microscope, 16
Eyes
    of ammocoete larva, 190
    of centipedes, 146
    of clamworm, 113
    of crayfish, 135, 141
    of dogfish shark, 193
    of frogs, 208
    of grasshopper, 149
    of honey bee, 150
    of horseshoe crabs, 129
    of lamprey, 191
    of land snails, 107
    of spider, 131
    of squid, 111
    of turtle, 225
    of yellow perch, 198
Eyespots. *See* Ocelli

F

Face
    muscles of, 239–242
Facial region, 236
Fairy shrimp, 141, 142f, 143
Fallopian tube, 253
Family, 7
Fang, 131, 146
Fascia, 237
Fasciculi, 237
Fasciola, 73
Fat body, 158, 218
Feather, 229
Feather stars, 9, 165–166
Feeding habits
    of amebas, 27, 27f
    of ammocoete larva, 189
    of amphioxus, 186
    of *Aurelia,* 68
    of brittle star, 171–172
    of crayfish, 134
    of euglenoids, 35
    of *Hydra,* 59
    of *Paramecium,* 45
    of sea cucumber, 177
    of sea star, 169–170
    of spiders, 131
    of squids, 110–111
    of yellow perch, 198
Feeding polyps, 64
Felis domestica, 7
Female pores, 116, 123

Femoral artery, 221, 261
Femoral veins, 220, 259
Femur, 131, 149, 231, 236
Fetal circulation, 255, 258–259, 259f
Fibula, 231, 236
Filaments, 104, 104f
Filoplume feathers, 229
Filopodia, 25
Fine–adjustment knob, 17–18, 17f, 19
Fin rays, 186, 198
Fins
  anal, 198
  caudal, 185, 189, 191, 193, 198
  dorsal, 185, 186, 189, 191, 193, 198
  lateral, 111
  pectoral, 193, 198
  pelvic, 193, 198
  ventral, 185
Fishes, 188–206
  bony, 181, 198–202, 199f–201f
  jawed, 11, 181, 192–198, 193f–195f, 197f
  jawless, 181, 188–192
Fishing polyps, 64
Flagella, 25, 33, 34, 35f, 37
Flagellated flame cells, 75
Flat surface of mirror, 17
Flatworms. See Platyhelminthes
Flexion, 238
Flexor muscles, 140, 238
Flight feathers, 229
Flight muscles, 231, 231f
"Floaters," 19
Florometra, 166
Floscularia, 94, 95f
Flour beetle, 4–5, 5f
Flukes
  digenetic, 10, 72–73, 78–81
  monogenetic, 72
Food vacuoles, 27, 45, 47, 61
Foot
  of clam, 100, 103
  of land snail, 107, 108
  of molluscs, 98
  of squid, 110f, 111
Foot gills, 138
Foot protractor muscle, 101, 103
Foot retractor muscles, 101, 103
Foramen magnum, 236
Foramen ovale, 258, 261
Foraminiferans, 25, 28
Forebrain, 190
Foreleg
  of honey bee, 151
  of pig, muscles of, 241t, 243
Forelimbs
  of frog, 208, 210
  of pig, 236, 257–258
  of pigeon, 231
Forequarter, muscles of, 239–242, 240t–241t, 241f, 242f

Forewings, 149
Fossa ovalis, 259
Free–living nematodes, 92
Free–swimming planula, 62, 68
Freshwater clam, 99–107, 100f–106f
Freshwater lamprey, 188
Freshwater leeches, 123
Freshwater segmented worms, 10
Freshwater sponges, 49, 54
Freshwater triclads, 73
Frogs, 11, 181, 207–223
Fundus, 250
Funnel, 109, 111
Funnel retractor muscles, 111
Furcula, 231

G

Gallbladder
  of ammocoete larva, 189, 190–191
  of dogfish shark, 194
  of frog, 217
  of pig, 249, 251
  of turtle, 227
  of yellow perch, 201
Ganglia, 106
Garden centipedes, 128
Garden slug, 107f
Garden spider, 130–132, 131f, 132f
Gas gland, 202
Gastric filaments, 68
Gastric ligaments, 169
Gastric mill, 141
Gastric muscles, 140
Gastric pouches, 68
Gastrocnemius, 211, 212t, 213, 214
Gastrodermis
  of Gonionemus, 64
  of Hydra, 60, 61f
  of Obelia, 63
Gastrohepatic ligament, 251
Gastroliths, 140
Gastropoda, 10, 99, 107–109, 107f–109f
Gastrotricha, 94–95, 95f
Gastrovascular cavity, 57
  of Gonionemus, 64
  of Hydra, 59, 60f
  of Metridium, 70
  of Obelia, 63
  of planarians, 75
Gastrozooids, 64
Gemmules, 52, 53f
Genera, 7
Genital aperture, 108
Genital artery, 259
Genital bulb, 158
Genital openings, 136
Genital operculum, 130
Genital papilla, 254
Genital plates, 176

Genital pores, 78, 83, 130, 138, 170, 178
Genital vein, 259
Geotaxis, in Paramecium, 46
Germ layers, 57
"Ghosts," 19
Giant featherduster worm, 115f
Giant fibers, 120
Giant squid, 109
Giardia intestinalis, 37
Giardia lamblia, 36f, 37
Gill arches, 197
Gill bailer, 138
Gill bars, 186, 187, 189
Gill chamber, 135, 197
Gill filaments, 104, 104f, 138, 197, 200
Gill lamellae, 189, 191
Gill pouches, 189, 191, 197
Gill rakers, 197, 200
Gill rays, 197
Gills
  of ammocoete larva, 191
  of arthropods, 127
  of bivalvia, 103, 104
  of crayfish, 135, 138
  of echinodermata, 165
  of horseshoe crab, 130
  of molluscs, 98
  of sea stars, 167, 171
  of sea urchin, 175
  of squid, 111
  of yellow perch, 198, 200
Gill slits
  of ammocoete larva, 189
  of amphioxus, 186, 187
  of dogfish shark, 193
  external, 189, 191, 197
  internal, 189, 197
  pharyngeal, 180, 185
  of yellow perch, 200
Gizzard, 118
Gland cells, 60, 61f
"Glassy rind," 27
Glenoid fossa, 236
Globigerina, 25, 28, 29f
Glochidium, 106, 106f
Glossina, 36
Glottis, 216, 217, 226, 247
Glutathione, 59
Gluteus, 212t
Gluteus medius, 243, 244t
Gnathobases, 129
Gnathostomata, 11, 181
Gnathostomulida, 72
Goldfish, 204–206
Gonads. See also Ovary; Testes
  of Aurelia, 68
  of clam, 106
  of crayfish, 140
  of cricket, 158
  of Metridium, 70
  of sea star, 170
Gonangia, 62, 63
Gonionemus, 58, 63–64, 64f
Gonium, 37f

Gonoducts. See Oviducts; Vas deferens
Gonopore, 63, 146, 148
Gonotheca, 63
Gonyaulax, 37
Gooseneck barnacle, 142
Gordius, 95
Gorgonia, 58
Gorgonocephalus, 166
Gracilis, 211, 243, 244t
Gracilis major, 212t, 213
Gracilis minor, 212t, 213
Gradual metamorphosis, 155
Granuloreticulosans, 25
Grasshopper, 128, 148–150, 149f, 150f
Gravid proglottids, 83, 84
Gray cricket, 155–158, 156f, 157f
Greater curvature, 250
Green glands, 141
Green hydra, 59
Green sea urchin, 173–174, 174f
Gregarina, 41, 41f
Gregarinea, 25
Growth lines, 100, 108, 199, 225
Gubernaculum, 253
Gullet
  of Aurelia, 68
  of Gonionemus, 64
  of Paramecium, 44
Gynecophoric canal, 80
Gyrodactylus, 72

H

Hagfishes, 11, 181, 188
Halichondria, 53
Haliclona, 53
Hard palate, 247
Head
  arteries of, 257–258
  of clamworm, 113
  of cricket, 155
  of dogfish shark, 193
  of earthworm, 116
  of frog, 208
  of grasshopper, 149, 149f
  of honey bee, 150–151, 150f
  of land snail, 107, 108
  of pig, 245–246, 246f, 257–258
  of planarians, 73
  of squid, 110f, 111
  veins of, 257
  of yellow perch, 198
Head rectractor muscles, 111
Heads, of muscles, 238
Heart
  of ammocoete larva, 189, 190
  of clam, 105
  of crayfish, 140
  of cricket, 155
  of dogfish shark, 196

of frog, 219–221, 220*f*
of molluscs, 98
of pig, 255–256, 261–262, 261*f*
of sea squirt, 182
of turtle, 228
ventral, 185
Heartbeat
of crayfish, 143
of *Daphnia,* 143
of frog, 221
Heart urchins, 166
Hectocotyly, 111
Heliozoans, 28
Helix, 99, 107, 107*f*
Hemiazygous vein, 257
Hemichordata, 11
Hemiptera, 162, 162*f,* 163, 163*f*
Hemocoel, 127
Hemocyanin, 130
Hemolymph, 127, 140
Hepatic arteries, 140
Hepatic cecum, 186
Hepatic duct, 251
Hepatic portal system, 220, 259–260
Hepatic portal vein, 186, 196, 220, 257, 259, 260
Hepatic veins, 186, 196, 220, 260
Hepatopancreas, 140
Hermaphroditism, in sea squirt, 184
Heterocercal fin, 193
Heterodera, 92
Heterokont flagella, 25
Hexactinellida, 49
High–power objective, 16, 18
Hilus, 252
Hindbrain, 190
Hindlegs
arteries of, 260–261
muscles of, 243–245, 244*t,* 245*f*
Hindlimbs, 208, 210, 231
Hindquarter, muscles of, 243–245, 243*f,* 244*t,* 245*f*
Hindwings, 149
Hinge ligament, 100, 101
Hip, muscles of, 243, 243*f,* 244*t*
Hirudinea, 10, 113, 123, 123*f,* 124*f,* 125
Hirudo, 113, 123, 123*f,* 124*f,* 125
Hirudo medicinalis, 123, 123*f,* 124*f,* 125
Holobranch, 198
Holophytic nutrition, 35
Holothuroidea, 9, 166, 176–179, 176*f,* 178*f*
Holozoic nutrition, 45
Homarus, 128, 134
Homarus americanus, 134
Homocercal fin, 198
Homoptera, 161, 161*f,* 162, 162*f,* 163, 163*f*
Honey bee, 150–152, 150*f*–152*f*

Honey bee sting, 151–152, 152*f*
Hooks, 81, 83, 83*f*
Hookworm, 92–93, 93*f*
Horns of uterus, 253
Horsehair worms, 10, 95, 95*f*
Horseshoe crabs, 11, 127, 128–130, 129*f,* 130*f*
House cricket, 155–158, 156*f,* 157*f*
Human(s)
blood fluke of, 78–81
liver fluke of, 78, 79*f*
skeleton of, 235*f*
Humerus, 231, 236
Hyaline cap, 27
Hyaline cortex, 27
Hyalonema, 49
Hydra, 58–61
Hydra littoralis, 58
Hydranths, 62
Hydrocaulus, 62, 63
Hydroid colonies, 58, 61–63
Hydroids, 9, 57, 62
Hydromedusae, 9, 63–64, 64*f*
Hydrorhiza, 62
Hydrostatic skeleton, 57
Hydrotheca, 63
Hydrozoa, 9, 58–65
Hymenoptera, 160, 160*f,* 161, 161*f*
Hyoid apparatus, 226
Hypaxial muscles, 199
Hyperbranchial groove, 186, 187
Hypnotoxin, 60
Hypopharynx, 149, 155
Hypostome, 59, 60*f,* 63
Hypothesis formulation, in scientific method, 2, 3–4

I

Ichthyomyzon, 188
Ileocecal valve, 251
Ileum, 251
Ilia, 226
Iliac arteries, 197, 221, 259
Iliacus, 244*t,* 245
Ilium, 210, 230, 236
Illuminator, 17
Image–forming optics, 16
Immature proglottids, 83
Incisors, 246
Incomplete septa, 70
Incurrent aperture, 102
Incurrent canals, of *Sycon,* 50, 51, 51*f*
Incurrent siphon, 182
Inferior jugular veins, 196
Infraspinatus, 241*t,* 242
Infundibulum, 254
Inguinal canal, 252, 253
Ink sac, 111
Innominate artery, 228
Innominate bones, 236
Innominate veins, 220, 257

Insecta, 11, 128, 148–163
Insertion, of muscles, 238
Interlamellar junctions, 104, 104*f*
Internal carotid artery, 220, 257
Internal gill slits, 189, 197
Internal iliac arteries, 261
Internal iliac veins, 259
Internal jugular vein, 220, 257
Internal maxillary vein, 257
Internal nares, 216, 226
Internal oblique, 243, 244*t*
Internal thoracic artery, 250, 258
Internal thoracic vein, 250, 257
Interstitial cells, 60, 61*f*
Intestinal roundworm, 88–92, 90*f,* 91*f*
Intestine
of ammocoete larva, 189
of amphioxus, 186
of *Ascaris,* 89, 92
of clam, 106
of crayfish, 140
of dogfish shark, 195
of earthworm, 118, 120
of frog, 216–217
of pig, 249, 251
of sea cucumber, 177
of sea squirt, 183
of sea star, 169
of snail, 108
of turtle, 227
of yellow perch, 201
Intracellular digestion, 59, 75
Ischium, 210, 226, 236
Isopoda, 142
Isoptera, 160, 160*f,* 162, 162*f*

J

Jawed fishes, 11, 181, 192–198, 193*f*–195*f,* 197*f*
Jawless fishes, 181, 188–192
Jaws, 111, 114, 171, 226
Jaw worms, 72
Jejunum, 251
Jellyfish, 9, 57, 62
"Jet propulsion," 63, 64*f*
Jointed appendages, 127
Joint gills, 138
Joints, 236–237
Jugular veins, 196, 220, 257
Julus, 148

K

Katharina, 98
Kerona, 59
Kidney
of ammocoete larva, 189, 190
of clam, 105
of dogfish shark, 196
of frog, 218
of pig, 250, 251–252
of yellow perch, 201

Kinetoplasta, 25
"Kissing bug," 37

L

Labial palps, 103, 106, 150–151
Labium, 149, 150
Labrum, 146, 149, 150
Ladder type of nervous system, 75
Lamellae, 104, 104*f,* 130, 197
Lampetra, 188
Lampreys, 11, 181, 188–192, 192*f*
Lamp shells, 10
Lampsilis, 99
Land snail, 107–109, 107*f*–109*f*
Languets, 183
Lappet, 67
Large intestine
of frog, 217
of pig, 249, 251
Larva
ammocoete, 188–191, 189*f*
amphiblastula, 51, 52
of brine shrimp, 143
of *Clonorchis,* 78
parenchymula, 52
planula, 62, 68
of sea squirt, 184, 184*f*
of *Sycon,* 51, 52
of *Tribolium confusum,* 5*f*
trilobite, 130, 130*f*
Larvacea, 181
Laryngeal pharynx, 247
Larynx
of frog, 216, 217
of pig, 247
of turtle, 227
Lateral canal, 170, 171, 177
Lateral fins, 111
Lateral groove, 189, 191
Lateral lines, 89, 191, 193, 198
Lateral nerves, 75, 119
Lateral neurals, 120
Lateral teeth, 102
Lateroneural vessel, 118
Latissimus dorsi, 240*t,* 242
Leeches, 10, 113, 123, 123*f,* 124*f,* 125
Left aorta, 228
Left atrium, 219, 221, 256, 261
Left carotid artery, 228
Left common carotid artery, 257
Left–handed shells, 108
Left lung, 227
Left pulmonary arteries, 258
Left subclavian artery, 228, 257
Left systemic arch, 228
Left ventricle, 256, 261–262
Leg muscles, of frog, 212–213, 212*t,* 213*f,* 214*f*
Leishmania, 25
Lenses, of microscope, 16
Lens paper, 18–19
Leopard frog, 207

Lepas, 142, 142*f*
Lepidoptera, 159, 159*f*
Lesser curvature, 250
Leuconoid sponge, 50*f,* 53
Leucosolenia, 49, 50*f,* 53
Levator, 238
Life cycle
    of *Aurelia,* 67*f*
    of *Clonorchis,* 78
    of *Gregarina,* 41
    of *Obelia,* 62*f*
    of *Plasmodium falciparum,*
        42*f,* 43*f*
    of *Plasmodium vivax,* 43*f*
    of *Schistosoma,* 79–80, 80*f*
    of *Tribolium confusum,* 4
    of *Volvox,* 34*f*
Light
    optical path of, through
        microscope, 18*f*
    reflected, 22
    transmitted, 22
Limax, 107*f*
Limifossor, 98
Limulus, 128–130, 129*f,* 130*f*
Limulus polyphemus, 128
Linea alba, 212
Lines of growth, 100, 108,
    199, 225
"Lion's mane jellyfish," 66
Lips
    of *Ascaris,* 89
    of sea urchin, 174
Lithobius, 128, 146
Liver
    of ammocoete larva, 189, 190
    of crayfish, 140
    of dogfish shark, 194
    of frog, 217
    of pig, 249, 251
    of turtle, 227
    of yellow perch, 201
Liver fluke, 78, 79*f*
Lizards, 11, 181
Lobopodia, 25
Lobster, 134–141
Locomotion
    of amebas, 25, 27, 29, 29*f*
    of annelids, 112
    of brittle star, 171
    of clamworm, 113
    of earthworms, 115
    of euglenoids, 34, 35*f*
    of frogs, 207, 210
    of leech, 123
    of millipede, 148
    of *Paramecium,* 43–44
    of pigeon, 231
    of planarians, 73
    of sea urchin, 174–175
    of squid, 109–110
    of *Volvox,* 33
    of yellow perch, 199
Logistic population growth, 3
Loligo, 99, 109–111, 110*f*
Long bone, 236, 237*f*
Longissimus dorsi, 243, 244*t*

Longitudinal fission, in
    euglenoids, 35
Longitudinal lines, 89, 91
Longitudinal muscle cells, 89
Longitudinal muscle layer, 120
Longitudinal muscles, 76, 91
Long monaxons, 52
Longus, 211
Lorica, 94
Lower mandible, 231
Low–power objective, 16, 18
Lubber grasshopper, 128,
    148–150, 149*f,* 150*f*
Lumbar arteries, 221
Lumbar vertebrae, 230, 236
Lumbricus, 112, 115–122,
    117*f*–120*f*
Lumbricus terrestris, 115
Lungs
    of frog, 217
    of land snail, 108
    of turtle, 227–228
Lymnaea stagnalis, 108
Lysosomes, 27
Lytechinus, 166, 174

M

Macracanthorhynchus, 95
Macracanthorhynchus
    hirudinaceus, 95
Macrogametes, 33
Macrogametocytes, 43
Macronucleus, 47, 48
Macropodus opercularis,
    202–203, 203*f*
Madreporite plate, 167, 170,
    171, 176
Magnification, in microscope, 20
Magnus, 211
Major, 211
Malacostraca, 142, 142*f*
Malaria, 42
Male pores, 116, 118, 123
Mallophaga, 161, 161*f*
Malpighian tubules, 158
Mammae, 232
Mammalia, 11, 181, 232–262
Mandibles, 134, 138, 146, 149,
    150, 231
Mandibular gland, 246
Mandibular muscles, 140
Mandibulata, 134
Mantle, 98, 99, 100*f,* 102–103,
    108, 111, 182
Mantle cavity, 98, 103–104,
    108, 111
Manubrium, 64, 236
Marine leeches, 123
Marine mussels, 99
Marine segmented worms, 10
Marine sponges, 49
Marrow, 236
Marrow cavity, 236
Masseter, 240*t,* 242
Mastax, 94

Mastigophora, 9
Mature proglottids, 83
Maxillae, 134, 138, 146, 149, 151
Maxillary teeth, 216
Maxillary veins, 257
Maxillipeds, 136, 138, 146
Maxillopoda, 142
Mealworms, 41
Meconium, 250–251
Mecoptera, 160, 160*f*
Median dorsal aorta, 186
Mediastinal pleurae, 250
Mediastinal septum, 250
Mediastinum, 250
Medicinal leech, 123, 123*f,*
    124*f,* 125
Medulla, 252
Medusa, 57, 62
Medusa buds, 63
Mehlis gland, 78, 84
Mellita, 166
Mercenaria, 99
Merostomata, 11, 127
Merozoites, 42
Mesenchyme, 51
Mesenteric arteries, 197
Mesenteric veins, 260
Mesenteries, 194, 249
Mesoglea
    of *Aurelia,* 66
    of *Gonionemus,* 64
    of *Hydra,* 60, 61*f*
    of *Obelia,* 63
    of *Sycon,* 51
Mesohyl, 51–52
Mesonephric kidney, 191, 218
Mesorchium, 195
Mesosternum, 210
Mesothorax, 149, 151, 155
Mesovarium, 227, 253
Metacarpals, 236
Metamerism, 112, 127, 185
Metamorphosis, gradual, 155
Metanauplius, 143
Metapleural folds, 185, 186
Metatarsals, 236
Metatarsus, 131
Metathorax, 149, 151, 155
Metazoan organisms, 9
Metridium, 58, 68–70, 69*f*
Metridium senile, 68
Microaquariums, 48
Microciona, 53
Microgametes, 33
Microgametocytes, 43
Micronucleus, 47
Microscope, 16–23
    compound light, 16–22
    getting acquainted with,
        18–19
    magnification in, 20
    measuring size of objects,
        20–21
    optical path of light
        through, 18*f*
    parts and operation of,
        16–18, 17*f*

    stereoscopic dissecting,
        22–23, 22*f*
    taking control of, 19–20
Midbrain, 190
Middle leg, 151
Millipedes, 11, 128, 147*f,* 148
Minippe, 142*f*
Miracidia, 79
Molgula, 180, 182
Mollusca, 10, 98–111
Molting, 135, 155
Monaxons, 52, 53*f*
Monhystera, 92
Monochus, 92
Monocular microscope, 18–19
Monoecism
    in *Clonorchis,* 78
    in earthworms, 116, 118
    in *Hydra,* 61
    in leech, 123
    in planarians, 75
    in *Sycon,* 52
Monogenea, 72
Monogenetic flukes, 72
Monoplacophora, 98
Monostyla, 94
"Moon jelly," 66
Mopalia, 98
Motor nerves, of amphioxus, 187
Mouth
    of amphioxus, 186
    of *Ascaris,* 89
    of *Aurelia,* 68
    of centipede, 147*f*
    of clam, 103, 106
    of clamworm, 113
    of *Clonorchis,* 78
    of dogfish shark, 193
    of earthworm, 116
    of frog, 216, 216*f*
    of *Gonionemus,* 64
    of grasshopper, 149*f*
    of honey bee, 150*f*
    of horseshoe crabs, 129
    of *Hydra,* 59, 60*f*
    of lamprey, 191
    of land snail, 107, 108
    of leech, 123
    of *Metridium,* 69, 70
    of *Obelia,* 63
    of *Paramecium,* 44
    of pig, 232, 246–248
    of planarians, 73, 74*f,* 75
    of sea cucumber, 176
    of sea squirt, 183
    of sea star, 166, 168
    of sea urchin, 174
    of squid, 111
    of turtle, 224, 226, 226*f*
    of *Vorticella,* 47
    of yellow perch, 198, 200
Mucins, 245
Mucosa lining, 251
Mucous gland, 108
Müllerian duct, 196
Multiple hemoglobin system,
    204–206

Muscle bands, 177
Muscle layers, 78, 120
Muscles/muscular system
    of amphioxus, 186
    circular, 76
    of clam, 103, 103f
    of crayfish, 140
    of earthworm, 120
    of frog, 211–215, 212t, 213f,
        214f
    longitudinal, 76, 91
    of pig, 237–245, 239f,
        240t–241t, 241f–243f,
        244t, 245f
    of pigeon, 231
    of planarians, 76
    of sea cucumber, 177
    of sea squirt, 182
    of sea star, 170
    of squid, 111
    of yellow perch, 199
Mussels. See Freshwater clam
Mylohyoid, 240t, 242
Myomeres, 199
Myosepta, 187
Myotomes, 186, 187, 189, 190
Myriapods, 128, 146–148, 147f
Mysis, 143
Mytilus, 99
Myxini, 11, 181

N

Nacre, 108
Nacreous layer, 100
Naked amebas, 25–28
Nasal fossae, 210
Nasohypophyseal canal, 190
Nasopharynx, 247
Nauplius, 143
Nautiloids, 99
Nautiluses, 10
Necator americanus, 92
Neck, muscles of, 239–242
Nematocysts, 57
    of Astrangia, 71
    of Aurelia, 68
    of Hydra, 59, 60
    of Metridium, 68–69
    of Obelia, 63
Nematoda, 10, 88–94
Nematomorpha, 10, 95, 95f
Nemertea, 10, 72
Neomenia, 99
Neopilina, 98
Nephridia, of earthworm,
    117, 119
Nephridial tubules, 187
Nephridiopores, 116, 119, 123
Nephrostome, 119
Nereis, 112, 113–114, 114f
Nereis virens, 113
Nerve cords, 75, 84, 158, 187,
    189, 190
    dorsal tubular, 180, 185, 186
    ventral, 91, 119, 120

Nerve net, 60
Nerve ring, 170
Nervous system
    of ammocoete larva, 190
    of amphioxus, 187
    of clam, 106
    of crayfish, 141
    of cricket, 158
    of earthworm, 119
    ladder type of, 75
    of planarians, 75
    of sea squirt, 182
    of sea star, 170
Neural canal, 187, 190
Neurocoel, 190
Neuropodium, 114
Neuroptera, 161, 161f
Nictitating membrane, 208, 225
Nipples, 232
Noctiluca, 37, 37f
Nondirectional illumination,
    74–75
Nostril
    of dogfish shark, 193
    of lamprey, 190, 191
    of pig, 232
    of turtle, 224–225, 226
    of yellow perch, 198
Notochord, 180, 181f, 185, 186,
    187, 189, 190
Notochordal sheath, 187
Notopodium, 114
Nucleus
    of amebas, 27–28
    of Paramecium, 44–45
    of Vorticella, 47
Nudibranchs, 10
Nutrition
    in amphioxus, 186
    in Aurelia, 68
    in crayfish, 134
    in Euglena, 35
    holophytic, 35
    holozoic, 45
    osmotrophic, 36, 41
    in Paramecium, 45
    in sea star, 169–170
    in sporozoans, 41
    in squids, 111
    in Trypanosoma, 36
Nutritive–muscular cells, 60, 61f

O

Obelia, 58, 61–63
Objectives, of microscope, 16
Obligate heterotrophs, 36
Oblique muscle, 212, 212t, 243,
    244t, 249
Observation, in scientific
    method, 2, 3
Ocelli
    of centipedes, 146
    of grasshopper, 149
    of honey bee, 150
    of planarians, 74f, 75

of sea star, 167, 170
of spider, 131
Octocorallia, 58
Octopus, 99
Octopuses, 10, 99
Ocular micrometer, 21, 21f
Oculars, of microscope, 16, 17f
Odonata, 160, 160f
Oil–immersion objective, 16, 19
Olecranon process, 236
Olfactory lobe, 190
Oligochaeta, 10, 112, 115–122,
    117f–120f
Ommochromes, 110
Omosternum, 210
Onchosphere, 84
Ootype, 78
Open circulatory system
    in clam, 105, 105f
    in crayfish, 140
Opercular cavity, 200
Opercular pump, 200
Operculum, 198
Ophioderma, 171
Ophiopholis aculeata, 172f
Ophiura, 166
Ophiuroidea, 9, 166, 171–173
Ophthalmic artery, 140
Opisthobranchia, 107
Opisthosoma, 130
Oral–aboral flattening, of sea
    star, 166
Oral arms, 68
Oral cavity, 200
Oral disc, 47, 69, 191
Oral groove, 44, 45
Oral hood, 185, 186, 189, 190
Oral lobes, 64
Oral papillae, 189, 190
Oral plates, 171
Oral pump, 200
Oral shields, 171
Oral siphon, 182
Oral sucker, 78, 123
Oral surface
    of brittle star, 171
    of sea star, 167–168
    of sea urchin, 174
Oral tentacles, 186
Oral valves, 200
Orbital fossae, 210
Orconectes, 134
Order, 7
Organelles, 24
Organs, 72
Organ–system level of
    organization, 88
Origin, of muscles, 238
Orthoptera, 155, 163, 163f
Osculum, 49, 50
Osmoregulation
    in amebas, 27, 30
    in Ascaris, 89
    in euglenoids, 35
    in Paramecium, 44
    in planarians, 75
Osmotrophic nutrition, 36, 41

Ossicles
    of brittle stars, 171
    of sea stars, 168, 170
Ossification, 236
Osteichthyes, 198–202,
    199f–201f
Ostia, 104, 140, 218
Ostium tubae, 196
Ostracoda, 142, 142f
Otic vesicles, 190
Oval, 202
Ovarian arteries, 260
Ovarian veins, 220, 259
Ovary
    of amphioxus, 187
    of Ascaris, 90, 92
    of clam, 106
    of Clonorchis, 78
    of crayfish, 140
    of cricket, 158
    of dogfish shark, 196
    of earthworm, 118
    of frog, 218
    of Hydra, 61, 61f
    of pig, 253
    of sea squirt, 184
    of sea star, 170
    of tapeworms, 84
    of turtle, 227
Oviductal gland, 196
Oviducts
    of Ascaris, 90, 92
    of crayfish, 136, 140
    of cricket, 158
    of dogfish shark, 196
    of earthworm, 118
    of frog, 218
    of grasshopper, 150
    of pig, 253
    of tapeworms, 84
Ovipositor, 150, 158
Ovoviviparity, in dogfish shark,
    192–193, 196
Oxygen transport, 204

P

Pacifastacus, 134
Painted turtle, 224–228,
    225f–227f
Palatal folds, 226
Palate, 247
Pallial cartilages, 111
Pallial line, 102, 103
Pallium. See Mantle
Palps, 113
Pancreas
    of dogfish shark, 195
    of frog, 217
    of pig, 251
    of turtle, 227
    of yellow perch, 201
Pancreatic duct, 251
Panulirus, 134
Papillae, 252
Papillary muscles, 261

Papulae, of sea stars, 167
Paradise fish, 202–203, 203*f*
Paragordius, 95
Parallel muscle, 237
Paramecium, 25, 43–46, 44*f*
Paramylon, 35
Parapodia, 113–114
Parasites, 36, 38, 41, 42
Parasitic amebas, 28, 28*f*
Parasitic nematodes, 92–94
Parastichopus, 166, 176, 177
Parastichopus californicus, 176
Paraxial rod, flagella with, 25
Parenchyma, 72, 76, 78
Parenchymula larva, 52
Parfocal lenses, 18
Parietal arteries, 197, 260
Parietal pericardium, 250
Parietal peritoneum, 120, 194,
    227, 249
Parietal pleura, 250
Parietal vein, 259
Parotid duct, 246
Parotid gland, 245
Patella, 131
Pauropoda, 128
Pauropus, 128
Pearl, 103
Pecten, 106
Pecten (pollen rake), 151
Pectineus, 244*t*, 245
Pectoral, 240*t*, 242
Pectoral fin, 193, 198
Pectoral girdle, 210, 226, 236
Pectoralis, 212*t*
Pectoralis muscle, 231, 231*f*
Pedal ganglia, 106
Pedicel, 131
Pedicellariae, 165, 167, 174
Pedipalps, 129, 131
Pellicle, 34, 35*f*, 44, 47
Pelmatohydra pseudoligactis, 58
Pelomyxa, 25
Pelvic canal, 236
Pelvic fin, 193, 198
Pelvic girdle, 226, 230, 231, 236
Pelvis, 252
Penis, 253
Pentaradial symmetry, 165, 166
Peranema, 34
Perca, 198–202, 199*f*–201*f*
Perca flavescens, 198
Pereiopods, 138
Pericardial cavity, 196, 201
Pericardial membrane, 103
Pericardial sac, 105, 227
Pericardial sinus, 140
Pericardium, 103, 140, 219, 250
Periderm, 233
Periosteum, 236
Periostracum, 100, 102, 108
Periproct, 176
Perisarc, 62
Peristalsis, 158
Peristaltic contraction, 115
Peristome, 47, 70, 174
Peristomial membrane, 111, 168

Peristomial tentacles, 113
Peristomium, 113, 116
Peritoneum, 120, 120*f*, 168, 170,
    194, 249
Peroneus, 212*t*, 213
Perophora, 184
Petromyzon marinus, 188
Phagocytes, 27
Phagocytosis, 27, 27*f*, 61
Phalanges, 236
Pharyngeal chamber, 76
Pharyngeal gill slits, 180, 185
Pharyngeal sheath, 75
Pharynx
    of ammocoete larva, 189
    of amphioxus, 186, 187
    of *Ascaris,* 89
    of clamworm, 113, 114
    of *Clonorchis,* 78
    of dogfish shark, 193
    of frog, 216
    of *Metridium,* 70
    of pig, 247
    of planarians, 73, 74*f*, 75, 76
    of rotifers, 94
    of sea cucumber, 177
    of sea squirt, 182, 183
    of turtle, 226
    of yellow perch, 200
Philodina, 94, 95*f*
Photoreceptor cells, 186
Photosynthesis, 35
Phototaxis
    in *Daphnia,* 144–145
    in *Paramecium,* 46
    in planarians, 73, 74–75
Phylum (phyla), 7
Physa, 99
Physalia, 58
Physalia pelagica, 64
Phytoflagellates, 33, 37, 37*f*
Pig, 232–262
    circulatory system of,
        255–262, 256*f,*
        258*f*–261*f*
    digestive system of,
        245–251, 246*f*–249*f*
    external structure of,
        232–233
    muscular system of,
        237–245, 239*f,*
        240*t*–241*t,* 241*f*–243*f,*
        244*t,* 245*f*
    reproductive system of,
        252–255, 254*f*
    skeleton of, 233–237, 234*f*
    urinary system of, 251–252,
        252*f*
Pigeon, 229–231
Pigment cells, 110
Pinacocytes, 52, 52*f*
Pineal organ, 191
Pinocytosis, 27, 29
Pinworm, 93–94, 94*f*
Placenta, 254
Placobdella, 113
Placoid scales, 193, 193*f*

Planarians, 10, 72, 73–77,
    74*f,* 76*f*
Plane (flat) surface of mirror, 17
"Plant–flagellates," 33
Planula, 62, 68
Plasmalemma, 26
Plasma membrane, 44
Plasmodium, 42–43
Plasmodium falciparum, 42*f*
Plasmodium vivax, 43*f*
Plastron, 224
Plates, 171
Platyhelminthes, 10, 72–87
Platyias, 94
Platysma, 239
Plecoptera, 160, 160*f*
Plectus, 92
Pleopods, 138
Pleura, 250
Pleural cavity, 250
Pleural fluid, 250
Pleuron, 155
Pleuropericardial membrane,
    228
Pleuroperitoneal cavity, 227
Pneumatophore, 64
Pneumostome, 108
Podia. *See* Tube feet
Poison fang, 146
Poison gland, 131, 151, 152*f*
Polian vesicles, 177
Pollen basket, 151
Pollen brush, 151
Pollen combs, 151
Pollen packer, 151
Pollen rake, 151
Polyaxons, 52
Polychaeta, 10, 112,
    113–114, 114*f*
Polygyra, 107
Polymorphism, 57, 64
Polyp, 57, 62, 64
Polyplacophora, 10, 98
Polystoma, 72
Pond snail, 108
Population growth, 3–4
Pores, 49, 50, 51, 170
Porifera, 9, 49–56
Portal system, of frog, 220
Portal veins, 196
Portuguese man–of–war, 64
Postanal tail, 180, 185
Postcardial vein, 186
Postcava, 256, 259
Postcaval veins, 219, 220, 261
Postcaval venous circulation,
    259, 260*f*
Posterior adductor muscles, 101
Posterior aorta, 105
Posterior cardinal sinuses, 196
Posterior cardinal veins, 190
Posterior deep pectoral,
    240*t,* 242
Posterior foot retractor
    muscles, 101
Posterior mesenteric artery, 260
Posterior mesenteric vein, 260

Posterior trunk, 75
Posterior vena cava, 220
Precardial vein, 186
Precava, 256
Precaval veins, 219, 220,
    257, 261
Prediction, in scientific method,
    2, 4
Prehallux, 208
Premolars, 246
Prepollux, 208
Primary lamellae, 197
Primary septa, 70
Prismatic layer, 100–101, 108
Proboscis, of nemertea, 72
Procambarus, 134
Processus vaginalis, 252
Proglottids, 81
Pronephric kidney, 189, 190
Prosobranchia, 107
Prosoma, 129
Prosopyles, of *Sycon,* 50,
    51, 51*f*
Prostate gland, 253
Prostomium, 113, 116
Prothorax, 149, 151, 155
Protochordata, 180
Protomerite, 41
Protonephridia, 75, 78
Proton pumps, 27
Protoplasmic flow, 25
Protoplasmic level of
    organization, 24
Protoplasmic strands, 33
Protopod, 138
Protozoa, 24–48
Protozoea, 143
Protractor muscles, 238
Pscoptera, 161
Pseudocardinal teeth, 102
Pseudocoel, 88, 89, 91
Pseudocoelomates, 88–97
Pseudopodia, 25, 27, 61
Psoas major, 244*t,* 245
Psocoptera, 162
Ptychodiscus, 37, 37*f*
Pubis, 210, 226, 236
Pulmocutaneous arch, 221
Pulmonary arch, 228
Pulmonary arteries, 219, 221,
    256, 258
Pulmonary circuit, 219, 255, 258
Pulmonary trunk, 258, 261
Pulmonary veins, 219, 220, 228,
    256, 258, 261
Pulmonata, 107–109, 107*f*–109*f*
Pulsation rate, of *Paramecium,*
    temperature and, 44
Pupa, of *Tribolium confusum,*
    5*f*
Purple sea urchin, 173
Pycnogonida, 127
Pygostyle, 230
Pyloric ceca, 169, 171, 201
Pyloric chamber, 140
Pyloric duct, 169
Pyloric end, 250

Pyloric portion of stomach, 201
Pyloric stomach, 169
Pyloric valve, 195, 201, 216, 251

Q

Quadriceps femoris, 243, 244*t*
Quadrula, 99
Quahogs, 99
Queen, 152
Quill, 229

R

Rachis, 229
Radial artery, 257
Radial canal
    of *Aurelia,* 68
    of *Gonionemus,* 64
    of sea cucumber, 177
    of sea star, 170, 171
    of *Sycon,* 50, 51, 51*f*
Radial chambers, 70
Radial nerve, 170, 171
Radial symmetry, tetramerous,
    66
Radiate animals. *See* Cnidaria;
    Ctenophora
Radiating canals, 44
Radiolarians, 25, 29
Radius, 231, 236
Radula, 98, 107, 108
Ragworm. *See* Clamworm
Rana, 207–223
Rana catesbeiana, 207
Rana pipiens, 207
Ray–finned fishes, 11
Rays, 11, 181
    of sea star, 166, 167
Rectal ceca, 169
Rectal gland, 195
Rectum
    of cricket, 158
    of dogfish shark, 195
    of pig, 251
    of squid, 111
Rectus, 211
Rectus abdominis, 212, 212*t,*
    243, 244*t*
Rectus femoris, 243, 244*t*
"Red tides," 37
Reflected light, 22
Reflecting mirror, 17
Reflexes
    crossed, 182
    direct, 182
Regeneration, of planarians,
    84–85, 85*f*
Renal arteries, 197, 260
Renal opening, 138
Renal portal system, 220
Renal portal veins, 196, 220
Renal pyramids, 252
Renal veins, 220, 259
Renilla, 58

Replicates, 5
Reproduction. *See* Asexual
    reproduction; Sexual
    reproduction
Reproductive cells, 33
Reproductive duct, 170
Reptilia, 11, 181, 224–228
Reservoir, 34
Respiration/respiratory system
    in amphioxus, 187
    in arthropods, 127
    in bivalvia, 103, 104
    in clam, 104
    in clamworm, 114
    in crayfish, 138
    in cricket, 158
    cutaneous, 217
    in dogfish shark, 197–198
    in echinodermata, 165
    in frog, 217
    in horseshoe crab, 130
    in land snail, 108
    in molluscs, 98
    in pig, 247, 247*f*
    in sea cucumber, 177
    in sea squirt, 182
    in sea stars, 167, 171
    in sea urchin, 175
    in spider, 131–132
    in squid, 111
    in turtle, 227–228
    in yellow perch, 200
Respiratory epithelium, 187
Respiratory trees, 177
Rete mirabile, 202
Reticulopodia, 25
Retractile muscles, 177
Retractor muscles, 238
Revolving nosepiece, of
    microscope, 16
Rhabdites, 76
Rhabditis, 92
Rhabdodermella, 50
Rhizopodans, 25
Rhizostoma, 58
Rhomboideus, 240*t,* 242
Rhomboideus capitis, 240*t,* 242
Rhopalium, 67
Ribbon worms. *See* Nemertea
Ribs, 230, 236
Right aorta, 228
Right atrium, 219, 221, 228,
    256, 261
Right carotid artery, 228
Right common carotid
    artery, 257
Right–handed shells, 108
Right pulmonary arteries, 258
Right subclavian artery,
    228, 257
Right systemic arch, 228
Right ventricle, 256, 261
Ring canal
    of *Aurelia,* 68
    of *Gonionemus,* 64
    of sea cucumber, 177
    of sea star, 170

River lamprey, 188
Rock dove, 229–231
Rock lobster, 134
Romalea, 128, 148–150,
    149*f,* 150*f*
Rostellum, 83, 83*f*
Rostrum, 135, 140, 185, 193
Rotation, 238
Rotifera, 94, 95*f*
Roundworms. *See* Nematoda
Rudimentary sixth toe, 208
Rudimentary thumb, 208
Rugae, 195, 251

S

Sacral hump, 208
Sacral vertebrae, 210, 226,
    230, 236
Salamanders, 11, 181
Salivary amylase, 245
Salivary glands, 155,
    245–246, 246*f*
Salpians, 181
Sand dollars, 9, 166
Sandworm. *See* Clamworm
Sarcodina, 9
Sarcomastigophora, 9, 24
Sarcopterygii, 181
Sartorius, 211, 212*t,* 213,
    243, 244*t*
Scales, 198–199
Scanning objective, 16
Scapula, 210, 226, 231, 236
Schistosoma, 73, 78–81
Schistosoma haematobium,
    79, 80
Schistosoma japonicum, 79, 80
Schistosoma mansoni, 79,
    80, 80*f*
Schistosomiasis, 79
Schizogony, 41, 42
Schizont, 43
Sciatic artery, 221
Sciatic vein, 220
Scientific method, 2–4
Sclerites, 149
Sclerosepta, 71
Sclerotic ring, 231
Scolex, 81, 83
Scolopendra, 146, 147*f*
Scorpions, 11, 127–128
Scrotum, 252
Scutes, 224
Scutigerella, 128
Scyphistomae, 68
Scyphomedusae, 63, 66–68
Scyphozoa, 9, 58, 66–68
Sea anemones, 9, 57, 58,
    68–70, 69*f*
Sea biscuits, 9
Sea clam, 99
Sea cucumbers, 9, 166,
    176–179, 176*f,* 178*f*
Sea lamprey, 188

Sea lilies, 9, 165–166
Sea spiders, 127
Sea squirt, 180, 181–184,
    182*f*–184*f*
Sea stars, 9, 166–171, 167*f*–169*f*
Sea urchins, 9, 166, 173–176,
    174*f,* 175*f*
Sea walnuts, 9
Secondary lamellae, 197
Sedentaria, 114
Segmental ganglia, 158
Segmentation, 112
Segmented arteries, 186
Segmented worms, 10
Semicircular notch, 151
Semilunar valves, 261
Semimembranosus, 212*t,*
    244*t,* 245
Seminal grooves, 116
Seminal receptacles
    of crayfish, 136
    of earthworm, 116, 117, 118
    of liver fluke, 78
Seminal vesicles
    of *Ascaris,* 90, 92
    of dogfish shark, 196
    of earthworm, 117, 118
    of liver fluke, 78
    of pig, 253
Seminiferous tubules, 253
Semitendinosus, 244*t,* 245
Sense organs. *See also* Eyes
    of clam, 106
    of clamworm, 113
    of crayfish, 141
    of frog, 216
    of lamprey, 191
    of planarians, 75
    of sea star, 170
    of squid, 111
Sensillae, 123
Sensory cells, of *Hydra,* 61, 61*f*
Sensory hairs, 131
Sensory nerves, of amphioxus,
    187
Sensory papillae, 216
Sensory tentacles, 167
Sepia, 99
Septa, 70, 115
Septal filaments, 70
Septal perforations, 70
Serial homology, 136
Setae, 114, 120
Sex, development of, 33
Sex cells, 33
Sexual dimorphism, 80
Sexual reproduction
    in amphioxus, 187
    in *Ascaris,* 90
    in *Aurelia,* 68
    in clam, 106
    in *Clonorchis,* 78
    in cnidaria, 57
    in crayfish, 140
    in cricket, 158
    in dogfish shark,
        195–196, 195*f*

Sexual reproduction–*Cont.*
in earthworms, 116, 118–119, 118*f*
in frog, 218, 218*f*
in *Gonionemus,* 64
in horseshoe crabs, 130
in *Hydra,* 61
in land snail, 108
in leech, 123
in pig, 252–255, 254*f*
in planarians, 75
in *Plasmodium,* 43
in *Schistosoma,* 79
in sea cucumber, 178
in sea squirt, 184
in sea star, 170
in spiny–headed worms, 95
in squid, 111
in *Sycon,* 52
in *Volvox,* 33, 34*f*
Shaft
of bone, 236, 237*f*
of feather, 229
of rib, 236
Shank muscles, 212*t,* 213, 214*f*
Sharks, 11, 181, 192–198, 193*f*–195*f,* 197*f*
Shell
of clam, 99, 100–102
of land snail, 107
of molluscs, 98
of turtle, 224, 225, 225*f*
Shelled amebas, 28, 29*f*
Shell gland, 196
Short monaxons, 52
Shoulders
arteries of, 257–258
muscles of, 239–242
veins of, 257
Silk glands, 132
Simplex, 253
Sinistral shells, 108
Sinus venosus, 196, 219, 228
Siphon, 111, 182
Siphonaptera, 161, 161*f*
Siphonoglyph, 70
Siphons, 100
Skates, 11
Skeletal cups, 71
Skeleton. *See also*
Endoskeleton;
Exoskeleton
appendicular, 199, 210, 225, 231, 236
axial, 199, 210, 225, 230–231, 236
of cat, 234*f*
of frog, 210, 211*f*
of human, 235*f*
hydrostatic, 57
of pig, 233–237, 234*f*
of pigeon, 229–231, 230*f*
of sponges, 52–53
of turtle, 225–226, 225*f*
of yellow perch, 199
Skin. *See also* Dermis; Epidermis
of dogfish shark, 193, 193*f*

of frog, 217
of pig, 249
of planarians, 73
Skin gills
of echinodermata, 165
of sea stars, 167, 171
Skull
of cat, 233*f*
of frog, 210
of pig, 236
of pigeon, 230–231
of turtle, 225–226, 225*f*
Sleeping sickness, 36
Slime tube, 118, 118*f*
Slugs, 10, 99, 107*f*
Small dog tapeworm, 81, 82*f*
Small intestine
of frog, 216
of pig, 249, 251
of turtle, 227
Snails, 10, 99
aquatic, 108
land, 107–109, 107*f*–109*f*
pond, 108
Snakes, 11, 181
Social behavior, in honey bees, 152
Society of Protozoologists, 24
"Soft corals," 58
Soft palate, 247
Soil nematodes, 92
Sole, 177
Solenogastres, 98–99
Somatic cells, 33
Spawning, lampreys, 188
Species, 7
Spermatic arteries, 260
Spermatic cord, 253
Spermatic veins, 220, 259
Spermatozoa, 61
Sperm duct
of crayfish, 136, 140
of dogfish shark, 196
of tapeworms, 83
Sperm sacs, 196
Sphaeridia, 175
Spicules, 49, 50, 52–53, 89, 90
Spiders, 11, 127–128
Spider webs, 130, 131, 131*f*
Spine, 236
Spines
of sea star, 167
of sea urchin, 174
Spinnerets, 132
Spiny dogfishes, 192
Spiny–headed worms, 10, 95, 95*f*
Spiny lobster, 134
Spiracles, 146, 148, 149, 193
Spiral movements, 44
Spiral valve, 195
Spirobolus, 128, 148
Spirobranchus giganteus, 115*f*
Spirostomum, 48, 48*f*
Spirotrichonympha, 36*f*
Spisula, 99
Spleen
of dogfish shark, 195, 196

of pig, 250
of yellow perch, 201
Splenius, 240*t,* 242
Sponges, 9, 49–56
Spongia, 49
Spongilla, 49
Spongin, 49, 53, 54*f*
Spongocoel, 49, 50, 51, 51*f*
Spongy bone, 236
Sporocyst, 79
Sporogony, 41
Sporozoans, 41
Sporozoites, 41, 42
Spur, 151
Squalus, 192–198, 193*f*–195*f,* 197*f*
Squalus acanthias, 192
Squalus suckleyi, 192
Squids, 10, 99, 109–111, 110*f*
Stage, of microscope, 16, 17*f*
Stage micrometer, 21, 21*f*
Starch granules, 35
Statocysts, 63, 141
Stentor, 25, 46*f,* 47–48, 47*f*
Stentor coeruleus, 48
Stereoscopic dissecting microscope, 22–23, 22*f*
Sternal artery, 140, 141
Sternocephalic muscle, 240*t,* 242
Sternohyoid, 239, 240*t*
Sternothyroid, 239, 240*t*
Sternum, 155, 210, 230, 231, 236
Stigma, 33, 34
Stimuli, response to, in *Paramecium,* 45–46
Sting, of honey bee, 151–152, 152*f*
Stomach
of *Aurelia,* 68
of clam, 106
of crayfish, 140, 141
of dogfish shark, 195
of *Gonionemus,* 64
of pig, 249, 250
of sea cucumber, 177
of sea squirt, 183
of sea star, 168, 169
of turtle, 227
of yellow perch, 201
Stone canal, 170, 177
Stone crab, 142*f*
Stony corals, 58, 70–71, 71*f*
Striated muscles, 127
Strobila, 68, 81
Strobilation, 68
Strongylocentrotus, 166
Strongylocentrotus drobachiensis, 173–174, 174*f*
Strongylocentrotus purpuratus, 174, 175
Subclavian arteries, 197, 221, 228
Subclavian veins, 196, 220, 257
Subcutaneous lymph sacs, 211

Subesophageal ganglion, 141
Subgenital pit, 68
Subintestinal artery, 191
Subintestinal vein, 186
Sublingual glands, 246
Submaxillary gland, 246
Subneural vessel, 118, 120
Subpharyngeal ganglia, 119
Subpharyngeal gland, 191
Subscapular artery, 257
Subscapular vein, 220
Subspecies, 7
Substage condenser, 16
Substage illuminator, 17
Subumbrella, 63
Sucker(s), 81, 83, 83*f*
caudal, 123
oral, 78, 123
ventral, 78, 123
Superficial pectoral, 240*t,* 242
Suprabranchial chamber, 104
Supracoracoideus muscle, 231, 231*f*
Supraesophageal ganglia, 141
Suprascapula, 210
Supraspinatus, 241*t,* 242
Sus domesticus, 232–262
Swim bladder, of yellow perch, 201–202
Swimmerets, 135, 136, 138
Sycon, 49–55, 50*f*
Syconoid sponge, 50, 50*f*
Symbiotic mutualism, 37, 59
Symmetry. *See* Bilateral symmetry; Pentaradial symmetry; Radial symmetry
Symphyla, 128
Symphyses, 236
Synarthrosis, 236–237
Synsacrum, 230
Systemic arch, 221, 228
Systemic circuit, 219, 255
Systemic veins, 228

T

Tactile hairs, 141
Taenia, 73, 81–84, 81*f*–83*f*
Taenia pisiformis, 81, 81*f*
Taenia saginata, 83
Taenia solium, 81
Tagelus, 99
Tagmata, 127
Tail
of dogfish shark, 193
postanal, 180, 185
of yellow perch, 198
Tail bone, 236
Tail fan, 136
Tapeworms. *See* Cestoda
Tarantula, 131
Tarsometatarsus, 231
Tarsus, 131, 149, 236
Taxis, to stimuli, in *Paramecium,* 45–46

Taxonomic keys, 7–11
Taxonomy, definition of, 7
Tealia, 58
Tectibranchs, 10
Teeth
 of crayfish, 141
 of dogfish shark, 193
 of freshwater clam, 102
 of frog, 216
 of lamprey, 191
 of pig, 246
 of sea urchin, 174
 of yellow perch, 200
Tegument, 78, 81
Telson, 130, 136, 140
Temperature, and pulsation rate
  of *Paramecium,* 44
Tendon, 238
Tensor fasciae latae, 243, 244*t*
Tentacles
 of *Aurelia,* 67
 of clamworm, 113
 of *Hydra,* 59, 60*f*
 of land snail, 107, 108
 of *Metridium,* 69
 of *Obelia,* 63
 of sea cucumber, 176
 of sea squirt, 183
 sensory, 167
 of squid, 110*f,* 111
Tentacular bulbs, 63
Teredo, 99
Tergite, 155
Tergum, 155
Test, 182
Testes
 of amphioxus, 187
 of *Ascaris,* 90, 92
 of clam, 106
 of *Clonorchis,* 78
 of crayfish, 140
 of cricket, 158
 of dogfish shark, 195
 of earthworm, 118
 of frog, 218
 of *Hydra,* 61
 of pig, 253
 of sea squirt, 184
 of sea star, 170
 of tapeworms, 83
Testing predictions, 2–3, 4
Tetrahymena, 25, 46*f*
Tetramerous radial
  symmetry, 66
Tetrapoda, 11
Thaliacea, 181
Theca, 71
Thigh muscles, of frog,
  212–213, 212*t,* 213*f,* 214*f*
Thigmotaxis
 in *Paramecium,* 46
 in planarians, 73, 74
Thoracic vertebrae, 226,
  230, 236
Thorax
 of cricket, 155
 of grasshopper, 149

of honey bee, 151
of pig, 232, 250
Thorny corals, 58
Thread, 60
Threadworms, 10, 95, 95*f*
Thymus, 247
Thyone, 166, 176, 177
Thyrocervical artery, 258
Thyroid gland, 242, 247–248
Thysanoptera, 159, 159*f,*
  162, 162*f*
Thysanura, 162, 162*f*
Tibia, 131, 149, 231, 236
Tibialis anterior longus, 212*t*
Tibialis anticus longus, 213, 214
Tibialis posterior, 212*t*
Ticks, 11
Tissue level of organization, 57
Tissue–organ level of
  organization, 72
Toads, 11, 181, 207
Tongue
 of frog, 216
 of honey bee, 150
 of lamprey, 191
 of turtle, 226
 of yellow perch, 200
Tooth shells, 10
Trachea, 227, 247
Tracheal spiracle, 131–132
Tracheal system
 of arthropods, 127
 of spiders, 131–132
Tracheal tubes, 158
Transmitted light, 22
Transparent ruler calibration,
  20–21
Transverse abdominal muscle,
  243, 244*t*
Transverse canal, 84
Transverse nerves, 75
Transverse septum, 227
Transversus, 212, 212*t*
Trapezius, 240*t,* 242
Trematoda, 10, 72–73, 78–81
Triatoma, 37
Tribolium confusum, 4–5, 5*f*
Triceps, 211
Triceps brachii, 241*t,* 243
Triceps femoris, 212*t,* 213
Trichina worm, 93, 93*f*
Trichinella spiralis, 93, 93*f*
Trichocysts, 45
Trichodina, 59
Trichonympha, 36*f,* 37
Trichoptera, 159, 159*f,*
  160, 160*f*
Triclads, 73
Tricuspid valve, 261
Trilobita, 127
Trilobite larvae, 130, 130*f*
Trinomial nomenclature, 7
Triploblastic development,
  72, 127
Triradiates, 52
Trivium, 167
Trochanter, 131, 149

Trophozoites, 28, 28*f,* 41, 42–43
True coelom, 88, 98, 105, 127
True jellyfish, 9, 57, 58, 66–68
Truncus arteriosus, 219
Trunk
 of dogfish shark, 193
 of frog, 208
 of pig, arteries of, 260–261
 of yellow perch, 198
Trunk muscles, of frog, 212,
  212*t,* 213*f,* 214*f*
Trypanosoma, 25, 36–37, 36*f*
Trypanosoma brucei gambiense,
  36, 36*f*
Trypanosoma brucei
  rhodesiense, 36
Trypanosoma cruzi, 37
Trypanosomatidea, 25, 37
Tsetse flies, 36
Tube anemones, 58
Tube feet
 of brittle star, 171
 of sea cucumber, 177
 of sea star, 166, 167–168,
  169, 170, 171
 of sea urchin, 174
Tubercle, 174
"Tube within a tube"
  construction, 89
Tubeworms, 115*f*
Tubifex, 112
Tubularia, 58
Tunic, 182
Tunicates, 11, 180, 181–184,
  182*f*–184*f*
Turbellaria, 10, 72, 73–77,
  74*f,* 76*f*
Turtles, 11, 181, 224–228,
  225*f*–227*f*
Tylenchus, 92
Tympanic membrane, 208
Tympanum, 149
Typhlosole, 118, 191

U

Ulna, 231, 236
Ulnar artery, 257
Umbilical arteries, 250, 261
Umbilical cord, 232
Umbilical vein, 249, 257
Umbo, 100
Undulating membrane, 37
Unicellular organisms, 9
Uniramia, 11, 128, 146–163
Uniramous appendages, 138
Upper mandible, 231
Ureter
 of cricket, 158
 of frog, 218
 of pig, 252
Urethra, 251, 252, 253
Urinary bladder, 218, 251
Urinary system. *See*
  Excretion/excretory
  system

Urochordata, 11, 180, 181–184,
  182*f*–184*f*
Urogenital arteries, 221
Urogenital opening, 198, 253
Urogenital papilla, 196
Urogenital sinus, 192, 254
Uropods, 136
Urostyle, 210
Uterus
 of *Ascaris,* 90, 91
 of *Clonorchis,* 78
 of dogfish shark, 196
 of frog, 218
 of pig, 253
 of tapeworms, 84

V

Vagina
 of *Ascaris,* 90
 of pig, 253, 254
 of tapeworms, 84
Valves, 99, 100–102, 196
Valvular intestine, 195
Vane, 229
Vas deferens
 of *Ascaris,* 90, 92
 of *Clonorchis,* 78
 of cricket, 158
 of pig, 253
 of tapeworms, 84
Vas efferens, of *Clonorchis,* 78
Vastus intermedialis, 243, 244*t*
Vastus lateralis, 243, 244*t*
Vastus medialis, 243, 244*t*
Veins, 149
Velar flaps, 190
Velar tentacles, 186
Velum, 63, 151, 186, 189
Venous system
 of dogfish shark, 196, 197*f*
 of frog, 220
Ventral abdominal artery, 140
Ventral abdominal vein, 220
Ventral aorta, 186, 187, 191, 197
Ventral blood vessel, 118, 120
Ventral cirrus, 114
Ventral fin, 185
Ventral heart, 185
Ventral median lines, 89
Ventral nerve cords, 91, 119, 120
Ventral serratus, 240*t,* 242
Ventral sucker, 78, 123
Ventral thoracic artery, 140
Ventricles, 105, 196, 219,
  221, 261
Venus, 99
Vermiform appendix, 251
Vertebral column
 of frog, 210
 of pig, 236
 of turtle, 226
Vertebralima, 25
Vertebrata, 181
Vestigial oviduct, 218
Vibrissae, 232

Villi, 251
Visceral ganglia, 106
Visceral hump, 108
Visceral mass, 103
Visceral pericardium, 250
Visceral peritoneum, 120,
    194, 249
Visceral pleura, 250
Visceral skeleton, 210
Vitelline gland, 84
Viviparity, in dogfish shark, 196
Vocal sacs, 216
Volvox, 25, 33, 34*f*
Vomerine teeth, 216
Vorticella, 25, 46–47, 46*f,* 47*f*
Vulva
    of *Ascaris,* 89, 90
    of pig, 254

W

Walking legs
    of crayfish, 138
    of grasshopper, 149
    of honey bee, 151
    of horseshoe crab, 129
    of spider, 131
Water balance. *See*
    Osmoregulation
Water–expulsion vesicle. *See*
    Contractile vacuole
Water flea, 141–142, 142*f*
Water tubes, 104, 104*f*
Water–vascular system
    of echinodermata, 165
    of sea cucumber,
        176–177, 178*f*

of sea stars, 166, 167, 170
    of sea urchin, 176
Webbed toes, 208
Wet mount, 22, 22*f*
Wheel organ, 186
Whelks, 99
Whorl, 107
Wings
    of grasshopper, 149
    of pigeon, 231
Wolffian duct, 196
Workers, 152

X

Xiphisternum, 210, 236

Y

Yellow perch, 198–202,
    199*f*–201*f*
Yolk ducts, of *Clonorchis,* 78
Yolk glands
    of *Clonorchis,* 78
    of tapeworms, 84

Z

Zebra mussels, 99
Zoantharia, 58
Zoea, 143
Zoochlorellae, 59
Zooflagellates, 33
Zooids, 33, 64, 184
Zygotes, 33, 61, 106